INTERNATIONAL CENTRE FOR MECHANICAL SCIENCES

COURSES AND LECTURES - No. 216

CODING AND COMPLEXITY

EDITED BY

G. LONGO
UNIVERSITY OF TRIESTE

PREFACE BY

J. L. MASSEY
UNIVERSITY OF NOTRE DAME

SPRINGER-VERLAG WIEN GMBH

Originally published by Springer-Verlag Wien-New York in 1975

ISBN 978-3-211-81341-6 ISBN 978-3-7091-3008-7 (eBook)

DOI 10.1007/978-3-7091-3008-7

PREFACE

The picturesque northern Italian city of Udine, lying seventy kilometers to the north and west of Trieste, was the setting in July 1974 for a two week summer school on "Coding and Complexity". This school was held under the auspices of the Centre International des Sciences Mécaniques (CISM) in Udine and was the brainchild of Professor Giuseppe Longo of the University of Trieste, who doubles as a member of the CISM staff. It was Longo's intention that this summer school should probe the practicality of the "coding theorems" of information theory. In their original statement by Shannon, these theorems promised arbitrarily reliable communication at any rate less than channel capacity provided that "long enough" codes were used. But how long is "long enough" and how "complex" to instrument are such long codes? To answer these questions in the light of the quarter-century of information-theoretic research since the publication of Shannon's paper, Longo invited as lecturers a number of scholars who have been active in various areas of coding research. This volume is the collection of their written lectures which were in most instances further polished and amplified after the summer school in light of the many stimulating interchanges which followed their oral presentation. The summer school was a lively one that educated listener and lecturer alike.

Of the nine contributions in this volume, the first five deal with channel coding, or "error-correcting coding" to use the more common but less precise term. To talk of "correcting errors" already implies that the channel input and output alphabets coincide, something that information theorists have come to recognize is a wasteful feature in some communication situations where "soft decision" demodulation is substantially superior to "hard decision" demodulation.

The first contribution, "Error Bounds for Tree Codes, Trellis Codes, and Convolutional Codes with Encoding and Decoding Procedures", is that by this writer who took advantage of the ten lecture hours granted him by Professor Longo to give a rather complete exposition of these non-block forms of coding and to include a number of his previously unpublished results. This paper was written for the reader with no background in convolutional coding and was designed to bring him to the current state of knowledge in this field. The description of encoding procedures and decoding procedures (Viterbi decoding and sequential decoding) is complete enough that the reader can easily gauge the complexity of the necessary devices. Both Viterbi decoding and sequential decoding have the feature that they are as well adapted to soft-decision as to hard-decision decoding, a fact which largely accounts for the preponderance of convolutional codes over block codes in practical applications. The two asymptotic questions addressed directly in this paper are (1) the usual exponential error bounds on average error probability – which are simplified by the use of some natural, but new, ensembles and (2) the size of necessary ensembles to prove the error bounds – these ensemble "complexities" are measures of the necessary encoder complexity. One new result that may have practical significance is the demonstration of the desirability of choosing the memory of a convolutional encoder slightly greater than the tail length in sequential decoding applications.

The second contribution, "Threshold Decoding – A Tentative Survey", by Dr. J.M. Goethals of the MBLE Research Laboratory in Brussels is a masterful compilation and condensation of the most important research into the design of block error-correcting codes that can be decoded with majority logic gates. This writer, having been one of the early workers in this area and the first to use the term "threshold decoding", confesses his jealousy of the elegance and clarity of Goethals' presentation. The reader will see the fundamental role which the finite geometries play in block code constructions for conventional threshold decoding of orthogonal parity checks, and be introduced to the use of t-designs as a potentially powerful method for construction of non-orthogonal parity checks. The appeal of threshold decoding has been the attractively small decoder complexity; its drawback has been the lack of asymptotic results for codes with long block lengths.

The third contribution, "Some Current Research in Decoding Theory", by Professor L.D. Rudolph of Syracuse University has an interesting history. Rudolph came to the summer school in the guise of "listener", but was drafted by Longo to be a "lecturer" when, in the course of the many free-wheeling discussions among participants, it became clear that he had recently been engaged in some very promising research on decoding block codes. We already noted above that in some communication situations there is considerable merit in using soft-decision rather than hard-decision demodulation. Yet virtually all efficient decoding procedures known for block codes are of the "error-correcting" (or "error-and-erasure-correcting") type and hence are suitable only for hard (or almost-hard) decision modulation – the one exception being Forney's "generalized minimum distance" decoding and its later variants which, unfortunately, is effective only for high-rate codes. In his paper, Rudolph describes a natural mapping for real numbers, namely $[1 - \cos \pi(x + y)]/2$, *that reduces to modulo-two addition when x and y are restricted to the values 0 and 1 and that suggests a corresponding modification of threshold decoding circuits to handle real variables (i.e., soft decisions) rather than binary ones. Small (and thus manageable) examples indicate that the performance of a full correlation decoder might be achieved – an exciting prospect for the legion of block code devotees. Rudolph's paper also gives a particularly clear exposition of "sequential code reduction", a technique invented by Rudolph for reducing the complexity of a decoder by cleverly converting the decoding problem to that for successive subcodes of the original code. There are no asymptotic results in Rudolph's paper, but the ideas therein are pregnant with possibilities for the future.*

The fourth contribution, "Procedures of Sequential Decoding", by Dr. K. Sh. Zigangirov of the Institute for the Problems of Information Transmission in Moscow is a natural companion to the first paper where the "stack algorithm" was described in an introductory way. Zigangirov had given some lectures at the 1973 CISM school on information theory, also organized by Longo, who felt that the material covered by Zigangirov fitted very well into this school's subject matter. In this paper Zigangirov (who is the inventor of the stack algorithm which was later independently discovered, and slightly refined, by Jelinek) gives a careful analysis of the memory necessary to implement the stack algorithm, including a modified version of the algorithm employing a threshold on the likelihood values for nodes retained in the stack. He also considers the effect on the required storage when a limitation is imposed on the "back search limit" (i.e., the difference between the longest and shortest paths to nodes in the stack). His asymptotic results show in all cases that the probability of a decoding failure (i.e., a deletion of data due to memory overflow) decreases algebraically, rather than exponentially, with memory size – the asymptotic plague of all sequential decoding methods.

The fifth contribution, "Decoding Complexity and Concatenation Codes", by Dr. V.V. Ziablov of the Institute for the Problems of Transmission of Information in Moscow introduces an important generalization of Forney's concatenated codes in which an "inner" block code over a small alphabet is used together with an "outer" block code over a large alphabet. Ziablov's innovation is to allow both the outer and inner codes to vary in an interesting way. The "trick" in Ziablov's construction is that his inner codes are nested in the sense that his first inner code is a subcode of the second, which in turn is a subcode of the third, etc. Thus a codeword in the i-th inner code is also a codeword in the j-th inner code whenever $j > i$. *His paper gives an interesting second-order (i.e., two inner codes and two outer codes) example yielding an overall* $(n,k) = (63,24)$ *binary code with* $d_{min} = 16$ *– the BCH code at that length with* $d_{min} = 16$ *has only* $k = 23$. *Algebraic coding theorists might do well to try Ziablov's technique to search for interesting codes already at moderate lengths, but it is asymptotic results which are Ziablov's concern in this paper. He shows that the* d_{min}/n *lower bound demonstrable for his m-th order concatenated scheme is substantially better than that for Forney's original scheme (cf. Ziablov's Figure 7); and he shows that the product of decoder complexity and computation time is proportional to* $n^2 \log n$ *for a decoder correcting all the errors guaranteed correctable by this bound. Besides this new work on generalized concatenated codes, Ziablov's paper includes some additional results on the complexity of error-correcting and erasure-correcting algorithms for low-density parity check codes.*

The sixth contribution, "The Complexity of Decoders", by Professor J.E. Savage of Brown University is a natural culmination of the previous five papers on channel coding. In this very thorough paper, Savage introduces the reader to the general theory of complexity that he has himself pioneered. The basic concept in Savage's theory is that of "computational work", which roughly speaking is the product of the number of logical elements in a device with the number of cycles that the device executes. The basic inequality is that if the device implements some boolean function, then the computational work done must be at least the combinational complexity of this function, i.e., the fewest number of logical elements in a direct combinational circuit which implements the function. By bounding the combinational complexity of decoding functions that achieve some specified small error probability, Savage obtains lower bounds on decoder complexity. Upper bounds can be established by considering the complexity of known decoding procedures. The skepticism in the value of his lower bounds that might be generated by the apparent "looseness" of the bounding argument largely evaporates when Savage demonstrates that his bound is tight (within a constant factor) at least for low rate coding. Perhaps the most intriguing result in Savage's paper is his demonstration that, for a given very small probability of decoding failure, concatenated coding of block codes is inherently much less complex than either Viterbi or sequential decoding of convolutional codes. But Savage adds a caveat here – "very small" may mean very small indeed before the asymptotic superiority wins out. This writer would add a second caveat – decoding failure for sequential decoding means "deletion of data", not "incorrect data" as it does in block coding. Nonetheless, Savage's results should heighten interest in concatenated coding for possible future applications such as inter-computer communications where extremely high reliability may be mandatory.

The last three papers in this collection differ in nature from the previous six which confine their attention to channel coding with one sender and one receiver. The first of these three, "Complexity and Information Theory", by Professor A. de Luca of the University of Salerno, while it shares some concepts with that of Savage described above, is not at all about coding complexity in the usual sense. Rather, it considers the use of "complexity" as a replacement for "probability" in the axiomatization of information theory. Although probabilistic concepts are well accepted as valid in the communications context, there are many other areas where information-theoretic notions are appropriate but the usage of probabilities is either suspect or clearly invalid. Kolmogorov was the first to suggest the use of complexity theory as a substitute for probability, the amount of "information" in a binary sequence being defined as the size of the smallest program that generates the sequence. In this paper, de Luca gives a careful and scholarly survey of Kolmogorov's notions and later related work, as well as the necessary background from formal logic required to make these notions precise. This is not a paper for the faint-hearted reader, it is both long and deep – its mastery will require considerable computational work in the sense of Savage – but it is remarkably well-written and thorough. It concludes with de Luca's own contribution of a "logical" approach to information theory in which the information which x gives about y is defined as the difference in the number of steps required to prove y when x is added as an axiom to a specified logical system. If that sounds esoteric, it was meant to. But de Luca expounded these ideas in Udine with such clarity and enthusiasm that this writer was caught up in the intellectual excitement and dragged to a reasonable understanding of work far from his own field. So, it is hoped, will be the reader's fate.

The next contribution, "Introduction to Multiple Users Communication

Systems", by Professor P.P. Bergmans of the University of Ghent deals with the generalization to the situation where many senders share a channel which jointly reaches to their respective users, perhaps the area of most intense theoretical interest today in information theory. After a lucid general introduction to such "multi-way" communication situations, Bergmans restricts his attention to the "degraded broadcast channel" originally introduced by Cover. This channel can be viewed as a succession of noisy channels with a different user at the output of each channel, the most disadvantaged user being that at the very end. Each sender can take his turn at sending his message to his respective user, but such "time-sharing" is inherently inefficient as asymptotic arguments show. Bergmans then introduces the concept of "superposition coding" in which the senders collaborate by cleverly "adding" their codeword to that from the other users. A simple but illuminating example is given which shows the superiority over time-sharing in a very "non-asymptotic" case. Bergmans then goes on to show how Slepian's permutation codes for the Gaussian channel can be used as the basis for simple, yet effective, superposition coding schemes for a Gaussian degraded broadcast channel.

The last contribution, "Algorithms for Source Coding", jointly by Dr. F. Jelinek (who was unable to attend the summer school) of the IBM Watson Research Center in Yorktown Heights, New York, and Professor Longo, is the only one which deals with source coding, or "data compression" as it is often called. After reviewing the usual block-to-variable length codes and the optimum Huffman codes, the authors go on to consider the much less well known variable length-to-block codes. Run length codes are a prototype of these latter codes which offer many practical advantages, such as ease of synchronization, over the former codes. One of the real services of this paper is its presentation of the optimum variable-to-block codes for memoryless sources that were found by A. Tunstall in his unpublished 1968 Ph.D. thesis at the Georgia Institute of Technology. Both the Huffman and Tunstall algorithms for optimum coding require a computational effort that grows exponentially with sequence length. The authors then go on to consider two source coding algorithms that are efficient in the sense of asymptotically optimum compression, but whose complexity grows only linearly with sequence length. The first such algorithm described is one due to Elias in unpublished work which takes advantage of the interpretation of a binary sequence as a fraction (with the "binary point" to the left of the sequence) in the interval [0,1). The second such algorithm is that proposed by Schalkwijk for implementing the Lynch-Davisson "universal" coding scheme in which a binary n-tuple is transmitted by sending first its weight (in radix-two form) followed by its index (also in radix-two form) in the lexicographical ordering of all the binary n-tuples of that weight. Both the novice in source coding and the expert will find this paper of interest.

James L. Massey
Freimann Professor of Electrical Engineering
University of Notre Dame
Notre Dame, Indiana 46556 USA

October 11, 1976

CONTENTS

ERROR BOUNDS FOR TREE CODES, TRELLIS CODES, AND CONVOLUTIONAL CODES WITH ENCODING AND DECODING PROCEDURES *

James L. Massey

Freimann Professor of Electrical Engineering

University of Notre Dame

Notre Dame, Indiana U.S.A.

* The original work reported herein was supported in part by the U.S.A. National Aeronautics and Space Administration under NASA Grant NGL 15-004-026 at the University of Notre Dame in liaison with the Goddard Space Flight Center.

1. Introduction

One of the anomalies of coding theory has been that while block parity-check codes form the subject for the overwhelming majority of theoretical studies, convolutional codes have been used in the majority of practical applications of "error-correcting codes." There are many reasons for this, not the least being the elegant algebraic characterizations that have been formulated for block codes. But while we may for aesthetic reasons prefer to speculate about linear block codes rather than convolutional codes, it seems to me that an information-theorist can no longer be inculpably ignorant of non-block codes. It is the purpose of these lectures to provide a reasonably complete and self-contained treatment of non-block codes for a reader having some general familiarity with block codes.

In order to make our presentation cohesive, we have borrowed heavily from other authors (hopefully with appropriate credit) and have included some of our own work that was previously unpublished. In the latter category are the formulations of the general ensembles of tree codes and trellis codes, given in Sections 3 and 4 respectively, together with the random coding bound for the ensemble of tree codes in Section 3. The random coding bound of Section 4 for trellis codes was, of course, given earlier by Viterbi for convolutional codes (which are the linear special case of trellis codes as we define the latter) for the case $T = M$ but we have somewhat generalized this bound.

Also in the category of our previously unpublished work is the low "complexity" ensemble of convolutional codes which is shown in Section 7 to meet the same upper bound on decoding error probability as for the entire ensemble of convolutional codes. There are a number of other places in our treatment where the reader familiar with past work in convolutional coding will recognize some novel treatments of the subject matter and we hope that these have contributed to the cohesiveness of these lectures.

2. The Two-Codeword-Exponent for Discrete Memoryless Channels

The discrete memoryless channel (or DMC) is a channel with a finite input alphabet and a finite output alphabet that acts independently on each input digit and whose statistics do not vary with time. Letting $A = \{a_1, a_2, \ldots, a_q\}$ and

$B = \{b_1, b_2, \ldots b_{q'}\}$ be the input and output alphabets respectively, we can specify a DMC by stating the conditional probability $P(b_j \mid a_i)$ of receiving b_j when a_i is transmitted for $j = 1,2,\ldots q'$ and $i = 1,2,\ldots q$. A DMC is often shown by a directed graph in which the edge from node a_i to node b_j is labelled with $P(b_j \mid a_i)$ as shown in Fig. 2.1 for the binary symmetric channel (BSC) for which $A = B = \{0,1\}$. The quantity ϵ is called the "crossover probability" of the BSC.

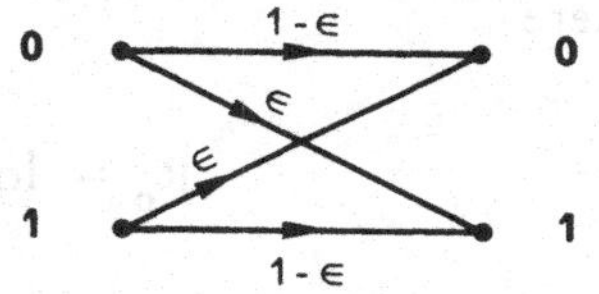

Fig. 2.1 The Binary Symmetric Channel (BSC)

An (N,R) **block code** for a DMC is an ordered set of $m = 2^{NR}$ N-tuples over the input alphabet A of the DMC, $(\mathbf{x}_1, \mathbf{x}_2, \ldots \mathbf{x}_m)$. We shall write $\mathbf{x}_j = [x_{j1}, x_{j2}, \ldots x_{jN}]$. The parameters N and R are the code **length** and **rate** respectively. We say that R is the rate in "bits per channel use" because, when the m codewords are equally likely, we send $\log m = NR$ bits of information in N uses of the DMC. (Here and hereafter, the base 2 is used for all logarithms.)

A maximum likelihood decoder (MLD), when $\mathbf{y} = [y_1, y_2, \ldots y_N]$ is received over the DMC, chooses as its estimate of the index of the transmitted codeword (one of) the index (es) j which maximizes

$$P(\mathbf{y} \mid \mathbf{x}_j) = \prod_{n=1}^{N} P(y_n \mid x_{jn}).$$

Consider now the simplest interesting codes, i.e. codes with only $m = 2$ codewords or $(N, R = 1/N)$ codes. Let Q be a probability distribution over the channel input alphabet A. To each code $(\mathbf{x}_1, \mathbf{x}_2)$, we assign the probability

$$P(\mathbf{x}_1, \mathbf{x}_2) = \prod_{n=1}^{N} Q(x_{1n}) Q(x_{2n})$$

which is the probability of selecting that code by choosing each code digit independently according to Q. Let $P_e(\mathbf{x}_1, \mathbf{x}_2)$ be the decoding error probability with a MLD for the code $(\mathbf{x}_1, \mathbf{x}_2)$ and a given probability assignment on the codewords. Then

$$\overline{P}_e = \sum_{\mathbf{x}_1 \epsilon A^N} \sum_{\mathbf{x}_2 \epsilon A^N} P_e(\mathbf{x}_1, \mathbf{x}_2)\, P(\mathbf{x}_1, \mathbf{x}_2)$$

is the average error probability with MLD for the ensemble of codes with $m = 2$

codewords of length N.

It is not difficult to show (Cf. Gallager [1]) that, regardless of the probability assignment on the two codewords,

$$\bar{P}_e \leqq 2^{-NR_o} \tag{2.1}$$

where

$$R_o = -\log\{\min_Q \sum_{y \in B} [\sum_{x \in A} Q(x)\sqrt{P(y|x)}]^2\} \tag{2.2}$$

when Q is taken as the minimizing distribution in (2.2). Since there must be at least one code whose $P_e(x_1, x_2)$ is no worse than average, it follows from (2.1) that we can find, for increasing N, a sequence of codes whose decoding error probability with MLD remains below the decreasing exponential 2^{-NR_o}. The exponent R_o, as defined by (2.2), is called the **two-codeword-exponent** for the DMC.

For DMC's with a binary input alphabet A = { 0,1 }, the calculation of R_o is simplified by the fact that Q(0) = Q(1) = 1/2 is always the minimizing distribution in (2.2). For instance, for the BSC one easily finds that

$$R_o \text{ (for BSC)} = 1 - \log[1 + 2\sqrt{\epsilon(1-\epsilon)}]. \tag{2.3}$$

In particular, we find from (2.3) that R_o is .45, .50 and .55 when ϵ is .057, .045 and .033 respectively.

In the following sections, we shall use R_o to find bounds on $\bar{P}_e$ with MLD for much more interesting code ensembles than that of codes with two codewords.

3. Tree Codes

We define an (L,T) m-ary tree to be a rooted tree such that (1) m branches diverge from each node at depth less than L from the root node and (2) one branch diverges from each node at depth less than L +T but at least L from the root. Here, L, T and m are integers such that $L \geqslant 1$, $T \geqslant 0$ and $m \geqslant 2$. We call L and T the **dividing length** and **tail length** respectively of the tree. In Fig. 3.1, we show the (L = 3, T = 2) binary tree. We note in general that the (L,T) m-ary tree has m^L terminal nodes and these terminal nodes are at depth L + T from the root.

We now define an (N,R,L,T) **tree code** for a DMC as the assignment of N channel input letters to each branch of an (L,T) (2^{NR})-ary tree. Note that $m = 2^{NR}$ is the number of branches diverging from each node in the "dividing part" of the tree. We call R the **rate** of the tree code and we define the **constraint length** of the tree code as

$$N_t = (T + 1)N \qquad (3.1)$$

which is just the number of digits on the "long paths" from the last dividing nodes to the terminal nodes.

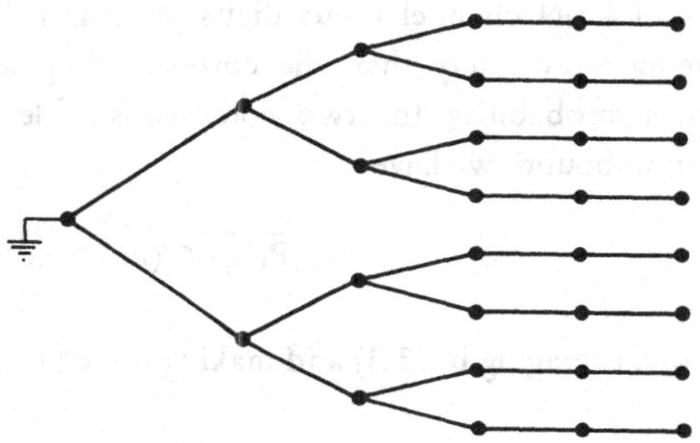

Fig. 3.1 The (L= 3, T = 2) Binary Tree

An (N,R,L,T) tree code is a special type of block code for the DMC. As a block code, it has $m_B = m^L$ codewords of length $N_B = (L+T)N$, namely the m_B N_B-tuples which label the paths from the root to each terminal node. As a block code, its rate R_B is given by

$$R_B = \log(m_B)/N_B = \frac{L}{L+T} R \qquad (3.2)$$

which, for the usual case in practice where $L \gg T$, is approximately the same as its rate R as a tree code.

We now derive an upper bound on the average decoding error probability, $\overline{P}_e$, for MLD of the ensemble of (N,R,L,T) tree codes for a given DMC when each code is assigned the probability equal to selecting that code when each channel input digit placed in the (L,T) m-ary tree is selected independently according to the minimizing distribution Q in (2.2).

Let E_i, $1 \leqslant i \leqslant L$, be the event that some path to a terminal node diverging from the correct path at the node at depth L-i from the root is at least as likely to produce the received sequence as the corresponding segment of the correct path. Then

$$P_e \leqslant P(E_1 \cup E_2 \cup \ldots \cup E_L)$$

where the inequality is needed because we might decode correctly when the correct path is "tied" with another path as most probable. By the union bound,

$$P_e \leqslant P(E_1) + P(E_2) + \ldots P(E_L). \qquad (3.3)$$

There are $(m\text{-}1)m^{i-1}$ paths to terminal nodes which diverge from the correct path from the node at depth L-i from the root and each of these paths is T + i branches, or $(T+i)N$ channel input digits in length. The probability of one of these paths being more likely than the corresponding segment of the correct path is just the error probability for two codewords of length $(T+i)N$. Thus, using (2.1) and a union bound, we have

$$\overline{P(E_i)} \leqslant (m-1)m^{i-1}\, 2^{-(T+i)NR_o}. \tag{3.4}$$

Now, averaging in (3.3) and making use of (3.4), we obtain

$$\overline{P}_e \leqslant (m-1)2^{-(T+1)NR_o} \sum_{i=1}^{L} m^{i-1}\, 2^{-(i-1)NR_o} .$$

We recognize the summation above as a geometric series which, when $m < 2^{NR_o}$, converges upon letting $L \to \infty$ to give

$$\overline{P}_e \leqslant \frac{m-1}{1-m\,2^{-NR_o}}\, 2^{-(T+1)NR_o},\quad m < 2^{NR_o}.$$

Finally, overbounding $m-1$ by m in the numerator and using $m = 2^{NR}$ and $N_t = (T+1)N$, we obtain the result that the average decoding error probability for the ensemble of (N,R,L,T) tree codes with MLD on a DMC satisfies

$$\overline{P}_e < c_t\, 2^{-N_t R_o} \text{ when } R < R_o \tag{3.5}$$

where

$$c_t = \frac{1}{2^{-NR} - 2^{-NR_o}} \tag{3.6}$$

is a relatively unimportant constant for any fixed rate R, $R < R_o$.

The bound (3.5) is quite remarkable in several respects. First, it shows that we can operate with tree codes at rates very close to R_o and still obtain R_o as the exponent of error probability just as if there were only two codewords. There is no need to use small information rates R to get a rapid decrease of error probability with constraint length. Second, the bound is completely independent of the dividing length L of the tree and depends only on the tail length T through $N_t = (T+1)N$.

Thus we can take $L \gg T$ to ensure, by (3.2), that the "nominal" rate R of the tree code is very close to the true information rate in bits per channel use. Third and finally, our derivation of the bound was extremely simple.

It is possible to extend the arguments used here to obtain a bound on $\overline{P}_e$ for tree codes in the region $R_0 \leqslant R < C$ where C is channel capacity for the DMC. For details of this argument, the reader is referred to Johannesson [2]. The case of greatest practical interest, however, is for R slightly less than R_o for which we have no need of the more general bound.

We have already noted that an (N,R,L,T) tree code for a DMC is a special type of ($N_B = (L+T)N$, $R_B = RL/(L+T)$) block code. It is also interesting to note that we can consider an (N,R) block code for a DMC to be the special case of an (N,R, L = 1, T = 0) tree code. Whether block codes or tree codes are more general is really just a matter of taste. To our taste, we prefer to think of the block code as the special case of a tree code.

4. Trellis Codes

By introducing "memory" into tree codes, we shall be led naturally to a very interesting class of codes which we call "trellis" codes for reasons that will become obvious.

Consider first how one might encode an (N,R,L,T) tree code. Since $m = 2^{NR}$ branches diverge from each node at depth less than L from the root, we can use a sequence $i_0, i_1, \ldots, i_{L-1}$ of L m-ary information digits (whose alphabet for the moment we shall consider to be $\{0,1,\ldots,m-1\}$) to select which branch to follow as we move away from the root. In Fig. 4.1, we show our rule for choosing the appropriate branch.

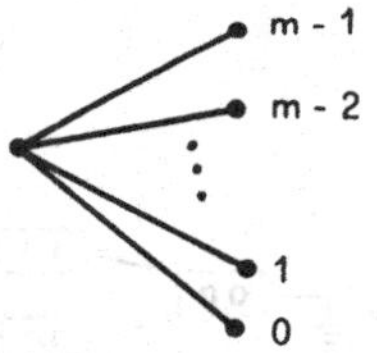

Fig. 4.1 The Convention by Which an Information Digit Selects a Branch to Follow

Thus, for instance, the information sequence $i_0, i_1, i_2 = 0,0,0$ would cause transmission of the channel input letters on the lowermost path of the tree as in Fig. 3.1 for an (N,R = 1/N, L = 3, T = 2) tree code.

Suppose now for some integer M, $T \leqslant M \leqslant L + T$, we label each node in an (L,T) m-ary tree with the previous M information digits leading to that node. By way of convention, we take information digits previous to the origin as 0's and, similarly, we take 0 as the "information digit" corresponding to the unique path from each node at depth less than L + M but L or more from the root. We then obtain what we shall call an (L,T,M) **node-labelled** m-ary **tree.** In Figure 4.2, we show the (L= 3, T= 2, M= 2) node-labelled binary tree.

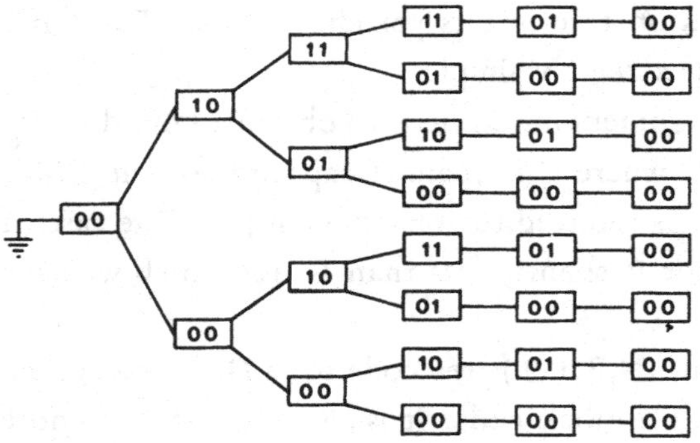

Fig. 4.2 The (L = 3, T = 2, M = 2) Node-Labelled Binary Tree

Suppose now that the tree code has "memory" M in the sense that for any two nodes at the same depth with the same label (i.e. having the same M previous information digits) the same remaining encoded sequence will result when the same sequence of information digits is applied beginning at either node. We then need to retain only one node with each label at each depth and all branches previously entering nodes with this label can be made to enter this single node with the label. We define an (L,T,M) m-ary **trellis** to be the result of such coalescing of similarly labelled nodes in an (L,T,M) node-labelled m-ary tree. In Figure 4.3, we show (a) the (L= 3, T= 2, M = 2) binary trellis and (b) the (L = 3, T= 1, M= 2) binary trellis. In general, the (L,T,M) trellis will have m^{M-T} terminal nodes.

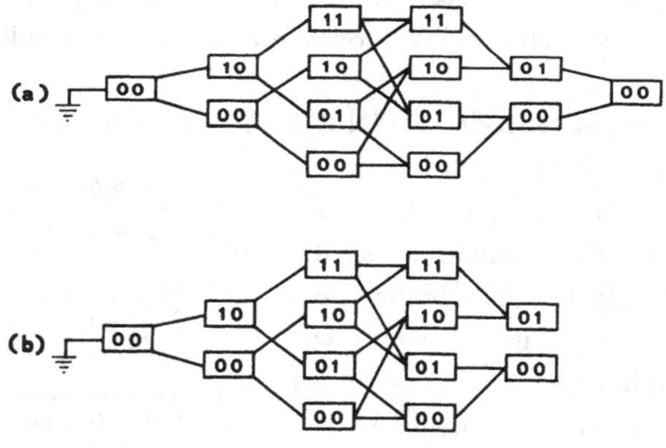

Fig. 4.3 (a) The (L = 3, T = 2, M = 2) Binary Trellis and (b) The (L = 3, T = 1, M = 2) Binary Trellis

We now define an (N,R,L,T,M) **trellis code** for a DMC as the assignment of N channel input letters to each branch of an (L,T,M) (2^{NR})-ary trellis.

As our discussion has made clear, an (N,R,L,T,M) trellis code for a DMC is a special type of (N,R,L,T) tree code so we still continue to call

$$N_t = (T + 1)N$$

the **constraint length** of the trellis code and, moreover, (3.2) still relates the "nominal" rate R of the trellis code to its true rate R_B in bits transmitted per use of the channel.

We now proceed to derive an upper bound on the average error probability for MLD on a DMC over the ensemble of trellis codes when we assign to each code the probability of its selection when each channel input digit placed on the (L,T,M) m-ary trellis is selected independently according to the minimizing distribution Q in (2.2). We consider first the special case M = T from which we shall later infer the general result with very little additional effort. Note that, when M = T, there is just one terminal node in the trellis, namely the node labelled 00...0. Thus, every path through the trellis starts at the root and ends at this node. (See Fig. 4.3(a).) Again we let E_i be the event that some path to the terminal node which diverges at depth L - i from the correct path is at least as likely to produce the received sequence as the corresponding segment of the correct path. Then the bound (3.3) again applies for MLD, i.e.

$$P_e \leq \sum_{i=1}^{L} P(E_i) \tag{4.1}$$

Next, we further decompose the events E_i. A path diverging at depth L -i can first remerge with the correct path at depth L+M+1 - i since its information digits must agree with those on the correct path for M successive branches for remergence to occur. Let A_{ij}, $1 \leq j \leq i$, be the event that some path diverging at depth L - i and first remerging at depth L + M + j - i is at least as likely to produce the received sequence as the corresponding M +j branch segment of the correct path. There are at most m^{j-1} (m - 1) such paths since M of their information digits agree with those on the correct path and their first information digit disagrees with that on the correct path. The probability of one of these paths being more likely than the correct path is just the error probability for two codewords of length (M +j)N. Thus, using (2.1) and a union bound, we have

$$\overline{P(A_{ij})} \leq (m-1)m^{j-1}\, 2^{-(M+j)NR_0}. \tag{4.2}$$

Moreover, we note that

$$E_i = A_{i1} \cup A_{i2} \cup \ldots \cup A_{ii}$$

so that a union bound gives

$$P(E_i) \leq \sum_{j=1}^{i} P(A_{ij}).$$

Taking averages and using (4.2), we obtain

$$\overline{P(E_i)} \leqslant (m-1)\, 2^{-(M+1)NR_o} \sum_{j=1}^{i} m^{j-1}\, 2^{-(j-1)R_o}.$$

Overbounding by letting $i \to \infty$, we recognize the above expression to be precisely the same as that leading to (3.5) with T replaced by M. Thus, we have

$$\overline{P(E_i)} < c_t\, 2^{-(M+1)NR_o}, \quad R < R_o \tag{4.3}$$

where c_t is defined in (3.6). Finally, taking averages in (4.1) and using (4.3), we obtain

$$\overline{P}_e \leqslant c_t\, L\, 2^{-(M+1)NR_o}, \quad R < R_o \tag{4.4}$$

as our basic bound for MLD of the ensemble of (N,R,L,T,M = T) trellis codes. Unlike the case for tree codes, this bound depends linearly on the dividing length L of the trellis and one can easily convince himself that this dependence on L is real and not merely the result of a weakness in our bounding arguments.

We now extend the above arguments to apply to the ensemble of (N,R,L,T,M) trellis codes without the restriction that M = T. We have already pointed out that the trellis will have m^{M-T} terminal nodes in the general case and that $T \leqslant M \leqslant L + T$.

Suppose that some path to a terminal node diverging from the correct path at depth L - i is more likely than the corresponding segment of the correct path. If this path remerges with the correct path, then such an error event is included in the event E_i for a trellis code with T = M. If this path does not remerge with the correct path, then such an error event is included in the event E_i for a tree code with tail length T. Thus $\overline{P}_e$ for MLD of the ensemble of (N,R,L,T,M) trellis codes is overbounded by the sum of $\overline{P}_e$ for the ensemble of (N,R,L,T = M,M) trellis codes and $\overline{P}_e$ for the ensemble of (N,R,L,T) tree codes, that is

$$\overline{P}_e < c_t L\, 2^{-(M+1)NR_o} + c_t\, 2^{-(T+1)NR_o}$$

which we may rewrite as

$$\overline{P}_e < c_t\,[1 + L2^{-(M-T)NR_o}]\,2^{-N_t R_o} \tag{4.5}$$

when $R < R_o$. Inequality (4.5) is our basic bound for MLD of the ensemble of (N,R,L,T,M) trellis codes.

Inequality (4.5) provides the clue for avoiding the undesirable dependence on L as in (4.4) for trellis codes with M = T. For suppose we choose

$$L2^{-(M-T)NR_o} \leqslant 1$$

or, equivalently

$$M - T \geqslant \frac{\log L}{NR_o}\quad, \tag{4.6}$$

then (4.5) becomes

$$\overline{P}_e \leqslant 2\,c_t\,2^{-N_t R_o}. \tag{4.7}$$

We conclude that making M only slightly greater than T according to (4.6) is sufficient to reduce P_e for a trellis code very close to that for a full tree code.

We have already noted that an (N,R,L,T,M) trellis code for a DMC is a special type of (N,R,L,T) tree code. It is interesting to note that we can alternatively think of the (N,R,L,T) tree code as the special case of an (N,R,L,T,M = L +T) trellis code since, when M = L +T, the trellis degenerates to a tree with $m^{M-T} = m^L$ terminal nodes.

The term "trellis" was first coined by Forney to describe graphs of the type in Fig. 4.3(a) which he employed to study convolutional codes. As we shall see, convolutional codes are a special subclass of the class of trellis codes which we defined here. The bounds (4.5) and (4.7) have been extended to rates R in the region $R_o < R < C$ by Johannesson [2] who also reported simulation results showing that the necessary excess of memory over tail length to make P_e independent of L is very closely given by the righthand side of (4.6).

5. Tree and Trellis Encoders–Convolutional Encoders

The observation above that the ensemble of (N,R,L,T) tree codes for a DMC coincides with the ensemble of (N,R,L,T,M = L+ T) trellis codes permits us to consider encoders for trellis codes alone without loss of generality.

Proceeding as in the previous section, we suppose that $i_0, i_1,...i_{L-1}$ are the m-ary "information digits" to be encoded, each of which controls the branch to be followed by the encoder at the nodes within the dividing length of the trellis. We shall associate a "time unit" with each encoded branch, and hence we may write the encoded sequence as $\mathbf{t}_0, \mathbf{t}_1, ..., \mathbf{t}_{L-1}, \mathbf{t}_L, ..., \mathbf{t}_{L+T-1}$ where $\mathbf{t}_u$, the encoded branch at time u, is an N-tuple of q-ary channel input digits. Similarly, $\mathbf{i}_u$ is the "information digit" at time u for $0 \leq u < L$. We suppose that the sequence $i_0, i_1,...i_{L-1}$ is augmented by a "tail" of T 0's where 0 is just some designated digit in the m-ary alphabet. For convenience, we shall write the entire sequence as $i_0, i_1, ..., i_{L+T-1}$ where $\mathbf{i}_u = \mathbf{0}$ for $L \leq u < L+T$. We shall also use the convention that $\mathbf{i}_u = \mathbf{0}$ for $u < 0$.

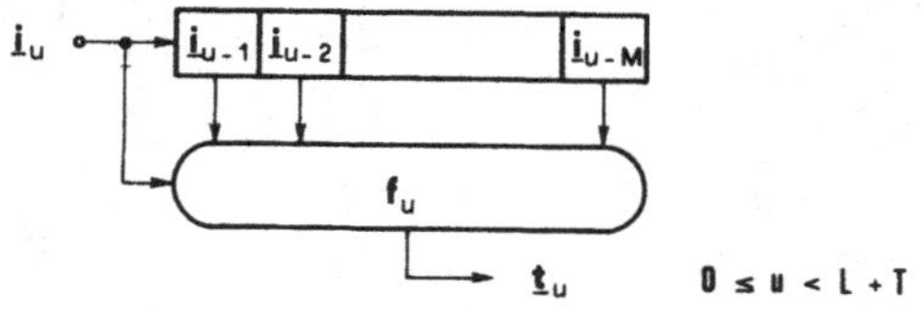

Fig. 5.1 A General (n,R,L,T,M) Trellis Encoder

With the above conventions, a general (N,R,L,T,M) trellis encoder may be represented as shown in Figure 5.1. Each square box denotes a delay of one time unit and the oval box denotes a general function from the m-ary information sequence to the set of channel input N-tuples. It should be noted that this function f_u is in general time-dependent. When $f_u = f$ for all u, $0 \leq u < L+T$, we call the encoder **fixed** (or "time-invariant").

A convolutional encoder is just a "linear" trellis encoder. To make this a meaningful statement, we must of course have an appropriate algebraic structure for the "information digits" and the encoded digits. To accomplish this, we suppose that the channel input alphabet is GF(q), the finite field of q elements (which requires that q be some power of a prime.) We then require $m = q^K$ so that each i_u may be taken as a K-tuple over GF(q). Notice that

$$R = \frac{\log m}{N} = \frac{K}{N} \log q \tag{5.1}$$

which is the rate in bits per channel use. It is common in coding theory to speak of K/N as the code rate but, more properly, this is the "dimensionless rate" of the code. For all our examples, we shall use q = 2 (the rules of GF(q) then being modulo-two arithmetic) for which case R = K/N.

For brevity, we hereafter write simply F rather than GF(q). A general (N,R,L,T,M) **convolutional encoder** is then representable as in Fig. 5.1 where we require $\mathbf{i}_u \in F^K$, $\mathbf{t}_u \in F^N$, and where we require the functions f_u, $0 \leq u < L+T$, to be linear functions from $F^{(M+1)K}$ to F^N. We can then represent such functions f_u as

$$t_u = i_u G_0(u) + i_{u-1} G_1(u) + ... + i_{u-M} G_M(u) \qquad (5.2)$$

where each $G_i(u)$ is a K x N matrix over F. For a fixed convolutional encoder (FCE), these matrices do not depend on u so that (5.2) becomes

$$t_u = i_u G_0 + i_{u-1} G_1 + ... + i_{u-M} G_M \qquad (5.3)$$

where each G_j is a K x N matrix over F. In Figs. 5.2 (a) and (b), we show a general

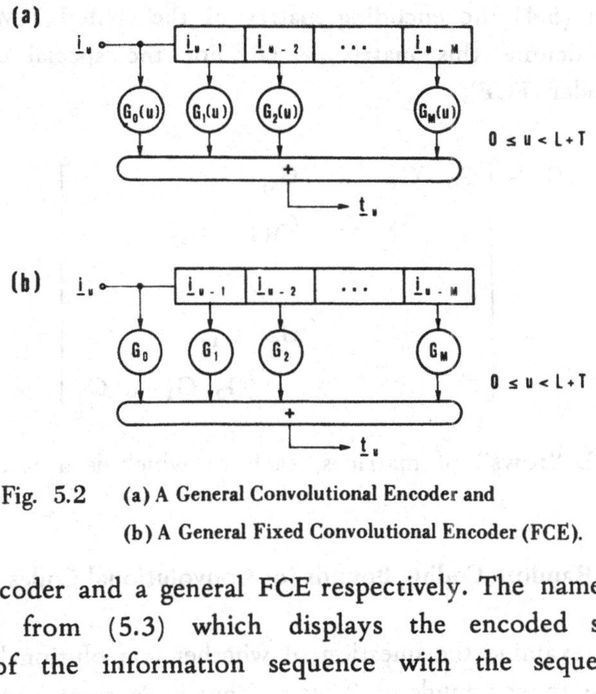

Fig. 5.2 **(a) A General Convolutional Encoder and**
(b) A General Fixed Convolutional Encoder (FCE).

convolutional encoder and a general FCE respectively. The name "convolutional encoder" stems from (5.3) which displays the encoded sequence as the "convolution" of the information sequence with the sequence of matrices $G_0, G_1, ... G_M$.

We shall find it convenient to write $i_{[u,v]}$ and $i_{[u,v)}$ for the sequences $i_u, i_{u+1}, ... i_v$ and $i_u, i_{u+1}, ... i_{v-1}$ respectively and similarly for $t_{[u,v]}$ and $t_{[u,v)}$. In this notation, the operation of a general convolutional encoder may, according to (5.2), be written as

$$\mathbf{t}_{[0,L+T)} = \mathbf{i}_{[0,L)} \begin{bmatrix} G_0(0) & G_1(1) & \cdots & G_M(M) & & & \\ & G_0(1) & \cdots & G_{M-1}(M) & G_M(M+1) & & \\ & & \ddots & \vdots & \vdots & & \\ & & & G_0(M) & G_1(M+1) & \cdots & \\ & & & & \ddots & & \\ & & & & & G_0(L-1) \ldots G_T(L-1) & \end{bmatrix} \tag{5.4}$$

where the blank portions of this matrix are assumed to be filled with zeroes. We call the matrix in (5.4) the **encoding matrix** of the (N,R,L,T,M) convolutional encoder and we denote this matrix as $\mathbf{G}$. For the special case of a fixed convolutional encoder (FCE),

$$\mathbf{G} = \begin{bmatrix} G_0 & G_1 & \cdots & G_M & & & \\ & G_0 & \cdots & G_{M-1} & G_M & & \\ & & \ddots & & & & \\ & & & G_0 & G_1 \cdots & & \\ & & & & G_0 & G_1 & \cdots & G_T \end{bmatrix} \tag{5.5}$$

where there are L "rows" of matrices, each of which is a K x N matrix over $F = GF(q)$.

6. Random Coding Bounds for Convolutional Codes

We now examine the question of whether convolutional tree and trellis codes can meet the upper bounds on P_e established in Sections 3 and 4 respectively for the general classes of tree and trellis codes. We note that the symmetry of linear codes requires at once the restriction to channels which are **symmetric** in the sense that $Q(x) = 1/q$ for all x is the minimizing distribution in (2.2). Fortunately, this includes all binary input channels which are the cases of greatest practical interest. (One can handle channels for which $Q(x) = 1/q$ is not the appropriate distribution by considering convolutional codes over alphabets larger than q letters and mapping the requisite number of letters in the larger alphabet to each q-ary letter to approximate the desired probability. In fact, q need not then be a power of a prime. We omit the awkward details of such a generalization.)

There is an artifice required with convolutional codes, as indeed with linear codes of every type, to show that some ensemble can achieve some random coding bound which arises from the fact that the all-zero information sequence is always encoded as the all-zero codeword. Thus, there is no way to pick an ensemble of convolutional codes so that, over the ensemble, the codeword assigned to the all-zero information sequence would appear to be "randomly selected" according to the given Q(x). To obviate this difficulty, one considers adding a random sequence to each codeword before transmission over the channel, i.e. the information sequence $\mathbf{i}_{[0,L+T)}$ is transmitted as the sequence $\tau_{[0,L+T)} = \mathbf{t}_{[0,L+T)} \mathbf{r}_{[0,L+T)}$ where the digits in $\mathbf{r}_{[0,L+T)}$ are independently selected according to the distribution $Q(x) = 1/q$ for all x. We can write this as

$$\tau_{[0,L+T)} = \mathbf{i}_{[0,L)}\,\mathbf{G} + \mathbf{r}_{[0,L+T)} \tag{6.1}$$

which makes it clear that the same random sequence is added to each codeword for a specific code **G**. Regardless of the ensemble of convolutional codes, i.e. regardless of the probability distribution over the matrices **G**, it follows from (6.1) that any given $\mathbf{i}_{[0,L)}$ is equally likely to be encoded into any given $\tau_{[0,L+T)}$. Thus, the artifice of the "added-random-sequence" results, over the ensemble of codes, in making the encoded sequence for any given path in the tree or trellis appear to have its digits selected independently according to the distribution $Q(x) = 1/q$ for all x.

We next observe that in deriving bounds on $\overline{P}_e$ in Sections 3 and 4, we used only the independence between two "unmerged" paths in the tree or trellis. In other words, the only property our code ensembles need to ensure that these bounds on $\overline{P}_e$ hold is that the digits on any two paths diverging from some node are mutually independent over the span to the next node (if any) where the two paths join again. We call an ensemble of codes **pairwise independent** if it enjoys this property. It follows that, when we employ the added-random-sequence artifice, an ensemble of tree or trellis codes over $F = GF(q)$ will be pairwise independent if and only if the **difference** of the encoded sequences on any two paths diverging from some node is, over the ensemble of codes, a sequence in which each digit is indepently selected according to $Q(x) = 1/q$ up to the node (if any) where the paths first merge again. Now suppose that $\mathbf{i}_{[0,v)}$ and $\mathbf{i}'_{[0,v)}$ are the information sequences describing such a pair of paths which diverge at depth u and remain unmerged at least to depth v. Then we see from (6.1) that the difference of their encoded sequences from node u to node v is the same as the encoded sequence $\mathbf{t}''_{[u,v)}$ resulting from encoding the difference sequence $\mathbf{i}''_{[0,v)} = \mathbf{i}_{[0,v)} - \mathbf{i}'_{[0,v)}$ since the random sequence cancels in the subtraction. Moreover, we note that $\mathbf{i}''_{[0,u)} = 0$ and $\mathbf{i}''_u \neq \mathbf{0}$ since the two paths diverge at node u. Further $\mathbf{i}''_{[u,v)} = [\mathbf{i}''_u, \mathbf{i}''_{u+1}, \ldots, \mathbf{i}''_{v-1}]$ can contain no run of M consecutive $\mathbf{0}$ information branches except possibly the last M of these information branches or else the paths would merge before node v. We summarize these observations as:

Lemma 6.1: An ensemble of (N,R,L,T,M) convolutional codes with an added-random-sequence is pairwise independent if and only if for every choice of u and v such that $0 \leqslant u < L$ and $u < v \leqslant L+T$ and for every choice of $\mathbf{i}_{[0,v)}$ such that $\mathbf{i}_j = \mathbf{0}$ for $j < u$, $\mathbf{i}_u \neq \mathbf{0}$, and $\mathbf{i}_{[u,v-1)}$ contains no internal run of M consecutive information K-tuples, the probability assignment on the codes in the ensemble is such that the resultant $\mathbf{t}_{[u,v)}$ is a sequence whose component digits are independently distributed and each has the distribution $Q(x) = 1/q$ all x.

As a first application of this lemma, we prove:

Theorem 6.1: The ensemble of (N,R,L,T,M = L + T) fixed convolutional codes with an added-random-sequence such that each digit in each of the K x N matrices G_i, $0 \leqslant i < L+T$, is independently selected according to $Q(x) = 1/q$ is pairwise independent. Consequently, the bound (3.5) for random tree codes holds also for this ensemble when $Q(x) = 1/q$ is the minimizing distribution in (2.2).

To prove this theorem, we first use (5.4) and (5.5) to write

$$\mathbf{t}_{[u,L+T)} = [\mathbf{i}_u, \mathbf{i}_{u+1}, \ldots \mathbf{i}_{L-1}] \begin{bmatrix} G_0 & G_1 & G_2 & \cdots & G_{L+T-u-1} \\ & G_0 & G_1 & \cdots & G_{L+T-u-2} \\ & & \ddots & & \vdots \\ & & & G_0 & \cdots\ G_T \end{bmatrix} \tag{6.2}$$

when $\mathbf{i}_j = \mathbf{0}$ for $j < u$. By Lemma 6.1, it suffices to show that $\mathbf{t}_{u+j}$ is equally likely to be any of the q^N q-ary N-tuples regardless of the values of $\mathbf{t}_u, \mathbf{t}_{u+1}, \ldots \mathbf{t}_{u+j-1}$ when $\mathbf{i}_u \neq \mathbf{0}$. (Since $M = L+T$ it is impossible to have a run of M consecutive $\mathbf{0}$ branches in $\mathbf{i}_{[u,L+T-1]}$ so this hypothesis of Lemma 6.1 will not explicitly be needed.) We see from (6.2) that $\mathbf{i}_u G_j$ will be a component of the sum for $\mathbf{t}_{u+j}$ and, moreover, G_j has played no part in determining $t_{[u,u+j)}$. Since over the ensemble of codes for any fixed choice of G_0, ... G_{j-1} and hence of $t_{[u,u+j)}$, G_j is equally likely to be any of the q^{KN} K x N q-ary matrices; thus, since $i_u \neq 0$, $i_u G_j$ is equally likely to be any q-ary N-tuple. Thus t_{u+j}, whose value is $i_u G_j$ plus a fixed vector determined by G_0, ... G_{j-1}, is also equally likely to be any q-ary N-tuple and the theorem is proved.

We see from Theorem 6.1 that **fixed** convolutional codes with M = L+ T are as "good" as general tree codes for transmitting information through a "symmetric" DMC, i.e. a DMC for which Q(x) = 1/q is the minimizing distribution in (2.2). One might naturally expect that fixed convolutional codes for any M are as good as general (N,R,L,T,M) trellis codes but, surprisingly, this has not yet been demonstrated. We show now that "time-varying" convolutional codes are as good as general trellis codes and, in the next section, explore the reasons why fixed codes pose problems that are still unsolved.

Theorem 6.2: The ensemble of (N,R,L,T,M) convolutional codes with an added-random-sequence such that each digit in each of the matrices $G_i(u)$ for $0 \leqslant i \leqslant M$ and $0 \leqslant u < L + T$ is independently selected according to Q(x) = 1/q is pairwise independent. Consequently, the bounds (4.3) and (4.5) for general random trellis codes, with parameters (N,R,L,T, M + T) and (N,R,L,T,M) respectively, hold also for this ensemble when Q(x) = 1/q is the minimizing distribution in (2.2).

To prove this theorem, we first write

$$t_{[u,v)} = [0,\ldots 0, i_u, \ldots i_{v-1}] \begin{bmatrix} G_M(u) & & & \\ G_{M-1}(u) & G_M(u+1) & \ddots & \\ \vdots & \vdots & & G_M(v-1) \\ \vdots & \vdots & & \\ G_1(u) & G_2(u+1) & \ddots & \\ G_0(u) & G_1(u+1) & \ddots & G_2(v-1) \\ & G_0(u+1) & \ddots & G_1(v-1) \\ & & \ddots & G_0(v-1) \end{bmatrix} \tag{6.3}$$

where there are M 0's preceeding i_u as applying when $i_j = 0$ for $j < u$. Suppose further that $i_u \neq 0$ and that $i_{[u,v-1)}$ contains no internal span of M consecutive 0 branches. By Lemma 6.1, it suffices to show that t_{u+j} is equally likely to be any q-ary N-tuple regardless of the values of $t_u, t_{u+1}, \ldots t_{u+j-1}$. But the only matrices $G_i(k)$ affecting t_{u+j} are $G_0(u+j), G_1(u+j), \ldots G_M(u+j)$, none of which affect any of the previous transmitted branches. Moreover

$$t_{u+j} = i_{u+j}G_0(u+j) + i_{u+j-1}\, G_1(u+j) + \ldots + i_{u+j-M}\, G_M(u+j)$$

and at least one of the information branches in this expression must be non-zero, say; $i_{u+j-k} \neq 0$. Then regardless of the choice of the other matrices affecting $t_{[u,u+j]}$, the vector $i_{u+j-k}\, G_k(u+j)$ over all choices of $G_k(u+j)$ will take on the value of every q-ary N-tuple the same number of times. Thus, the theorem is proved.

It is interesting to compare the "sizes" of the ensembles in Theorems 6.1 and 6.2 where by "size" we mean the number of digits which must be chosen to specify a code.

To specify a code in the ensemble of fixed (N,R,L,T, M = L + T) convolutional codes of Theorem 6.1, we must select (L + T)KN digits to specify the matrices G_i and another (L + T)N digits to specify the added-random-sequence. Hence, this ensemble has size (L +T) (K + 1)N. Since there are (L + T)N encoded digits, we can say the "per-digit size" or "complexity" of the ensemble is K + 1. The "complexity" in this sense is just the number of code parameters that must be selected for each digit in a codeword.

To specify a code in the ensemble of "time-varying" (N,R,L,T,M) convolutional codes of Theorem 6.2, we must select (M + 1) (L + T)KN digits to specify the matrices $G_i(u)$ and another (L +T)N digits to specify the added-random-sequence. Hence, this ensemble has "size" [(M + 1)K +1] (L + T)N and its "complexity" is (M + 1)K + 1.

One wonders whether good trellis codes are really more complex than good tree codes or whether we have not yet found the right bounding arguments for trellis codes. The next section will give some support to the latter view. Further support comes from noticing that our ensemble of trellis codes reaches its maximum "complexity" (L + T)K + 1 when its memory M is its maximum value of L + T - 1. But it is precisely at this point that the trellis codes become tree codes and, thus Theorem 6.1 assures us that "complexity" only K + 1 suffices.

7. A Class of Good, Low "Complexity Convolutional Trellis Codes

In this section we demonstrate a class of (N,R,L,T,M) convolutional codes which have the pairwise independence property and have also much smaller "complexity" than the ensemble in Theorem 6.2. In Fig. 7.1, we show the encoder

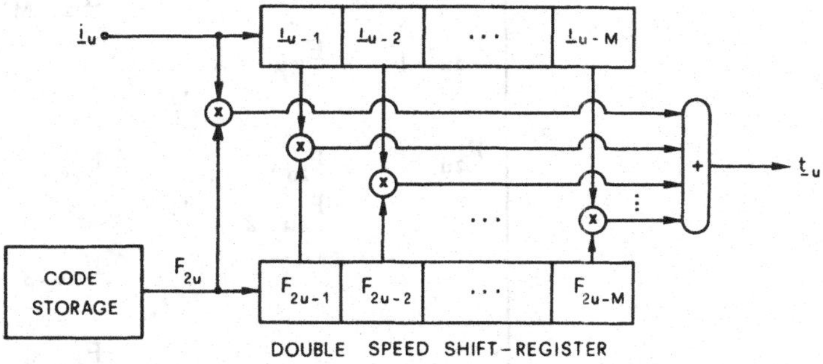

Fig. 7.1 Canonical Encoder for Low "Complexity" Trellis Codes

structure for this new class of convolutional codes. Each F_j is a K x N matrix over GF(q). The code is specified by the matrices F_j, $0 \leqslant j \leqslant 2(L+T-1)$, which are initially all in storage and fed at twice the rate of the information branches into a shift-register where the previous M of these matrices are available for multiplying the previous M information branches. In formal terms, these matrices F_j specify a time-varying convolutional code in the manner

$$G_i(u) = F_{2u-i} \tag{7.1}$$

but this formal relationship disguises the encoder structure.

For this new class of codes, we now prove:

Theorem 7.1: The ensemble of (N,R,L,T,M) convolutional codes with an added-random-sequence for which $G_i(u) = F_{2u-i}$ such that each of the digits in each of the matrices F_j, $0 \leqslant j \leqslant 2(L+T-1)$ is independently selected according to $Q(x) = 1/q$ is pairwise independent. Consequently, the bounds (4.3) and (4.5) for random trellis codes hold also for this ensemble when $Q(x) = 1/q$ is the minimizing distribution in (2.2)

To prove this theorem, we first note that, for our new class of codes, (6.3) becomes

$$
\mathbf{t}_{[u,v)} = [0,\ldots,0,\mathbf{i}_u,\mathbf{i}_{u+1},\ldots,\mathbf{i}_{v-1}]
\begin{bmatrix}
F_{2u-M} & & \\
F_{2u-M+1} & F_{2u-M+2} & \\
\vdots & \vdots & \ddots \quad F_{2v-M-2} \\
F_{2u-1} & F_{2u} & \ddots \\
F_{2u} & F_{2u+1} & \ddots \quad F_{2v-4} \\
 & F_{2u+2} & \ddots \\
 & & \ddots \quad F_{2v-3} \\
 & & \ddots \quad F_{2v-2}
\end{bmatrix}
$$

Again suppose that $\mathbf{i}_u \neq \mathbf{0}$ and that $\mathbf{i}_{[u,v-1)}$ contains no internal span of M consecutive $\mathbf{0}$ branches. This assumption ensures that the M +1 information branches multiplying the M + 1 non-blank matrices in each "column" of the righthand supermatrix are not all $\mathbf{0}$. Note also that the matrices F_i move diagonally upward as one moves to the right in the supermatrix. The encoded digit $\mathbf{t}_{u+j}$ is controlled by the span $\mathbf{i}_{[u+j-M,\, u+j]}$ of information branches. Let $\mathbf{i}_{u+k-M}$ be the rightmost of these branches which is not $\mathbf{0}$. This information branch multiplies a matrix F_i in the summation for $\mathbf{t}_{u+j}$ which has not affected any of the previous encoded branches since this must be the first time in its movement upward from the right that this matrix F_i has encountered a non-zero information branch. Thus, by the repeat our earlier arguments, $\mathbf{t}_{u+j}$ will be equally likely to be any q-ary N-tuple regardless of the values of $\mathbf{t}_u$, $\mathbf{t}_{u+1}$, ... $\mathbf{t}_{u+j-1}$. The pairwise independence of this ensemble of codes now follows from Lemma 6.1 and the theorem is proved.

To specify a code in this new ensemble, we must first specify the 2(L +T)KN digits in the matrices F_i. A further (L + T)N digits are required to specify the added random sequence. Hence, the ensemble has size (2K +1) (L +T)N or, equivalently, "complexity" 2K +1. This is a considerable improvement over the "complexity" (M+1)K+1 of the full ensemble of time-varying (N,R,L,T,M) codes which we showed in the previous section were "good" in the sense of meeting the bounds on P_e of Section 4. Still, the new ensemble has a "complexity" about twice that, K +1, for the ensemble of (N,R,L,T, M = L + T) fixed convolutional codes which are "good" tree codes.

One rather mysterious aspect of this new class of "good" (N,R,L,T,M) trellis codes is that it does not include the fixed codes as a subclass. By comparison of (5.5) and (7.1), we see that the only fixed codes in the new class are those for which F_i is the same matrix for all i and these happen to be very poor codes.

The basic difficulty in trying to prove that the ensemble of fixed convolutional codes is "good" in the sense of meeting the bounds on P_e of Section 4 is that the only known way of proving such bounds for various code ensembles is via pairwise independence. Unfortunately, the ensemble of fixed convolutional codes is not a pairwise independent ensemble when $M < L + T$ so the "goodness" of fixed codes cannot be proved in the standard way. There is, however, nothing that says an ensemble must be pairwise independent to be "good" - in fact the ensemble in which one picked only the time-varying convolutional code with smallest P_e for each set of parameters N,R,L,T and M would not be pairwise independent but would surely be "good". We suspect that the ensemble of fixed (N,R,L,T,M) convolutional codes is indeed "good." The verification of this conjecture, or a proof that it is false, is an interesting open problem not without practical significance. All of the convolutionally-coded systems used in practice to date have employed fixed codes. If fixed codes are truly not as good as time-varying codes, improvement of these practical systems might be accomplished or significantly better systems might be built in the future.

8. Signal Flowcharts for Fixed Convolutional Encoders

Viterbi [3] has made clever use of signal flowchart techniques to analyze fixed convolutional encoders (FCE's). Their use is perhaps best explained through an example.

In Fig. 8.1 (a), we show an N = 2, K = 1 (R = 1/2) binary FCE with memory M = 2. We have not yet specified L or T and, in fact, the versatility of convolutional codes arises partly from the fact that these parameters can be chosen to fit the practical situation at hand. The signal flowchart for this encoder is shown in Fig. 8.1 (b) and is constructed as follows. There is a node for every "state" $[i_{u-1}, i_{u-2}]$ which is shown with the state as its label. From each state, there is a directed edge to each of the possible successor states, $[0, i_{u-1}]$ and $[1, i_{u-1}]$ for $i_u = 0$ and

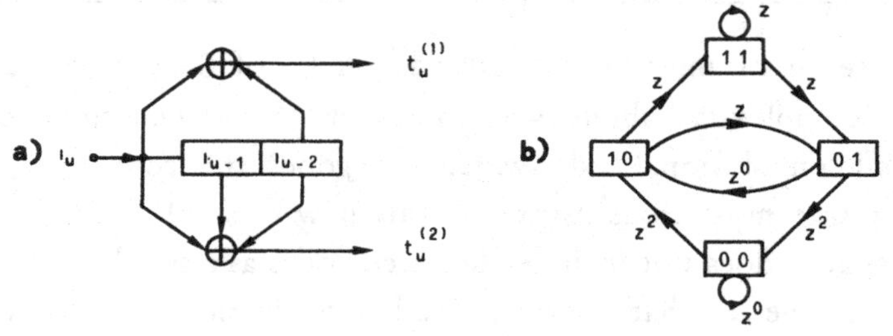

Fig. 8.1 (a) A binary FCE and (b) Its Corresponding Signal Flowchart

$i_u = 1$ respectively, which is labelled with z^w where w is the Hamming weight of the encoded branch $[t_u^{(1)}, t_u^{(2)}]$ that results from the indicated transition. For instance, when the state is $[i_{u-1}, i_{u-2}] = [1,1]$, the input $i_u = 0$ causes the encoded branch $[t_u^{(1)}, t_u^{(2)}] = [1, 0]$ and causes the next state to be $[0, 1]$. Since the Hamming weight (i.e. the number of non-zero digits) of the encoded branch is 1, the transition from state $[1, 1]$ to state $[0, 1]$ is labelled with z^1.

It should be clear from this example how one constructs the corresponding signal flowchart for any FCE. In general, since there are $(q^K)^M = q^{KM}$ states in the FCE, there will also be q^{KM} nodes in the signal flowchart so that even in the binary case ($q = 2$) the construction of the flowchart is impractical unless KM is quite small. Since the input $\mathbf{i}_u = \mathbf{0}$ applied to the state $[\mathbf{i}_{u-1}, \ldots, \mathbf{i}_{u-M}] = [\mathbf{0}, \ldots \mathbf{0}]$ always results in $\mathbf{t}_u = \mathbf{0}$ for any FCE, the flowchart will always have a self-loop labelled $z^0 = 1$ at the zero state. The FCE is said to be **catastrophic** if there is any other directed loop in the flowchart whose branches are labelled $z^0 = 1$, i.e. whose "loop gain" is 1. The encoder of Fig. 8.1 (a) is non-catastrophic. The encoder of Fig. 8.2 (a) is, however, catastrophic since besides the self-loop at

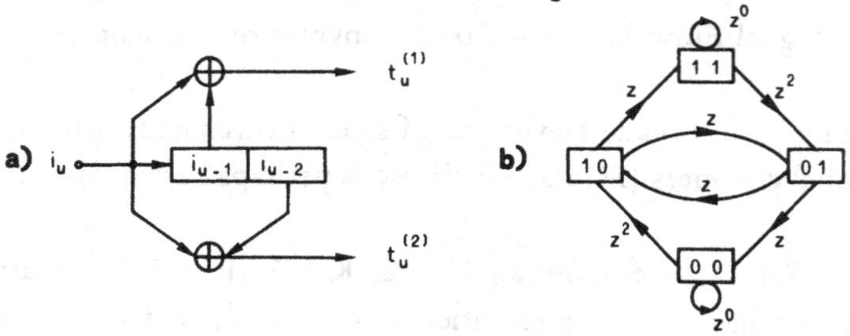

Fig. 8.2 (a) A Catastrophic Binary FCE and (b) Its Corresponding Flowchart

state $[0, 0]$ there is another directed loop with unity gain, namely the self-loop at state $[1, 1]$. (We shall later give an equivalent characterization of "catastrophic" which will make it more clear why such a pejorative term is used for this kind of FCE).

When the zero state and its self-loop are removed from the signal flowchart for an FCE, one obtains a signal flowchart with an input node (at the tails of the undeleted edges leaving the zero state) and an output node (at the points of the undeleted edges entering the zero state). There is a well-defined transmission gain, A(z), from this input node to this output node if and only if the FCE is non-catastrophic since then and only then will there be no closed loops with unity gain. The transmission gain, A(z), can be found by standard signal flowchart techniques. For example, for the flowchart in Fig. 8.1 (a), we find

$$A(z) = \frac{z^5}{1-2z} = \sum_{i=0}^{\infty} 2^i z^{5+i} \tag{8.2}$$

where the summation should be thought of as a formal power series in the indeterminate z.

We now show how to interpret the transmission gain

$$A(z) = a_0 + a_1 z + \ldots + a_i z^i + \ldots \tag{8.3}$$

obtained from the flowchart for FCE in terms of the trellis code with $L \to \infty$ specified by this FCE. For clarity, we show in Fig. 8.3 the initial portion of such a trellis for the FCE of Fig. 8.1 (a).

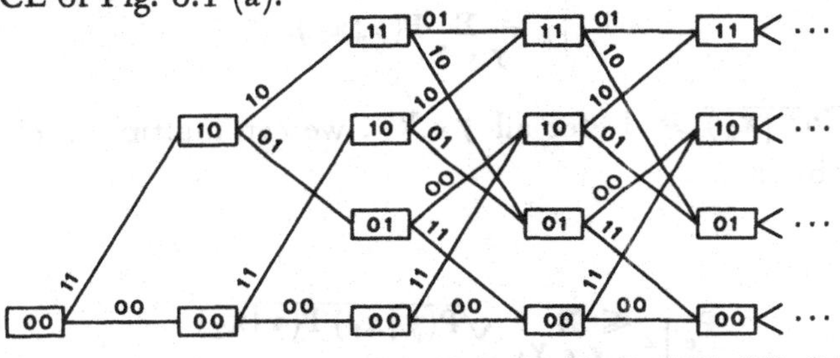

Fig. 8.3 A Portion of the $L \to \infty$ Trellis Code Defined by the FCE of Fig. 8.1 (a)

Consider any path from the input node to the output node of the flowchart with the zero state and its self-loop removed. The first branch corresponds to leaving the zero state and the last branch corresponds to the first return to the zero state. The gain z^w along this path is just the Hamming weight of this particular path through the trellis. It follows that the coefficient a_i in (8.3) is just the number of paths of Hamming weight i which depart from the all-zero lower path at the root node in the trellis and first remerge with the all-zero path at their termini. For example, from (8.2) it follows that there is only one such path of weight 5 in the trellis of Figure 8.1 as we can easily verify directly.

By the linearity of an FCE, it follows that a_i is also, for any given path through the entire trellis, the number of paths which depart from this path at the root node and first remerge with this path at their termini whose Hamming distance to the corresponding segment of the given path is exactly i. (The Hamming distance between two sequences is the number of positions in which they differ and, hence, is equal to the Hamming weight of their difference). We shall use this fact in the next section to obtain a useful upper bound on decoding error probability for a given specific FCE with MLD.

9. An Error Bound for Specific Convolutional Encoders

Consider a specific code $(\mathbf{x}_1, \mathbf{x}_2)$ containing two codewords of length N for use on a given DMC. Let $\mathbf{y} = [y_1, y_2, \ldots y_N]$ denote the received N-tuple. For a maximum likelihood decoder (MLD), the decoding region Y_1 for $\mathbf{x}_1$ is the set of all $\mathbf{y}$ such that $P(\mathbf{y}|\mathbf{x}_1) \geqslant P(\mathbf{y}|\mathbf{x}_2)$ [except that those $\mathbf{y}$ for which $P(\mathbf{y}|\mathbf{x}_1) = P(\mathbf{y}|\mathbf{x}_2)$ can be assigned arbitrarily either to Y_1 or to Y_2, the decoding region for $\mathbf{x}_2$.] Now consider the decoding error probability given that $\mathbf{x}_2$ is transmitted which we shall denote $P_e|_2$. We have

$$P_e|_2 = \sum_{\mathbf{y} \in Y_1} P(\mathbf{y}|\mathbf{x}_2). \tag{9.1}$$

Since $\sqrt{P(\mathbf{y}|\mathbf{x}_1)/P(\mathbf{y}|\mathbf{x}_2)} \geqslant 1$ for all $\mathbf{y} \in Y_1$, we can multiply each term in (9.1) by this factor to obtain

$$P_e|_2 \leqslant \sum_{\mathbf{y} \in Y_1} \sqrt{P(\mathbf{y}|\mathbf{x}_1)\, P(\mathbf{y}|\mathbf{x}_2)} \tag{9.2}$$

which we can then further overbound, by extending the summation to all $\mathbf{y}$, as

$$P_e|_2 \leqslant \sum_{\text{all } \mathbf{y}} \sqrt{P(\mathbf{y}|\mathbf{x}_1)\, P(\mathbf{y}|\mathbf{x}_2)}. \tag{9.3}$$

Since the righthand side of (9.3) is symmetric in $\mathbf{x}_1$ and $\mathbf{x}_2$, it follows that the same bound applies for $P_e|_1$, the probability of a decoding error given that $\mathbf{x}_1$ is transmitted. But then this must also be a bound on the decoding error probability P_e for MLD regardless of the probabilities assigned to the two codewords, i.e.

$$P_e \leqslant \sum_{\text{all } y} \sqrt{P(y|x_1)\,P(y|x_2)}. \tag{9.4}$$

Using the notation of Section 2, we can rewrite (9.4) as

$$P_e \leqslant \prod_{n=1}^{N} \sum_{y_n \in B} \sqrt{P(y_n|x_{1n})\,P(y_n|x_{2n})}.. \tag{9.5}$$

[The interested reader should now have no difficulty deriving the bound (2.1) by averaging the bound (9.5) over the ensemble of codes with two codewords.]

The bound (9.5) is surprisingly tight for MLD of a given code with two codewords on a given DMC. To illustrate its use, consider the binary symmetric channel (BSC) of Fig. 2.2. When $x_{1n} = x_{2n}$, the summation in (9.5) is unity. When $x_{1n} \neq x_{2n}$, the summation in (9.5) is $2\sqrt{\epsilon(1-\epsilon)}$ where ϵ is the crossover probability of the BSC. Hence, (9.5) becomes

$$P_e \leqslant [2\sqrt{\epsilon(1-\epsilon)}]^{d_H(x_1,x_2)} \tag{9.6}$$

where d_H denotes the Hamming distance between the indicated N-tuples. The simple bound (9.6) is often surprisingly close to the true P_e for MLD of the code (x_1, x_2) as the reader can verify by some detailed examples.

We note in general for any binary input channel (not just the BSC) that the bound (9.5) will take the form

$$P_e \leqslant \gamma^{d_H(x_1,\,x_2)} \tag{9.7}$$

where

$$\gamma = \sum_{y \in B} \sqrt{P(y|0)\,P(y|1)} \tag{9.8}$$

and $\gamma < 1$ unless the channel is "useless," i.e. unless $P(y|0) = P(y|1)$ for all y.

We now show, following Viterbi [3], how to combine the bound (9.7) with the transmission gain A(z) for a non-catastrophic FCE to obtain an upper bound on MLD of the trellis code defined by this particular FCE. More precisely, we shall upperbound the "first error probability" P_{e1} that the MLD decodes i_0

incorrectly, i.e. that the MLD chooses a path through the trellis which departs from the correct path at the root node. But an MLD will decode i_0 incorrectly if and only if there is some path departing from the correct path at the root node and first remerging with the correct path at its terminus such that the received sequence is more likely given that path than given the corresponding portion of the correct path. If this path is at Hamming distance i from the correct path then, by (9.7), the probability that it will cause i_0 to be decoded incorrectly is bounded above by γ^i. If there are a_i such paths at Hamming distance i, then, by the union bound, the probability that they will cause i_0 to be decoded incorrectly is bounded above by $a_i\gamma^i$. Taking into account all possible values of i with another union bound, we have finally

$$P_{el} \leqslant a_0 + a_1\gamma + a_2\gamma^2 + \ldots$$

which in light of (8.3) can be written

$$P_{el} \leqslant A(\gamma) \tag{9.9}$$

where γ is defined by (9.8). The bound (9.9) is our desired upper bound on the probability of incorrectly decoding i_0 when a MLD is used for the trellis code defined by the binary FCE associated with A(z). Since the bound applied for $L \to \infty$, it holds for any finite L **a fortiori** provided that T = M so that every path eventually remerges with the correct path.

As a specific example, consider the BSC with $\epsilon = .01$. For this channel

$$\gamma = 2\sqrt{\epsilon(1-\epsilon)} \cong .20.$$

Combining (9.9) and (8.2), we see that the probability of decoding i_0 incorrectly, when a MLD is used with a trellis code defined by the FCE of Fig. 8.1, is overbounded by

$$P_{el} \leqslant \frac{(.20)^5}{1-2(.20)} \cong 5 \times 10^{-4}$$

which is an improvement by at least a factor of 20 over the raw "error probability" of the channel.

It has recently been observed [21] that the bound on P_{el} given by (9.9) can be improved for the BSC as follows. The bound (9.6), although quite tight when $d_H(\mathbf{x}_1,\mathbf{x}_2)$ is even, can be replaced by

$$P_e \leqslant [2\sqrt{\epsilon(1-\epsilon)}]^{d_H(\mathbf{x}_1,\mathbf{x}_2)+1}, \quad d_H(\mathbf{x}_1,\mathbf{x}_2) \text{ odd}$$

since the same number of errors are required to cause a decoding error for the case of a given odd d_H as for the case when d_H is increased by 1. It follows then that, for the BSC,

$$P_{el} \leqslant a_0 + a_2\gamma^2 + a_4\gamma^4 + \ldots$$
$$+ a_1\gamma^2 + a_3\gamma^4 + \ldots$$

or

$$P_{el} \leqslant \frac{1}{2}[(1+\gamma)\,A(\gamma) + (1-\gamma)\,A(-\gamma)]. \tag{9.10}$$

For the BSC with $\epsilon = .01$ and the code with A(z) given by (8.2), we find from (9.10)

$$P_{el} \leqslant 2.3 \times 10^{-4}$$

which is an improvement on our earlier bound by a factor of more than 2. (In fact, Van de Meeberg [21] shows that (9.10) can be slightly further improved by taking $\gamma = \sqrt{2\epsilon}$).

Before closing this section, we should remark that P_{el} is also the probability of decoding i_u incorrectly given that $i_0, i_1, \ldots i_{u-1}$ have all been correctly decoded by the MLD in the case $L \to \infty$, We should also emphasize that the bound (9.9) applies for any binary input DMC and often gives a useful bound on the performance of a coding system employing a "short constraint length" convolutional code together with a MLD for such a binary input DMC.

10. Distance Measure for Fixed Convolutional Encoders

In this section, we describe some distance measures which have been proposed for fixed convolutional encoders (FCEs) and we indicate the relationship between these measures and the performance of various kinds of decoders for these codes.

The column distance of order i, denoted d_i, is defined to be the minimum

Hamming distance between two encoded paths $\mathbf{t}_{[0,i]}$ in the $L\to\infty$ trellis defined by the FCE resulting from information sequences $\mathbf{i}_{[0,i]}$ with differing values of $\mathbf{i}_0$. Equivalently, $\mathbf{d}_i$ is the minimum Hamming distance between two paths $i+1$ branches in length which diverge at the root node of the trellis. Because of the linearity of the FCE, d_i is equal to the minimum Hamming weight of an encoded path $\mathbf{t}_{[0,i]}$ resulting from an information sequence with $\mathbf{i}_0 \neq \mathbf{0}$. Letting W_H denote the Hamming weight of a sequence, we can make use of (5.4) and (5.5) to write

$$d_i = \min_{\mathbf{i}_0 \neq 0} W_H\left([\mathbf{i}_0, \mathbf{i}_1, \ldots \mathbf{i}_i] \begin{bmatrix} G_0 & G_1 & G_2 & \cdots & G_i \\ & G_0 & G_1 & \cdots & G_{i-1} \\ & & \ddots & & \vdots \\ & & & & G_0 \end{bmatrix}\right) \tag{10.1}$$

where we recall our earlier convention that $G_j \equiv 0$ when $j > M$ and M is the memory of the FCE.

To see the reason for the terminology "column distance", we consider the semi-infinite "supermatrix" G assumed by the righthand side of (5.5) as $L\to\infty$, i.e.

$$G = \begin{bmatrix} G_0 & G_1 & \cdots & G_M & & & & \\ & G_0 & \cdots & G_{M-1} & G_M & & & \\ & & \ddots & & & \ddots & & \\ & & & G_0 & G_1 & \cdots & G_M & \\ & & & & \ddots & & & \ddots \end{bmatrix} \tag{10.2}$$

When we speak of "superrows" and "supercolumns", we mean that the K x N matrices G_j are to be treated as single entries. For example, the second superrow of G is $[0, G_0\ G_1 \ldots G_M\ 0\ 0 \ldots]$. From (10.1), it follows that d_i is the minimum weight vector in the row space of the matrix formed by truncating G after $i+1$ supercolumns which includes a non-zero multiple of the first superrow in the truncated matrix.

It follows from (10.1) that d_i is a nondecreasing function of i. Moreover, every d_i is bounded above by the Hamming weight of any row in the finite matrix $[G_0\ G_1 \ldots G_M]$ so that the limit

$$d_\infty = \lim_{i\to\infty} d_i \tag{10.3}$$

always exists. It follows that

$$d_0 \leqslant d_1 \leqslant d_2 \leqslant \ldots d_i \leqslant \ldots \leqslant d_\infty . \qquad (10.4)$$

The distance, d_∞, is perhaps the single most important parameter of a convolutional code for reasons we shall soon make evident. The terminology "free distance," first used by this writer, has now been generally accepted for this distance measure and one will often find the symbols d_f or d_{free} in place of d_∞.

In terms of the signal flowchart for the FCE, d_∞ is the minimum weight of an infinitely long directed path (more precisely, the minimum W for the path gain Z^W) leaving the zero state. Since d_∞ is finite, such a minimum path must return to the zero state and remain in the self-loop thereafter when the FCE is non-catastrophic. From our definition of A(z) in Section 8 for a non-catastrophic encoder, it follows that

$$d_\infty = \min \{ i \,|\, a_i \neq 0 \} \qquad (10.5)$$

Hence, as the bound (9.9) indicates, d_∞ is the main determiner of P_e for maximum likelihood decoding (MLD) of the trellis code defined by the FCE and used on a given DMC. In general, for two FCE's of the same rate and memory, the one with larger d_∞ will give the smaller P_e. The same remark applies to "almost MLD" schemes such as sequential decoding which we shall discuss later. Because of its fundamental importance, considerable effort has gone into the search for FCE's which, for a given rate R and memory M, have the largest d_∞, cf. [4] and [5] for example.

The quantity d_M is called the **feedback-decoding minimum distance**, or simply the "minimum distance," of the FCE and one often sees the alternative notations d_{FD} and d_{min}. This distance is of importance for "algebraic decoders" of the type that estimate $\mathbf{i}_0$ from the "first constrain length" $\mathbf{r}_{[0,M]}$, then "subtract" the effect of $\mathbf{i}_0$ from the received sequence and use the same algorithm to estimate $\mathbf{i}_1$ from $\mathbf{r}_{[1,M+1]}$, etc. There exists such a decoder which correctly decodes $\mathbf{i}_0$, $\mathbf{i}_1$, ..., $\mathbf{i}_u$ whenever there are t or fewer errors in each constraint length $\mathbf{r}_{[j,M+j]}$, $0 \leqslant j \leqslant u$, when and only when

$$t \leqslant \frac{1}{2}(d_M - 1). \qquad (10.6)$$

This measure was first used by Wozencraft and Reiffen [6] who proved a "Gilbert lower bound" on d_M for codes of rate R = 1/N that was later generalized by this writer to all rates [7].

The (M + 1)-tuple $\mathbf{d} = [d_0, d_1, ..., d_M]$ has been recently introduced [5] and called the **distance profile** of the FCE. If $\mathbf{d}$ and $\mathbf{d}'$ are distance profiles for two FCE's of the same rate and memory, one says $\mathbf{d} > \mathbf{d}'$ if $d_j > d_j'$ for the smallest index j, $0 \leq j \leq M$, where $d_j \neq d_j'$. The code with the larger distance profile will have a minimal separation between paths diverging at the root node which grows more rapidly, at least initially, with depth into the trellis. For this reason, the code with the larger $\mathbf{d}$ will generally require less computation when sequential decoding is used than will the other code.

The **row distance of order** i, denoted r_i, is defined to be the minimum Hamming distance between two different paths in the (N, R, L + i + 1, T = M, M) trellis defined by the FCE. By the linearity of the FCE, it follows that

$$r_i = \min_{\mathbf{i}_{[0,i]} \neq 0} W_H\left([\mathbf{i}_0, \mathbf{i}_1, ..., \mathbf{i}_i] \begin{bmatrix} G_0 & G_1 & \cdots & G_M & & \\ & G_0 & G_1 & \cdots & G_M & \\ & & \ddots & \ddots & & \ddots \\ & & & G_0 & G_1 & \cdots & G_M \end{bmatrix}\right) \tag{10.7}$$

which is equivalent to saving that r_i is the minimum weight of non-trivial linear combinations of the rows in the matrix formed by truncating the super-matrix G after its first i + 1 superrows. In terms of the signal flowchart of the FCE, r_i is the minimum weight of a path of length M + i + 1 branches which diverges at some point from the zero state and terminates on the zero state.

It follows from (10.7) that r_i is non-increasing with i. Since every r_i is bounded below by 0, it follows that the limit

$$r_\infty = \lim_{i \to \infty} r_i$$

always exists. Thus, we have

$$r_0 \geq r_1 \geq r_2 \geq ... \geq r_i \geq ... \geq r_\infty \tag{10.8}$$

which should be contrasted with (10.4). Moreover, we see from (10.1) and (10.7) that

$$d_i \leq r_j \quad \text{all i and j.}$$

From (10.4) and (10.8), we then have the inequalities

$$d_0 \leqslant d_1 \leqslant \ldots \leqslant d_i \leqslant \ldots \leqslant d_\infty \leqslant r_\infty \leqslant \ldots \leqslant r_i \leqslant \ldots \leqslant r_1 \leqslant r_0 \qquad (10.9)$$

which lie at the heart of distance measures for FCE's. By the relationship we noted between the distance measures d_i and r_j and paths in the signal flowchart for the FCE, it follows that

$$d_\infty = r_\infty \qquad (10.10)$$

for a non-catastrophic FCE.

From its definition, we see that r_i specifies the error-correcting property of the $L = i + 1$, $T = M$ trellis generated by the FCE in that there exists a decoder that can correct all patterns of t or fewer errors in the full encoded sequence if and only if

$$t \leqslant \frac{1}{2}(r_i - 1). \qquad (10.11)$$

Nonetheless, the row distances r_i are not of much direct interest in convolutional coding for the reasons (1) that in practice the trellis length L is usually great enough that r_∞ is the appropriate distance, and (2) that in practice one always chooses a non-catastrophic FCE so that $r_\infty = d_\infty$. The main utility of the row distance is in assisting the calculation or bounding of the column distances via (10.9). For example, from the flowchart for the FCE in Fig. 8.1, we readily find in order that $d_0=2, d_1=3, d_2=3, d_3=4, d_4=4, d_5=5, r_0=5$. From (10.9) it then follows immediately that $d_i=5$ for all $i \geqslant 5$ (and $r_i=5$ for all $i \geqslant 0$). In particular, the free distance of this code is $d_\infty = 5$. It should be noted that (10.10) may not hold when the FCE is catastrophic. For example, from the flowchart for the FCE in Fig. 8.2, we find that $d_0 = 2$ and that $d_i = 3$ for all $i \geqslant 1$ so that $d_\infty = 3$. On the other hand, we find that $r_i = 4$ for all $i \geqslant 0$ so that $r_\infty = 4 > d_\infty = 3$.

For an intersting discussion of the distance measures considered here as well as their values for many specific FCE's, the reader is referred to the paper by Costello [8]. The paper by Johannesson [5] is a good source for excellent FCE's of rate $R = 1/2$ which is the rate most often used in practice.

11. Catastrophic Fixed Convolutional Encoders

In this section, we probe more deeply into the meaning of a catastrophic FCE as defined previously in Section 8 in terms of the signal flowchart of the FCE. We first note that our definition there is equivalent to saying that an FCE is catastrophic if and only if there is an information sequence $\mathbf{i}_{[0,\infty)}$ with $W_H(\mathbf{i}_{[0,\infty)}) = \infty$ that produces an encoded sequence $\mathbf{t}_{[0,\infty)}$ for which $W_H(\mathbf{t}_{[0,\infty)}) < \infty$. To see this equivalence, we note that if the signal flowgraph has a closed loop of weight 0 besides the self-loop at the zero state, then the information sequence $\mathbf{i}_{[0,\infty)}$ which drives the encoder to a state on the former loop and around the loop forever after is an infinite weight sequence whcih produces a finite weight encoded sequence. Conversely, if there is no zero weight loop besides the self-loop at the zero state, then an information sequence of infinite weight must drive the encoder through infinitely many loops besides the self-loop at the zero state and hence must produce an encoded sequence of infinite weight also.

We now give further characterizations of catastrophic FCE's arising from the fact that an FCE is a K-input, N-output linear sequential circuit (LSC). In fact, an FCE is an LSC with **input memory** M in the sense that its output depends only on the present input and the M preceeding inputs. An LSC with finite input memory is called feedforward (FF) and is distinguished by the fact that the entries in its transfer function matrix (written in the delay operator D) are all polynomials. For example, the K = 1, N = 2, M = 2 FCE in Fig. 8.1 has the K x N transfer function matrix

$$\mathbf{T}(D) = [1 + D^2 \quad 1 + D + D^2] \tag{11.1}$$

as can easily be read from its defining circuit.

A second LSC with transfer function $\mathbf{T}^*(D)$ is said to be a delay-Δ inverse of the LSC with transfer function T(D) whenever (when both are started in the zero state) the output sequence of the original LSC produces, when used as the input sequence to the inverse, an output sequence equal to the input sequence of the original except for a delay of Δ time units, that is when

$$\mathbf{T}(D)\, \mathbf{T}^*(D) = D^{\Delta}. \tag{11.2}$$

For instance, the simple LSC with transfer function matrix

$$T^*(D) = \begin{bmatrix} 1 \\ 1 \end{bmatrix} \tag{11.3}$$

is a delay $\Delta = 1$ inverse of the FCE of Fig. 8.1 whose transfer function is given in (11.1). Similarly the LSC with transfer function

$$T^*(D) = \begin{bmatrix} 1+D \\ D \end{bmatrix} \tag{11.4}$$

is a delay $\Delta = 0$ inverse, i.e. an **instantaneous inverse**, of the same FCE.

An LSC is called **feedforward-invertible** if it has a delay-Δ inverse, for some Δ, which is a feedforward LSC. As either of the above inverses for the FCE in Fig. 8.1 is feedforward-invertible. Forney [9] has shown that an LSC is feedforward-invertible if and only if there is no infinite weight input sequence which produces a finite weight output sequence. From our second characterization above of a catastrophic FCE, a third characterization immediately follows; namely that an **FCE is catastrophic if and only if it is not feedforward-invertible.**

Massey and Sain have shown that an FCE, or more generally any FF LSC, is feedforward-invertible if and only if the greatest common divisor of the K x K minors of its transfer function matrix is a monomial, i.e. of the form D^i for some i [10]. Hence, the FCE of Fig. 8.1 with $K = 1$ and $N = 2$ could be recognized as non-catastrophic from the fact that the polynomials in its transfer function matrix of (11.1) are relatively prime, that is, their greatest common divisor is 1. On the other hand, the transfer function matrix of the FCE of Fig. 8.2 is

$$T(D) = [1 + D \quad 1 + D^2] \tag{11.5}$$

and, since $(1 + D)^2 = 1 + D^2$, from the fact that the greatest common divisor of its entries is $1 + D$ one can conclude that this FCE is catastrophic. (Olson has extended this test for feedforward invertibility to arbitrary LSC's [11] and Forney [9] has given an elegant algebraic formulation of Olson's test.)

If one divides each entry of the transfer function matrix of (11.5) for the catastrophic FCE of Fig. 8.2 by the greatest common divisor $1 + D$ of its entries, one obtains the transfer function matrix for the simpler $M = 1$ FCE of Fig. 11.1. One readily finds that $d_0 = 2$ and $d_i = 3$ for $i \geqslant 1$ just as was the case for the catastrophic FCE from which the FCE of Fig. 11.1 was obtained. As the reader may

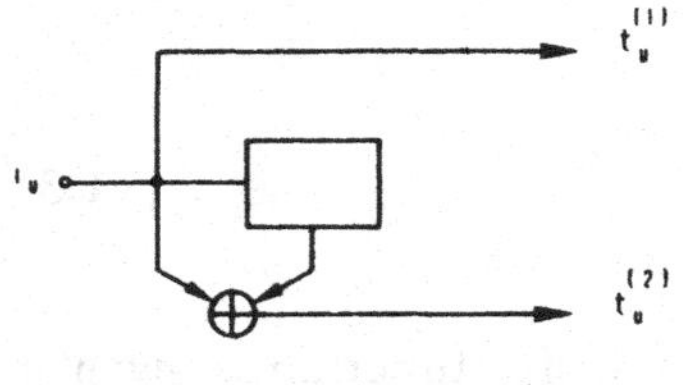

Fig. 11.1 The Non-Catastrophic FCE "Equivalent" to the Catastrophic FCE of Fig. 8.2.

suspect, this preservation of column distances was no accident – for any catastrophic FCE one can find an "equivalent" non-catastrophic FCE with the same column distances and perhaps less memory. This, in itself, is perhaps reason enough to eschew catastrophic FCE's.

While we have given three equivalent characterizations of catastrophic FCE's, the reader may not consider any of these ways of viewing the property as reason enough for the strong term "catastrophic." The fourth and last of our characterizations (which historically was the first characterization of the "catastrophic" property [10]) should make it clear why such a damning adjective is appropriate.

Consider the $L \to \infty$ trellis code generated by an FCE. A decoder is just a device for forming an estimate $\hat{\mathbf{i}}_{[0,\infty)}$ of the information sequence $\mathbf{i}_{[0,\infty)}$. The difference sequence

$$\boldsymbol{\delta}_{[0,\infty)} = \hat{\mathbf{i}}_{[0,\infty)} - \mathbf{i}_{[0,\infty)} \tag{11.6}$$

is just the sequence of **decoded information errors.** Now any decoder can be thought of as also making an estimate $\hat{\mathbf{t}}_{[0,\infty)}$ of the encoded sequence $\mathbf{t}_{[0,\infty)}$ transmitted over the channel simply by considering $\hat{\mathbf{t}}_{[0,\infty)}$ to be the result of encoding $\hat{\mathbf{i}}_{[0,\infty)}$. The difference sequence

$$\boldsymbol{\epsilon}_{[0,\infty)} = \hat{\mathbf{t}}_{[0,\infty)} - \mathbf{t}_{[0,\infty)} \tag{11.7}$$

in the sequence of **decoded transmission errors.** In fact, we can show the general situation as in Fig. 11.2 where it is clear that which encoder inverse is used is of no

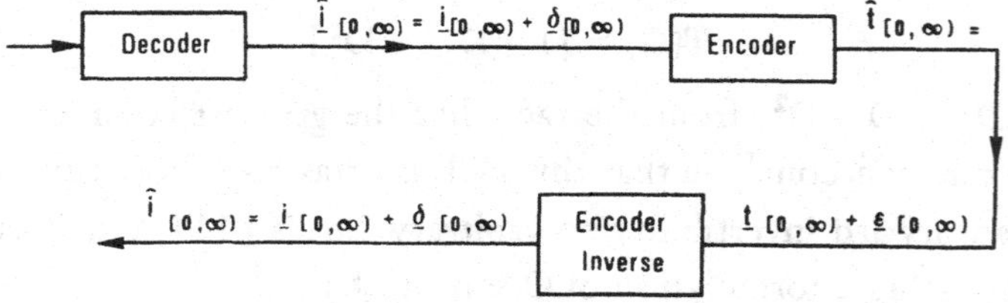

Fig. 11.2 A Canonic "Decomposition" of the Decoder for a FCE.

significance. By the linearity of the FCE and its inverse, it follows that the sequence $\boldsymbol{\delta}_{[0,\infty)}$ is the response of the encoder inverse to the sequence $\boldsymbol{\epsilon}_{[0,\infty)}$ – which fact we shall shortly use in a critical way. One can think of the decoder-encoder tandem in Fig. 11.2 as forming a "channel sequence estimator" whose output is inverted to yield $\hat{\mathbf{i}}_{[0,\infty)}$. In fact, as we shall see later, Viterbi decoders and sequential decoders truly perform this channel sequence estimation prior to estimating the information digits.

We think the reader would agree that it would be a "catastrophic" situation if the channel sequence estimator should make only a finite number of errors, i.e. $W_H(\boldsymbol{\epsilon}_{[0,\infty)}) < \infty$, but these resulted in an avalanche of infinitely many information decoding errors, i.e. $W_H(\boldsymbol{\delta}_{[0,\infty)}) = \infty$. But we now claim that an FCE is catastrophic if and only if every "realistic" channel-decoder pair are such that some channel behavior will cause $W_H(\boldsymbol{\epsilon}_{[0,\infty)}) < \infty$ but result in $W_H(\boldsymbol{\delta}_{[0,\infty)}) = \infty$. Conversely if an FCE is non-catastrophic, no channel-decoder pair can ever result in $W_H(\boldsymbol{\epsilon}_{[0,\infty)}) < \infty$ but $W_H(\boldsymbol{\delta}_{[0,\infty)}) = \infty$. By a "realistic" channel-decoder pair we mean a pair such that, regardless of what $\mathbf{i}_{[0,\infty)}$ is encoded, the channel can behave so as to cause the decoder to decide $\hat{\mathbf{i}}_{[0,\infty)} = \hat{\mathbf{0}}$. This rules out, for instance, the "noisless" BSC with 0 crossover probability and other "perfect" channels, and also rules out decoders that never estimate $\mathbf{0}$ but rules out no combination of a "real" channel and a "reasonable" decoder.

We prove the converse first because it is so simple. Suppose the FCE is non-catastrophic. It then has an FF inverse which we can assume is the inverse in Fig. 11.2 since it is immaterial which inverse is used. Suppose $W_H(\boldsymbol{\epsilon}_{[0,\infty)}) < \infty$ and recall that $\boldsymbol{\delta}_{[0,\infty)}$ is the response of the FF inverse to $\boldsymbol{\epsilon}_{[0,\infty)}$. Since the FF inverse has finite input memory, after a finite time it will be see only zero digits in the sequence $\boldsymbol{\epsilon}_{[0,\infty)}$ and its output must be zero thereafter. We conclude that $W_H(\boldsymbol{\delta}_{[0,\infty)}) < \infty$ which proves the converse.

Now suppose that the FCE is catastrophic. There is then an input sequence $\mathbf{i}_{[0,\infty)}$ with $W_H(\mathbf{i}_{[0,\infty)}) = \infty$ that causes a transmitted sequence $\mathbf{t}_{[0,\infty)}$ with $W_H(\mathbf{t}_{[0,\infty)}) < \infty$. Next, suppose the channel behaves so as to cause $\hat{\mathbf{i}}_{[0,\infty)} = \mathbf{0}$ which implies in turn that $\hat{\mathbf{t}}_{[0,\infty)} = \mathbf{0}$. We then have $W_H(\boldsymbol{\epsilon}) = W_H(\mathbf{0} - \mathbf{t}_{[0,\infty)}) = W_H(\mathbf{t}_{[0,\infty)}) < \infty$ but we also have $W_H(\boldsymbol{\delta}) = = W_H(\mathbf{0} - \mathbf{i}_{[0,\infty)}) = W_H(\mathbf{i}_{[0,\infty)}) = \infty$. Thus, a finite number of errors in estimating the channel sequence have been converted to an infinite number of errors in decoding the information digits and our claim is proved in full.

12. MLD (Viterbi Decoding) of Trellis Codes

In this section, we study how one might perform, in an efficient way, maximum likelihood decoding (MLD) for an arbitrary (N,R,L,T,M) trellis code used on an arbitrary discrete memoryless channel (DMC). The procedures that will be developed apply, of course, to convolutional codes which are the linear special case of trellis codes.

Suppose that $\mathbf{y}_{[0,L+T)}$ is the sequence received over the channel where (in what should now be familiar notation)

$$\mathbf{y}_{[0,L+T)} = [\mathbf{y}_0, \mathbf{y}_1, \ldots, \mathbf{y}_{L+T-1}]$$

and where each $\mathbf{y}_u$ is an N-tuple of channel output letters. Similarly, we write $\mathbf{x}_{[0,L+T)}$ for the encoded sequence, i.e. the sequence of channel input letters on the path through the trellis, resulting from the information sequence $\mathbf{i}_{[0,L+T)}$. A maximum likelihood decoder chooses as its estimate $\hat{\mathbf{i}}_{[0,L+T)}$ (one of) the sequence(s) which maximizes

$$P(\mathbf{y}_{[0,L+T)} | \mathbf{x}_{[0,L+T)}) = \prod_{u=0}^{L+T-1} P(\mathbf{y}_u | \mathbf{x}_u)$$

or, equivalently, which maximizes the statistic

$$\log P(\mathbf{y}_{[0,L+T)} | \mathbf{x}_{[0,L+T)}) = \sum_{u=0}^{L+T-1} \log P(\mathbf{y}_u | \mathbf{x}_u) . \qquad (12.1)$$

Since $\mathbf{y}$ will be fixed throughout our discussion, we shall write simply

$$L_0(\mathbf{x}_{[u,v]}) = \log P(\mathbf{y}_{[u,v]} | \mathbf{x}_{[u,v]}) \qquad (12.2)$$

and we note that

$$L_0(\mathbf{x}_{[u,v]}) = \sum_{i=u}^{v} L_0(\mathbf{x}_i) . \qquad (12.3)$$

In fact, we shall not insist that L_0 be the logarithmic function as in (12.2), but simply that it be a statistic whose maximization yields a MLD and which is additive according to (12.3). For instance, for the BSC, we can take

$$L_0(\mathbf{x}_i) = -\,d_H(\mathbf{x}_i, \mathbf{y}_i)$$

i.e. the negative of the number of channel crossovers or "errors" had $\mathbf{x}_i$ actually been transmitted.

The key to MLD for (N,R,L,T,M) trellis codes is the following:

Principle of Non-Optimality: If the paths $\mathbf{i}'_{[0,u]}$ and $\mathbf{i}''_{[0,u]}$ terminate at the same node of the trellis and

$$L_0(\mathbf{x}'_{[0,u]}) > L_0(\mathbf{x}''_{[0,u]}), \tag{12.4}$$

then $\mathbf{i}''_{[0,u]}$ cannot be the first u + 1 branches of (one of) the path(s) $\mathbf{i}_{[0,L+T)}$ which maximizes $L_0(\mathbf{x}_{[0,L+T)})$.

The proof of this property is very simple. Suppose on the contrary that $\mathbf{i}''_{[0,L+T)} = \mathbf{i}''_{[0,u]} * \mathbf{i}''_{(u,L+T)}$ [where here and after * denotes concatenation of sequences] maximizes the statistic L_0. Then consider the path $\mathbf{i}'_{[0,L+T)} = \mathbf{i}'_{[0,u]} * \mathbf{i}''_{(u,L+T)}$. The encoded sequence $\mathbf{x}'_{[0,L+T)} = \mathbf{x}'_{[0,u]} * \mathbf{x}_{(u,L+T)}$ must have $\mathbf{x}'_{(u,L+T)} = \mathbf{x}''_{(u,L+T)}$ since $\mathbf{i}'_{[0,u]}$ terminates on the same node as $\mathbf{i}''_{[0,u]}$ and we have used the same subsequent information sequence $\mathbf{i}''_{(u,L+T)}$ in both cases. Consequently,

$$\begin{aligned} L_0(\mathbf{x}'_{[0,L+T)}) &= L_0(\mathbf{x}'_{[0,u]} * \mathbf{x}''_{(u,L+T)}) \\ &= L_0(\mathbf{x}'_{[0,u]}) + L_0(\mathbf{x}''_{(u,L+T)}) \\ &> L_0(\mathbf{x}''_{[0,u]}) + L_0(\mathbf{x}''_{(u,L+T)}) \end{aligned}$$

by (12.4). Thus,

$$L_0(\mathbf{x}'_{[0,L+T)}) > L_0(\mathbf{x}''_{[0,L+T]})$$

which contradicts the assumption that $\mathbf{x}''_{[0,L+T)}$ maximized the statistic L_0.

We now show, in an example, how the principle of Non-Optimality specifies an efficient MLD procedure for trellis codes. In Fig. 12.1, we show the (n = 2, R = 1/2, L = 4, T = 2, M = 2) trellis code for the binary FCE of Fig. 8.1 and the received sequence $\mathbf{y}_{[0,6)}$ = [1,0,1,1,0,1,0,1,1,0,0,0] which has been received over a BSC with crossover probability $\epsilon < 1/2$. For our decoding statistic L_0, we use the negative of the number of errors between this received sequence and the

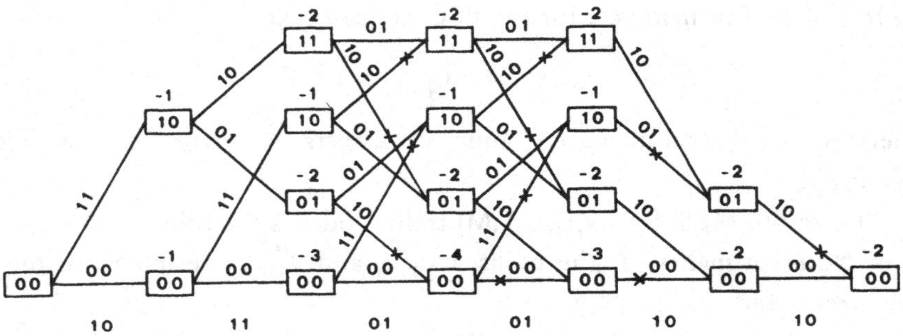

Fig. 12.1 An Example of MLD for the BSC

encoded sequence. As will soon be apparent, the number written above each node or "encoder state" is the statistic L_0 for the best path to that state.

For the nodes at depth 1 and 2 in the trellis, there is only one path to each node so there is no doubt about the best path to these nodes. At depth 3, the interesting situations begin. Of the two paths entering state [1, 1] we see that the upper has L_0=, - 2 and the lower has L_0 =- 3. By the principle of non-optimality, we can discard the lower path in our search for the best path through the trellis. We show this figuratively in Fig. 12.1 by a "cross" on the last branch of the lower path to indicate we have "cut" the path at this point to leave the upper path as the only remaining path into state [1, 1] . We repeat this same procedure for the other 3 states at depth 3, and above each state we show the value of L_0 for the one remaining path to that state. Since there is now exactly one path into each state at depth 3, there are exactly two paths now into each state at depth 4. We process each state at depth 4 by "cutting" the worse path into that state. We then have exactly one path into each state at depth 4 and hence exactly 2 paths again into each state at depth 5. We then process both states at depth 5 and thus create exactly two paths into the only state, [0, 0] at depth 6. We then process this state by "cutting" the worse path into it. We now find we are left with exactly one path through the entire trellis, namely the path (which is most easily read in reverse order from Fig. 12.1) $\mathbf{x}_{[0,\,6)}$ = [0,0,1,1,0,1,1,1,0,0,0,0] which is the encoding of $\mathbf{i}_{[0,6)}$ = [0,1,0,0,0,0]. 0,1,0,0,0,0 . Since we have been true to the principal of non-optimality and have discarded only paths which could not be optimal, we conclude that this remaining path is the optimal one. The MLD decision is thus $\hat{\mathbf{i}}_{[0,L)} = \hat{\mathbf{i}}_{[0,4)}$= [0,1,0,0].

It should be clear that we have now found a general method to do MLD for any (N,R,L,T,M) trellis code on any DMC. Up to depth M, there will be a unique path to each state and we can directly assign the statistic L_0. At depth M +1, there will be exactly $m = 2^{NR}$ paths into each state. We discard or "cut" all but the best one of these and assign the statistic of this best path to that state. We thus again create exactly m paths into each state at depth M + 2 and we process these states in the same manner. When we finally reach depth L + T of the tree, our final processing creates exactly one path into each of these m^{M-T}, terminal states (and in practice one would always use M = T with MLD.) That one of these m^{M-T} paths through the entire trellis which has the greatest statistic L must be the optimal path, i.e. the path to be chosen by a MLD.

This method of doing MLD for the trellis codes was introduced by Viterbi [12] and is now usually called "Viterbi Decoding." Viterbi at that time used this decoding procedure as a theoretical tool to prove the bound (4.4) for time-varying convolutional codes (which he also extended appropriately for rates R, $R_0 \leq R < C$,) but he was unaware at the time that his procedure was actually maximum-likelihood. That observation was first made by Omura [13] who also observed that Viterbi's algorithm was in fact the "dynamic programming" solution to the problem of optimal decoding. [What we have called the "Principle of Non-Optimality" is what Bellman calls the "Principle of Optimality" in dynamic programming but the former name strikes us as more illuminating in the present context.] The reader may wonder that these observations were made so late but he should know that Forney's use of "trellis diagrams" for convolutional codes, which makes these facts so transparent now, came considerably after Omura's work.

13. Implementing Viterbi Decoders

In the previous section, we gave the conceptual formulation of Viterbi decoding, i.e. MLD, for trellis codes. In this section, we consider how one might actually construct Viterbi decoders, either in hardware or computer software, for trellis codes generated by fixed convolutional encoders.

Because of their practical importance and conceptual simplicity, we shall consider only (N, R = 1/N, L, T = M, M) trellis codes for binary fixed convolutional encoders (FCE's). Although we shall be thus restricting ourselves to binary input channels, it is important to note that we have no corresponding restriction on the size of the channel output alphabet. "Binary output quantization" often severely degrades a channel in terms of its resultant R_0 or its resultant capacity. As will be apparent, Viterbi decoders can quite easily handle "high-order output quantization" of the physical channel.

Since $R = 1/N$, there are only $2^{RN} = 2$ branches entering each state in the trellis, at depth $M+1$ and greater, and there are a total of 2^M distinct states. A state at depth u in the trellis may be denoted $[i_{u-1}, i_{u-2}, \ldots i_{u-M}]$ where each i_j is a binary information digit.

Consider now the amount of "processing" done at each depth u, $M < u \leqslant L$ where all 2^M states are present in the trellis and each state has two branches leading into it. At such a given time u, the previous processing in the Viterbi algorithm will ensure that there are only two paths from the root node to each state at time u. For each such state, it is necessary to compare these two paths, discard the worse, compute the statistic L_0 on each of the two branches leading out of the state, and add this branch statistic to the statistic for the better of the two paths before sending these values on to the two successor states at depth u+ 1. To accomplish this task in hardware, it is natural to consider using a separate "microcomputer" for each of the 2^M states.

Let $M(b_1, b_2, \ldots, b_M)$ denote the "microcomputer" corresponding to state $[b_1, b_2, \ldots b_M]$. We can think of each microcomputer functioning at the u-th step of the computation, $u = 0, 1, 2, \ldots L+M$, according to the diagram shown in Fig. 13.2. At step u, the microcomputer $M(b_1, b_2, \ldots b_M)$ receives the output of two

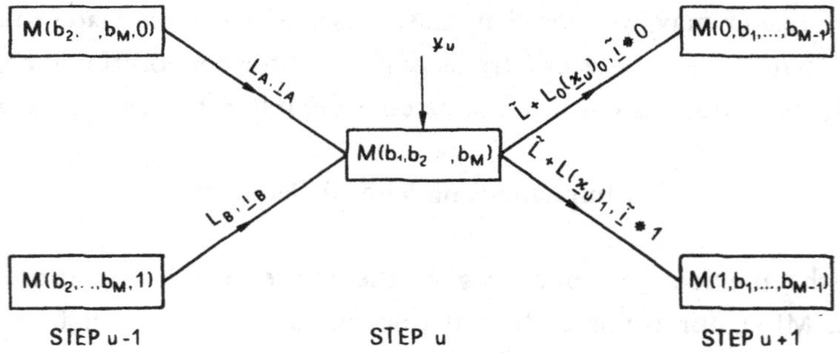

Fig. 13.2 **Functioning of Microcomputers in a Hardware Viterbi Decoder**

designated microcomputers from step u - 1 consisting of the statistic (L_A or L_B) and the information sequence ($\mathbf{i}_A$ or $\mathbf{i}_B$) for the two surviving paths into state $[b_1, b_2, \ldots b_M]$. The microcomputer $M(b_1, b_2, \ldots b_M)$ first compares these statistics and sets

$$\tilde{L} = L_A \text{ and } \tilde{\mathbf{i}} = \mathbf{i}_A \text{ if } L_A \geqslant L_B$$
$$\tilde{L} = L_B \text{ and } \tilde{\mathbf{i}} = \mathbf{i}_B \text{ if } L_A < L_B .$$

Since the encoded branch $\mathbf{x}_u$ is determined uniquely by the state and the information digit i_u, the labels on these branches, say $(\mathbf{x}_u)_0$ and $(\mathbf{x}_u)_1$ for $i_u = 0$ and $i_u = 1$ respectively, are always the same for a given microcomputer $M(b_1, b_2, \ldots b_M)$. Hence, given the received branch $\mathbf{y}_u$, this microcomputer can immediately determine (say by look-up in a stored table) the appropriate branch metrics, say $L_0(\mathbf{x}_u)_0$ and $L_0(\mathbf{x}_u)_1$ respectively. The microcomputer then adds these branch statistics to the statistic $\tilde{L}$, appends the appropriate information digit to $\tilde{i}$, and sends these results along to two designated microcomputers for their use in step u +1 of the computation. Thus each "microcomputer" is a rather simple digital device.

At the steps u, $u \leqslant M$ and $u > L$, we see that in principle not all microcomputers should be in operation since not all 2^M states are then present in the trellis or not all have 2 input branches. A simple trick avoids the complication of "activating" and "deactivating" certain microcomputers; at step 0 one initializes all microcomputers, except M(0,0,...0) which is initialized with $\tilde{L} = 0$, with sufficiently negagive L so that the path from these states can never be the final chosen path and, at $u = L + M$ when the computation is finished one ignores the output of all microcomputers except M(0,0,...0) whose output $\tilde{i}$ is taken as the decoded estimate. In this manner, all microcomputers may be left to function at all times which simplifies the controlling logic.

It should be pointed out that there are only 2^N different types of microcomputers, regardless of the value of M, for FCE's in which $(\mathbf{x}_u)_0$ is the complement of $(\mathbf{x}_u)_1$ [as is the case with all "good" R = 1/N binary encoders] since the number of different microcomputers is then the number of possible encoded branches $(\mathbf{x}_u)_0$ of N binary digits. For instance, for R = 1/2 which is the most important case in practice, there are only $2^2 = 4$ different types of microcomputers. Hence, a hardware Viterbi decoder for an M = 6, R = 1/2 FCE would have $2^6 = 64$ microcomputers but only 4 distinct types of microcomputers.

When a Viterbi decoder is implemented in software on a general-purpose digital computer, the usual method is to perform serially, at step u, the operations of each of the 2^M microcomputers of a hardware realization.

Whether implemented in hardware or software, it is clear that the "complexity" of a Viterbi decoder grows exponentially with M, the encoder memory. M = 6 or 7 appears to be about the limit of practicality. Rate 1/2 FCE's with memory $M \cong 6$ are, however, surprisingly good for their short constraint lengths $N_t = (M+1)\,N \cong 14$. In fact, at the present writing, if one is content with an error probabability of 10^{-4} or greater, Viterbi decoders operate at as small or smaller energy per information bit to noise power spectral density ratios on the "deep-space channel" as any other practical coding system and are rather widely used in space applications. The prospect is for even wider use of Viterbi decoders in future digital communication systems [3].

14. Decoding Tree Codes--The Fano Metric

We now take up the question of how to decode tree codes in a reasonable way. We can of course consider FCE's to generate a tree code as well as a trellis code so our results will apply, in particular, to convolutional codes. (Generally speaking, we choose to view FCE's as generating tree codes rather than trellis codes when M is so large that a Viterbi-type trellis decoder is impractical.)

When L is quite large, the number of paths 2^{NRL} in a tree code is so great that one cannot conceive of comparing each path with the received sequence to perform maximum-likelihood decoding (MLD). For instance, with $R = 1/N$ and $L = 100$ (which is "small" as practical tree codes go) there are $2^{100} \cong 10^{30}$ paths through the tree. A feeling for the size of this number can be obtained by reflecting that 10^{30} is about one million times greater than Avogadro's number! But since large numbers of paths stem from each of the early nodes in the tree, one suspects that it might be possible to discard all the paths stemming from such a node when the path to it is sufficiently bad without severely degrading performance from that of MLD. This, in fact, is the whole point of using tree codes. But to discard paths as "bad," we need some absolute measure of their quality which will take into account the fact that the paths we discard may of different lengths. To obtain such a measure of "quality metric" we are led naturally to consider the decoding problem for codes whose codewords (unlike those of block codes) have different lengths.

Let $(x_1, x_2, \dots x_S)$ be a code with S codewords,

$$\mathbf{x}_s = [x_{s1}, x_{s2}, \dots x_{sN_s}] \quad s = 1,2,\dots S$$

whose lengths $N_1, N_2, \dots N_s$ are in general different. Consider transmitting the message s by sending $\mathbf{x}_s$ over a DMC. The decoder must make its estimate $\hat{s}$ of which message was sent. Let

$$N = \max(N_1, N_2, \dots N_S)$$

and we require the decoder to estimate s from the received sequence $\mathbf{y} = [y_1, y_2, \dots y_N]$.

To avoid any information about s being sent over the DMC covertly after the codeword $\mathbf{x}_s$ has been transmitted and when $N_s < N$, we suppose that a random tail of $N - N_s$ digits obtained by independent selection from the channel input alphabet according to a probability distribution $Q(x)$ are appended to $\mathbf{x}_s$ for transmission over the DMC and that the decoder knows only the distribution $Q(x)$ used in this selection. We do not wish to assume the messages are equiprobable so we let P_s denote the probability that message s is sent. We then seek to find a decoding rule which minimizes the probability of a decoding error.

Let $P(s, \mathbf{y})$ denote the joint probability of sending message s and receiving $\mathbf{y}$. Since our channel is memoryless and the random digits following $\mathbf{x}_s$ are independently chosen, we have

$$P(s,y) = P_s \prod_{n=1}^{N_s} P(y_i \mid x_{si}) \prod_{t=N_s+1}^{N} P_Q(y_t) \tag{14.1}$$

where $P_Q(y)$ is the probability of receiving y given that a random digit is sent over the channel, i.e.

$$P_Q(y) = \sum_{x \in A} P(y \mid x)\, Q(x) \tag{14.2}$$

where A is the channel input alphabet. The probability of a decoding error is minimized by choosing $\hat{s}$ as the value of s which maximizes $P(s, \mathbf{y})$, or, equivalently, which maximizes

$$\frac{P(s,y)}{\prod_{i=1}^{N} P_Q(y_i)} = P_s \prod_{n=1}^{N_s} \frac{P(y_n \mid x_{sn})}{P_Q(y_n)} \tag{14.3}$$

as we have merely divided by a positive constant which is independent of s. Again equivalently, we may take logarithms in (14.3) to assert that the probability of a decoding error will be minimized by choosing $\hat{s}$ as that value of s which maximizes the statistic

$$L_f(s,y) = \sum_{n=1}^{N_s} \left[\log \frac{P(y_n \mid x_{sn})}{P_Q(y_n)} - \frac{1}{N_s} \log \frac{1}{P_s}\right] \tag{14.4}$$

We note the somewhat surprising but comforting fact that this "decoding metric" for message s depends only on that part of **y** of the same length as s.

Now consider how we can use the above metric to perform "almost MLD" of tree codes without exploring the entire code tree. As a specific instance, consider an $R = 1/N$ tree code with $N = 3$ for a binary input channel. Suppose that we have partially explored the tree as shown in Fig. 14.1. Which of the 4 terminal nodes should we extend to continue an efficient search for the same encoded path that a MLD would find at least "most of the time"? Recall that a MLD is equivalent to a minimum error probability decoder for equiprobable codewords. But assuming that the 2^{NRL} codewords in a tree code are equally likely is the same as assuming that, at any dividing node, the encoder follows any of the 2^{NR} branches with probability 2^{-NR}. Hence, we should assume that messages 1, 2, 3, and 4 in Fig. 14.1 have probabilities $P_1 = 2^{-1}$, $P_2 = P_3 = 2^{-3}$ and $P_4 = 2^{-2}$. We should then use these probabilities in (14.4) to find the decoding metric for each message. If we then extend the node with greatest metric and continue this process until we reach a terminal node in the full tree, we can be reasonably sure that we will have found the same path that a MLD would have found with its enormously greater searching. This is the basic concept behind **sequential decoding**, the two principal forms of which will be described in the next two sections.

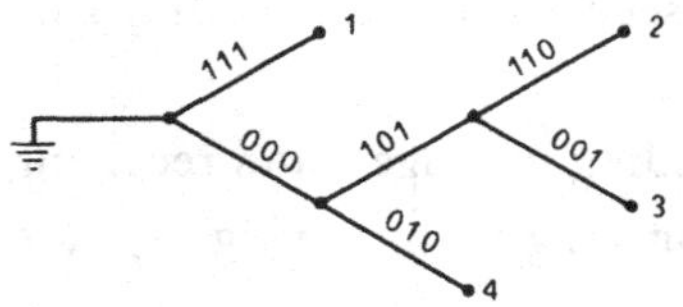

Fig. 14.1 **A Partially-Explored R = 1/3 Tree Code for a Binary Input Channel.**

Before launching into our discussion of sequential decoding, let us examine more closely the metric of (14.4) for a partially-explored tree code. If the node s is at depth u in the trellis, then $N_s = Nu$ (where N is the number of channel digits per branch) and the probability that the encoder reached this node in the tree in $P_s = 2^{-NRu}$. Substituting these values into (14.4) we find

$$L_f(s,\mathbf{y}) = \sum_{n=1}^{N_s} \left[\log \frac{P(y_n \mid x_{sn})}{P_Q(y_n)} - R\right] \qquad (14.5)$$

which is the appropriate decoding metric for tree codes.

The metric (14.5) was first used for tree codes by Fano [14] (in whose honor we have used the subscript f.) It is a remarkable tribute to Fano's intuition that he postulated this metric entirely on intuitive grounds. The analytical justification given here was formulated by this writer [15] almost ten years later!

A word should be said about the probability distribution Q(x) which should be used in (14.2). That this should be the minimizing distribution in (2.2) can be seen as follows. If our tree code is to give a decoding error probability P_e with MLD bounded as in (3.5), it should have the character of a tree code whose digits were selected independently according to this Q(x). Given any partially explored section, we should presume that the further digits in the unexplored section have the character as though they were selected independently according to Q(x) and this is the only knowledge we should presume of these digits until we explore further into the tree.

A numerical example may help the reader to obtain a better "feel" for the Fano metric. Consider a tree code for a BSC with crossover probability $\epsilon = .045$. For any binary input channel, and thus in particular for the BSC, the minimizing distribution in (2.2) is $Q(0) = Q(1) = 1/2$. For $\epsilon = .045$, one finds from (2.2) that $R_0 = .50$ so it is natural in our example to choose a tree code of rate $R = 1/2$. From (14.5) we see that the Fano metric for the n-th digit of the s path in the tree is given by

$$L_f(x_{sn}) = \log \frac{1 - .045}{.050} - .50 \cong + .50$$

when $x_{sn} = y_n$ whereas

$$L_f(x_{sn}) = \log \frac{.045}{.50} - .50 \cong - 3.5$$

when $x_{sn} \neq y_n$. In practice, one usually scales the metrics by a positive constant so that all metric values can be closely approximated by integers which then permits the use of integer arithmetic in the decoding apparatus. In the present example, one would choose a scale factor of 2 so that

$$L_f(x_{sn}) = +1 \text{ when } x_{sn} = y_n$$

$$-7 \text{ when } x_{sn} \neq y_n .$$

In the above example, we have changed our notation for the Fano metric to reflect the fact that the received vector **y** may be considered as fixed for a particular decoding situation and to reflect the fact that the Fano metric is an additive function of the digits along the assumed encoded path **x**. We shall continue henceforth with this simplified notation and write for instance

$$L_f(\mathbf{x}_s) = \sum_{n=1}^{N_s} L_f(x_{sn}) \tag{14.6}$$

where

$$L_f(x_{sn}) = \log \frac{P(y_n \mid x_{sn})}{P_Q(y_n)} - R. \tag{14.7}$$

15. Sequential Decoding--The Stack Algorithm

Sequential decoding is a generic name for any decoding procedure for a tree code which searches for the likely transmitted path for a given received sequence $\mathbf{y}_{[0\ L\ +\ T)}$ by successively exploring the encoded tree with the following constraints on the nature of the exploration :

(1) Any new nodes explored must be at the next depth beyond an already explored node, and

(2) No knowledge of the unexplored part of the tree is available except knowledge of the statistical distribution Q(x) which characterizes the ensemble of tree codes for which the code is use may be considered a "typical" member.

The discussion of the previous section has undoubtedly led the reader to deduce that the "obvious" sequential decoding algorithm is one which stores the Fano metric for the paths to all terminal nodes already explored and then extends to the next depth in the tree that path with the greatest Fano metric. This in fact is just the **stack algorithm** first considered by Zigangirov [16] and independently proposed later by Jelinek [17]. The adjective "stack" is used to indicate that the "natural" decoder operation is to store the previously explored nodes in a stack with decreasing metric down into the stack and, at each step, to extend the node at the top of the stack.

For R = 1/N tree codes where there are only two branches diverging from each node of the tree, we can state the strict **stack algorithm** as below where $\mathbf{x}_0$ denotes the encoded branch for the information digit 0 and $\mathbf{x}_1$ denotes the encoded branch for the information digit 1 on the two branches leading from the path i being extended. Each entry in the stack consists of a pair $[\mathbf{i}, L_f(\mathbf{x})]$ where **i** is the path through the tree, **x** the encoded digits on that path, and $L_f(\mathbf{x})$ the Fano metric for

x. The symbol Λ denotes the "empty string" which is the path to the root node of the tree. For $\mathbf{i} = \Lambda$, we have $\mathbf{x} = \Lambda$ and $L_f(\mathbf{x}) = 0$.

Step 0: Place $[\Lambda, 0]$ into the initially empty stack.

Step 1: Extend the top entry $[\mathbf{i}, L_f(\mathbf{x})]$ in the stack by forming

$$[i * 0, L_f(\mathbf{x}) + L_f(\mathbf{x}_0)]$$

and

$$[i * 1, L_f(\mathbf{x}) + L_f(\mathbf{x}_1)]$$

then deleting $[\mathbf{i}, L_f(\mathbf{x})]$ from the stack.

Step 2: Place the two newly-formed entries into the stack so that the stack remains ordered with entries with greater metric higher in the stack.

Step 3: If the top entry $[\mathbf{i}, L_f(\mathbf{x})]$ in the stack is a path through the entire tree, stop and choose $\hat{\mathbf{i}}_{[0,L+T)} = \mathbf{i}$. Otherwise, go to step 1.

It should be very plausible to the reader, in light of the discussion in Section 14, that the stack algorithm does essentially maximum likelihood decoding (MLD). In fact, one can show quite precisely that P_e for stack decoding of the ensemble of random tree codes satisfies the bound (2.2) except for a somewhat greater value of the "unimportant" constant c_t [17].

As it stands, however, the "strict" stack algorithm is not a very practical decoding procedure because of the increasing time required for step 2 as the size of the stack grows. Jelinek in his proposal of the stack algorithm also suggested a clever technique to speed up this searching step, at negligible cost in increased probability of error, by ignoring metric differences within a small specified **quantization parameter** Δ. More precisely, he considers a stack of "buckets" $\ldots B_{-2}, B_{-1}, B_0, B_1, \ldots$ such that B_j contains all entries $[i, L(x)]$ in the stack for which

$$j\Delta \leq L_f(\mathbf{x}) < (j + 1)\Delta \qquad (15.1)$$

and all "stores" and "fetches" from buckets are done on a "last in, first out" basis. Jelinek's **bucket stack algorithm** can be stated as follows:

Step 0: Place $[\Lambda, 0]$ into bucket B_0 of the initially empty stack.

Step 1: Fetch the most recent entry $[\mathbf{i}, L_f(\mathbf{x})]$ from the topmost non-empty bucket and form

$$[\mathbf{i} * 0, L_f(\mathbf{x}) + L_f(\mathbf{x}_0)]$$

and

$$[\mathbf{i} * 1, L_f(\mathbf{x}) + L_f(\mathbf{x}_1)].$$

Step 2: Store the two newly-formed entries into their appropriate buckets (as determined by (15.1).)

Step 3: If the most recent entry $[\mathbf{i}, L_f(\mathbf{x})]$ in the topmost non-empty bucket is a path through the entire tree, stop and choose $\hat{\mathbf{i}}_{[0, L+T)} = \mathbf{i}$. Otherwise, go to step 1.

Unlike the case for step 2 of the "strict" stack algorithm, the time required to execute step 2 of the "bucket" stack algorithm is independent of the number of entries already in the stack.

The practicality of the stack algorithm hinges on the amount of "computation" it requires before reaching a decoding decision. A computation is defined as the extension of one node, i.e. the performing of steps 1, 2 and 3 of the algorithm. Note that since one entry is deleted from the stack and two new entries are added in each computation, the number of computations performed at any time is equal to the number of entries in the stack. The quantity usually used to describe the computational effort is C_0, the number of computations required to decode divided by the number L of information bits decoded, i.e. the **per digit computation.** C_0 is of course a random variable whose value depends on how "noisy" is the received sequence $\mathbf{y}_{[0, L+T)}$. C_0 does not depend on the memory M of the convolutional code that might be used to generate the tree and depends very little on T. Hence, one usually chooses M and T so that the decoding error probability is negligibly small when the stack algorithm (or any other form of sequential decoding) is used. What limits performance is the decoding time available. If this time permits at most n_{max} computations, then the **deletion probability**

$$P_d = P(C_0 > \frac{n_{max}}{L}) \tag{15.2}$$

becomes the significant practical limitation. It turns out that C_0 is a particularly nasty kind of random variable, called a Pareto random variable (see [1] and [18]), for which $P(C_0 > n)$ decreases only as a small negative power of n. For this reason, P_d cannot be made extremely small with sequential decoding, $P_d \cong 10^{-3}$ being common in practice. Sequential decoding thus recommends itself to users who (1) have a feedback channel available on which to request repeats of deleted data, or (2) who do not mind deletion of small amounts of data, but who in either case demand a very small error probability in the undeleted data provided by the decoder.

The stack algorithm requires a substantial amount of storage but very little logical processing of its contents. Hence it is very well-matched to the characteristics of present-day mini-computers.

It is interesting to consider the nature of the path that will be selected by the stack algorithm (provided that it is granted sufficient time to complete its computation.) Let V_j , $0 \leq j \leq L+T$, denote the metric along a given path **i** out to depth j in the tree, that is

$$V_j = L_f(\mathbf{x}_{[0,j)}) \tag{15.3}$$

where $\mathbf{x}_{[0,j)}$ is the encoded path for $\mathbf{i}_{[0,j)}$ and similarly let V_j' denote the metric out of depth j along another path **i'**. We can then state:

The Non-Selection Principle for the Stack Algorithm: If the paths **i** and **i'** through the tree diverge at depth j and

$$\min(V_{j+1}, V_{j+2}, \dots V_{L+T}) > \min(V'_{j+1}, V'_{j+2}, \dots V'_{L+T}) \tag{15.4}$$

then **i'** cannot be the path at the top of the stack when the stack algorithm stops.

To prove this property, we first note that since **i** and **i'** diverge at depth j, we can write $\mathbf{i} = \mathbf{i}_{[0,j)} * \mathbf{i}_{[j,L+T)}$ and $\mathbf{i}' = \mathbf{i}_{[0,j)} * \mathbf{i}_{[j,L+T)}$. Hence, neither **i** nor **i'** can be the final path unless at some point the entry $[\mathbf{i}_{[0,j)}, V_j]$ reaches the top of the stack and is extended so that both $[\mathbf{i}_{[0,j+1)}, V_{j+1}]$ and $[\mathbf{i}'_{[0,j+1)}, V'_{j+1}]$ then enter the stack. At every subsequent point, $\mathbf{i}_{[0,j+1)}$ or one of its extensions must be in the stack so there will always be an entry in the stack whose metric V satisfies

$$V \geq \min(V_{j+1}, V_{j+2}, \dots V_{L+T}). \tag{15.5}$$

Now suppose that V'_{j+k} is minimum among $V'_{j+1}, \dots V'_{L+T}$. If **i'** reaches the top of the stack, then $\mathbf{i}'_{[0,j+k)}$ must have earlier reached the top of the stack since the complete path **i'** is an extension of this path. But by (15.4) and (15.5), $V'_{j+k} < V$ so that $\mathbf{i}'_{[0,j+k)}$ cannot reach the top of the stack as there is always in the stack some entry with a greater metric. By contradiction then, we conclude that **i'** cannot be the selected path.

Just as with the non-optimality principle for MLD, we can use the non-selection principle for stack decoding to find the path through the tree that will be found by a stack decoder. We would start at depth L - 1 and eliminate at each node that one of the 2 paths stemming therefrom with the smaller $\min(V_L, V_{L+1}, \dots V_{L+T})$. Then we would move to each node at depth L - 2 and eliminate that one of the two remaining paths stemming therefrom with the smaller $\min(V_{L-1}, V_L, \ldots V_{L+T})$, etc. When several centuries later (for L = 50, say) we reached depth 0, our elimination would leave us with one path, namely that one which would be selected finally by the stack algorithm. On the other hand, we might just apply the stack algorithm directly to the tree, starting at the root, and let it find its own path! The point is that (15.4) is only of importance conceptually in describing the nature of the finally selected path. Unlike the principal of non-optimality, it does not suggest directly a practical algorithm for finding the path of this nature.

It should be clear from the non-selection principle that P_e will be small for a stack decoder when and only when the metric tends to increase along the actual transmitted path through the tree but tends to decrease along every path diverging from this correct path. In fact, such reasoning is precisely what led Fano to the metric of (14.5). Fano reasoned that the first term in this metric is the "mutual information" between the received digit y_n and the hypothesized digit x_{sn}. If x_{sn} were the actual sent digit, the average of this mutual information should be channel capacity C so that the metric should on the average increase by the positive increment C - R at each digit on the correct path. On an incorrect path, the digit x_{sn} should be statistically independent of y_n so that the average mutual information should be 0 and thus the metric should on the average decrease by the negative increment - R at each digit along an incorrect path.

16. Sequential Decoding--The Fano Algorithm

The stack algorithm requires a substantial amount of storage in most applications. It uses this storage in order to perform an exploration of the tree that is "efficient" in the sense that each explored node is processed or "extended" only one time. The sequential decoding algorithm proposed by Fano [14] is a clever method for finding the path through the tree, as determined by the non-selection principle of the previous section, using almost no storage. The "trick" is to permit several extensions of the same node during the course of the search with all the necessary knowledge of previous searches reduced to a single stored number.

Perhaps Fano's algorithm can best be understood by an analogy. Consider the map of Fig. 16.1 which shows a tree of roads emanating from City A which we can consider the "root node" of the tree. (The "mysterious" city X whose elevation is very far below sea level is an artifice whose purpose will become clear but should

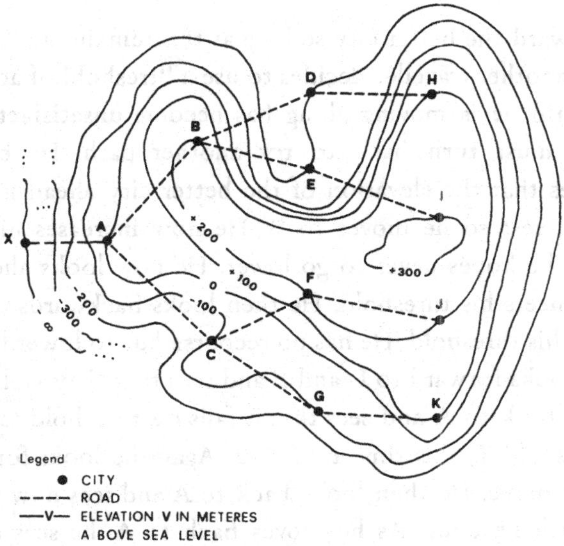

Fig. 16.1 Contour Map Used to Illustrate the Tree Search Performed by the Fano Algorithm.

be considered at depth - 1 from the root node A.) Suppose a traveller wishes to travel from A to that one of the terminal cities H, I, J, K that would be found by the stack algorithm, i.e. this path V_0, V_1, V_2, V_3 will win the inequality (15.4) against every path diverging from it. By applying the stack algorithm of the preceeding section, we find the stack contents after each computation to be as follows:

Computation No.	1	2	3	4	5
Contents of Stack	[A, 0]	[B, 200]	[C, −100]	[F, 200]	[J, 200]
		[C, −100]	[D, −200]	[G, −100]	[G, −100]
			[E, −200]	[D, −200]	[D, −200]
				[E, −200]	[E, −200]

Since J is a terminal city and is at the top of the stack, city J is the one that will be reached by a traveller who moves according to the stack algorithm. (J is not the highest terminal city, however; city I would be the one reached by a traveller who moved according to a MLD algorithm.) We see that it is necessary for the traveller to "remember" the elevation V of every city at the end of each of his explored paths in the course of his travelling in order to use the stack algorithm. On the other hand, the inequality (15.4) suggests that the traveller ought to be able, at each junction, to choose the road toward the better city so long as this remains an "acceptable" path.

Suppose another traveller decides to use a "treshold of acceptability" T to decide when the path he is moving along has become unsatisfactory according to (15.4) so that he must turn back to try another path. He begins at A with $T = V_A = 0$. He sees that the elevation of the better city ahead is $V_B = 200$ which meets his threshold test so he moves to B. He now increases his threshold T to $T = V_B = 200$ since he hopes never to go lower. He now looks ahead but sees that neither city ahead meets his threshold. He then looks backwards and sees that this city A is also below his threshold. He has no recourse but to lower his threshold T to $T = 100$. Again he looks forward to D and E and sees that they violate his threshold, then he again looks back to A and sees that it fails his threshold test also. Again, he must lower his threshold T, this time to $T = 0$. Again he looks forward to D and E but again he cannot move. He then looks back to A and sees now that he can move **backward** to A which he does. As he moves back to A, he sees that he is on the better path from A so his first action at A is to look forward on the other path to C but he then sees that he cannot move there. He then looks back to the mysterious city X and sees that he cannot move there. (He is trapped, there is no path through the tree from A that stays at 0 or above at all points!) He then reduces his threshold T to $T = -100$ and again looks forward. Again he moves to the better city B; but what should he now do to his threshold? (The traveller has forgotten by now, but we know that if he raises his threshold again to $V_B = 200$ he will be travelling forever in a loop.) The traveller knows however that **the threshold was not tight** at A on this move and he uses this as a signal not to meddle with T which he leaves at $T = -100$. He then looks forward and sees he cannot move. He looks back and sees that he can move backward to A which he does. Seeing that he has returned by the better road, he then looks forward along the worse road to C. He sees that he can move to C which he does and since his threshold $T = -100$ is already tight he need not worry about changing it. He then looks forward and sees that he can move which he does along the better path to F. He recalls that his threshold was tight at C for this move so he thus raises his threshold now to $T = V_F = 200$. He now looks forward and sees that he can move to J which he does. Since J is a terminal city he has completed his journey and ended up in the same city as the traveller who used the stack algorithm.

This second traveller has made use of the **Fano Algorithm** to guide his search of the tree [14]. We give a flowchart for this algorithm in Fig. 16.2. We

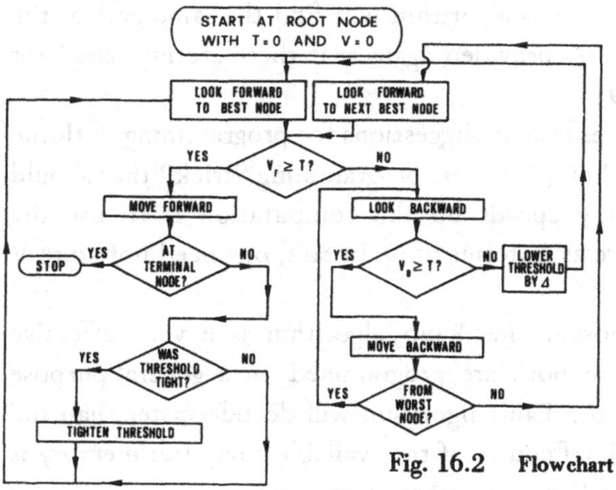

Fig. 16.2 Flowchart for the Fano Sequential Decoding Algorithm.

assume that there is a fictitious node, backwards from the root node, with metric $-\infty$ so that a look backward from the root node always results in lowering of the threshold by Δ. In this flowchart, V_f denotes the value of the metric for the path from the root node to the forward node at which one is looking, whereas V_B denotes the value of the metric for the path from the root node to the node being looked at which is behind the currently occupied node. By "tightening the threshold," we mean increasing T by the largest multiple of Δ such that the metric V from the root node to the currently occupied node still does not violate the threshold.

It is not hard to show that the Fano algorithm always finds the same path through the tree as the stack algorithm when (1) all node values are multiples of Δ and (2) there is no "tie" for the winning path, i.e. only one path through the tree is strictly not eliminated by the non-selection rule for the stack algorithm, cf. Geist [19]. Our traveller example suggests the key. The threshold will be lowered below 0 only at the root node and then only enough (if at all) to allow the path with greatest $\min(V_1, V_2, \ldots, V_{L+T})$ to be travelled. On the first arrival at any other node, the threshold will be tight before the next "look" since it was either tight on arrival or would be tightened by the outcome of the "was threshold tight" test. Hence, any other node has the same character search performed starting at first arrival there as does the root node. By iterating the root node argument, one finds that the path selected cannot be one that would have been eliminted by the non-selection rule. Since there is only one such path by assumption, one has verified that the Fano algorithm will find the same path as the stack algorithm. When condition (1) is not met, then, as might be expected, the Fano algorithm will find the same path as the bucket stack algorithm for the same Δ provided again that there are no "ties" for the winning path, again cf. Geist [19].

Geist has also given some excellent suggestions for programming both the stack algorithm and the Fano algorithm [20]. One programming "trick" that should always be used since it substantially speeds up the computation is to use the difference $Q = V - T$ and to do all tests in terms of Q. In fact, one need not store V and T separately.

When Δ is properly chosen, the Fano algorithm is a very effective sequential decoding procedure. When both are programmed on a general-purpose computer, simulations suggest that the Fano algorithm will decode faster than the stack algorithm for $R < .9\,R_0$ [20]. Of course, if the available computer memory is small, one would be forced to use the Fano algorithm in any case.

It is curious to note the history of sequential decoding. The order of topics in Sections 14, 15 and 16 (which seems the most logical ordering) is actually in the reverse order of their discovery! The first sequential decoding algorithm, due to Wozencraft [6], inspired all these subsequent discoveries but is much more complicated as well as less effective than the stack and Fano algorithms. It has taken a long time to reach the point where one understands why the early sequential decoding algrithms "worked" and what they were really doing.

17. Aknowledgment and References

It would have been impossible to have reached the view of non-block codes that we have shared with the reader of these lectures without the stimulation we have received from engaging in search in this subject with our students. In particular, we must acknowledge Dr. Daniel J. Costello and Dr. John M. Geist, both of whom I had the pleasure to supervise as graduate students, and Mr. Rolf Johannesson who is just completing a year of study with me on leave from his home university, the Technical University of Lund, Sweden. Among colleagues who have greatly influenced my thoughts on non-block codes are, preeminently, Dr. G. David Forney, Jr., of the Codex Corporation, Newton, Mass., U.S.A., and Prof. Robert G. Gallager of M.I.T., Cambridge, Mass., U.S.A. Not least, I must mention the skillful introduction to, and encouragement to work in, the field of non-block coding that I received as a graduate student at M.I.T. in 1961-1962 under the supervision of Prof.

John M. Wozencraft.

Finally, I wish to thank Prof. Giuseppe Longo of the University of Trieste, Italy, for his invitation to present these lectures at the Centre International des Sciences Mécaniques, Udine, Italy, and for his interest in and encouragement of their written form.

REFERENCES

[1] Gallager, R.G., *Information Theory and Reliable Communication*, John Wiley and Sons, New York, 1968.

[2] Johannesson, R., "On The Error Probability of General Tree and Trellis Codes with Applications to Sequential Decoding," Tech. Rpt. No. EE 7316, Dept. of Elec. Engr., Univ. of Notre Dame, Notre Dame, Indiana, U.S.A. December 1973.

[3] Viterbi, A.J., "Convolutional Codes and Their Performance in Communication Systems," *IEEE Trans. Comm Tech.*, Vol. COM-19, pp. 751-772, October, 1971.

[4] Larsen, K.J., "Short Convolutional Codes with Maximum Free Distance for Rates 1/2, 1/3, and 1/4", *IEEE Trans. Info. Th.*, Vol. IT-19, pp. 371-372, May, 1973.

[5] Johannesson, R., "Robustly-Optimal Rate One-Half Binary Convolutional Codes", *IEEE Trans. Info. Th.*, Vol. IT-21, pp. 464-468, July 1975.

[6] Wozencraft, J.M., and B. Reiffen, *Sequential Decoding*, M.I.T. Press, Cambridge, Mass., 1959.

[7] Massey, J.L., *Threshold Decoding*, M.I.T. Press, Cambridge, Mass., 1963.

[8] Costello, D.J., Jr., "A Construction Technique for Random-Error-Correcting Convolutional Codes," *IEEE Trans. Info. Th.*, Vol. IT-15, pp. 631-636, September 1969.

[9] Forney, G. David, Jr., "Convolutional Codes I: Algebraic Structure," *IEEE Trans. Info. Th.*, Vol. IT-16, pp. 720-738, November, 1970.

[10] Massey J.L., and M.K. Sain, "Inverses of Linear Sequential Circuits," *IEEE Trans. Computers*, Vol. C-17, pp. 330-337, April, 1968.

[11] Olson, R.R., "Note on Feedforward Inverses of Linear Sequential Circuits," *IEEE Trans. Computers*, Vol. C-19, pp. 1216-1221, December, 1970.

[12] Viterbi, A.J., "Error Bounds for Convolutional Codes and an Asymptotically Optimum Decoding Algorithm," *IEEE Trans. Info. Th.*, Vol. IT-13, pp. 260-269, April, 1967.

[13] Omura, J.K., "On the Viterbi Decoding Algorithm," *IEEE Trans. Info. Th.*, Vol. IT-15, pp. 177-179, January, 1969.

[14] Fano, R.M., "A Heuristic Discussion of Probabilistic Decoding," *IEEE Trans. Info. Th.*, Vol. IT-9, pp. 64-74, April 1963.

[15] Massey, J.L., "Variable-Length Codes and the Fano Metric," *IEEE Trans. Info. Th.*, Vol. IT-18, pp. 196-198, January, 1972.

[16] Zigangirov, K.Sh., "Some Sequential Decoding Procedures," *Prob. Pederachi Inform.*, Vol. 2. pp. 13-25, 1966.

[17] Jelinek, F., "A Fast Sequential Decoding Algorithm Using a Stack," *IBM J. Res. Dev.*, Vol. 13, pp. 675-685, November, 1969.

[18] Wozencraft, J.M., and I.M. Jacobs, *Principles of Communication Engineering*, John Wiley and Sons, New York, 1965.

[19] Geist, J.M., "Search Properties of Some Sequential Decoding Algorithms," *IEEE Trans. Info. Th.*, Vol IT-19, pp. 519-526, July, 1973.

[20] Geist, J.M., "A, Empirical Comparison of Two Sequential Decoding Algorithms," *IEEE Trans. Comm. Techn.*, Vol. COM-19, pp. 415-419, August, 1971.

[21] Van De Meeberg, L., " A Tightened Upper Bound on the Error Probability of Binary Convolutional Codes with Viterbi Decoding," *IEEE Trans. Info. Th.*, Vol. IT-20, pp. 389-391, May, 1974.

THRESHOLD DECODING
– A tentative survey –

J.M. Goethals
MBLE Research Laboratory
Avenue Van Becelaere 2,
B-1170 Brussels (Belgium)

Abstract

The content of these notes is a tentative survey of the results recently obtained concerning one of the most attractive methods known for decoding linear block codes. The material is divided into four sections, each followed by a list of references. The first serves as an introduction to the subject. The second section deals with geometric codes, which constitutes the largest class of threshold - decodable codes. The third section is concerned with the generalized Reed-Muller codes, which provide the setting for studying geometric codes. Finally, in section four, an example is given of the improvements of Rudolph's method which can be obtained from the combinatorial structure of a code.

1. Threshold Decoding of Linear Block Codes

1.1. Introduction: the Decoding Problem for Linear Block Codes '

A **block code** of length n is a collection of n-tuples $(a_1, a_2, \ldots a_n)$ from a given alphabet. The alphabet usually consists of the elements of a finite field GF(q). We shall mainly be concerned with binary codes, for which the alphabet simply consists of the two elements, 0 and 1, of the binary field GF(2). The set V(n,q) of all n-tuples from the field GF(q) has the structure of an n-dimensional space, and a block code over the alphabet GF(q) is **linear** when the set of codewords (= elements of the code) coincide with a k-dimensional subspace of V(n,q), for some k, $1 \leqslant k \leqslant n$. We shall refer to an (n,k) code in that case. Any such code is uniquely defined by a set of k linearly independent codewords, which form a **generating set**, in the sense that any codeword is uniquely expressible as a linear combination of these k generators. The code contains q^k codewords, corresponding to the q^k distinct linear combinations of the k generators. Hence, given a generating set $(\mathbf{u}_1, \mathbf{u}_2, \ldots, \mathbf{u}_k)$ for the code, the map

$$(a_1, a_2, \ldots, a_k) \rightarrow a_1\mathbf{u}_1 + a_2\mathbf{u}_2 + \ldots + a_k\mathbf{u}_k) \tag{1.1}$$

defines the **encoding process** which sets up a one-to-one correspondence between the set of k-tuples $(a_1, a_2, \ldots, a_k)$ from GF(q) and the set of codewords. A generating set is said to be in **systematic form** when the matrix formed with the k generators takes the form

$$\begin{bmatrix} \mathbf{u}_1 \\ \mathbf{u}_2 \\ \cdots \\ \mathbf{u}_k \end{bmatrix} = [I_k \mid P], \tag{1.2}$$

where the first columns form a unit matrix of order k. In that case, the first k components of the codeword defined by (1.1) coincide with the k-tuple $(a_1,a_2,...,a_k)$, while the remaining n-k components are linear combinations of these. More precisely, denoting by $p_{i,j}$, $1 \leqslant i \leqslant k$, $1 \leqslant j \leqslant n\text{-}k$, the entries of the matrix P appearing in (1.2), we have, for the codeword defined by (1.1),

$$a_1\mathbf{u}_1 + a_2\mathbf{u}_2 + ... + a_k\mathbf{u}_k = (a_1, a_2, ..., a_k, a_{k+1}, ..., a_n)$$
$$a_{k+j} = \sum_{i=1}^{k} a_i p_{i,j}, \quad j = 1,2,...,n\text{-}k. \tag{1.3}$$

Hence, any codeword is uniquely defined by its first k components, called **information symbols.** The remaining n-k symbols, defined by (1.3), are called the **parity symbols.** Up to a permutation of the coordinate places, any linear block code possesses a generating set in systematic form.

An (n,k) code, being a k-dimensional subspace of V(n,q), can be defined to be the set of n-tuples $(a_1,a_2,...a_n)$ from GF(q) satisfying a set of n-k linearly independent linear equations

$$\Sigma\, a_i b_{i,j} = 0, \quad j = 1,2,...,n\text{-}k. \tag{1.4}$$

We observe that the equations (1.3) form such a system, with coefficients

$$b_{i,j} = p_{i,j}, \quad 1 \leqslant i \leqslant k,$$

$$b_{k+i,j} = -\delta_{i,j}, \quad j = 1,2,...,n\text{-}k. \tag{1.5}$$

We assume now that after transmission through some noisy channel a received n-tuple $(r_1,r_2,...,r_n)$ is obtained which differs from the transmitted codeword $(a_1,a_2,...,a_n)$ by some **error sequence** $(e_1,e_2,...,e_n)$, that is

$$r_i = a_i + e_i, \quad i = 1,2,...,n, \tag{1.6}$$

where r_i and e_i are elements of GF(q), and addition is performed in that field. Since the set of equations (1.4) is satisfied by all codewords, we have, for j = 1,2,...,n-k,

$$\sum_{i=1}^{n} r_i b_{i,j} = \sum_{i=1}^{n} e_i b_{i,j} = s_j, \text{ say.} \tag{1.7}$$

The (n-k)-tuple $(s_1, s_2, \ldots s_{n-k})$ of values obtained in this way, which only depends on the error sequence, is called the **syndrome** of the error sequence. There are exactly q^k error sequences which produce a given syndrome. The **decoding problem** consists in finding that error sequence, producing a given syndrome, that is most probable for a given channel and a given transmitted codeword. For a symmetric memoryless channel (like the binary symmetric channel), the most probable error sequence is that one which has the minimum number of nonzero components, independently of the transmitted codeword. In that case, the decoding problem reduces to finding the solution $(e_1, e_2, \ldots, e_n)$ of the equations (1.7), with the minimum number of nonzero components. We observe that any two distinct solutions of (1.7) differ by a nonzero codeword. Hence, if every nonzero codeword has at least 2t+1 nonzero components, there is at most one solution of (1.7) with t or fewer nonzero components, and thus any error sequence with t or fewer nonzero components will be correctly decoded. For that reason, a linear code with the above property is said to be **t-error-correcting**. In practice, it is generally not feasible to find the most probable solution of (1.7), for an arbitrary syndrome, simply because of the enormous number of possibilities. An efficient soultion of the decoding problem depends on finding a simply instrumented method for determining the most probable solution of (1.7) for a high probability subset of the set of all syndromes. This subset usually consists of the syndromes corresponding to error sequences with e or fewer nonzero components, where $e \leqslant t$ for a t-error-correcting code. In that respect, threshold decoding appears to be a satisfactory method, when applicable, because of the simplicity of its implementation. The concept of a set of orthogonal parity cheks, and its generalizations, plays a central role in this method, and will first be discussed.

1.2 Orthogonal Parity Cheks and Majority Decoding

Let us first introduce some terminology. By a **parity chek rule** for a linear code, we mean a linear equation, like (1.4), satisfied by all codewords. For an (n,k) code, there are q^{n-k} parity check rules, each of which may be obtained by a linear combination of a set of n-k linearly independent ones. The value resulting from the application of a given parity check rule on a received n-tuple, and which only depends on the error sequence, will be called a **parity check.** Hence a parity check may be viewed as a linear combination

$$s_k = \sum_{i=1}^{n} e_i b_{i,k} \tag{1.8}$$

of the error digits e_i. We shall say that the i-th digit e_i is **checked** by the parity check (1.8) when $b_{i,k} \neq 0$.

Definition 1. A set of r parity checks (1.8) is said to be **orthogonal** on e_m if

$$b_{m,k} = 1, \text{ for all } s_k \text{ in the set,}$$

and if each of the remaining error digits e_i is checked by at most one parity check of the set.

Thus, e_m is able to affect all of the r patity checks in the set, but no other digit can affect more than a single parity check in the set. This observation leads to the following theorem.

Theorem 1. Provided that $[r/2]$ or fewer of the $\{e_i\}$ that are checked by a set of r parity checks s_k orthogonal on e_m are nonzero, then e_m is given correctly as that value of GF(q) which is assumed by the absolute majority of the set $\{s_k\} \cup \{0\}$ that is $e_m = \mathrm{Maj}(\{s_k\},0)$.

Proof. Assume that e among the $\{e_i\}$ that are checked are nonzero, and suppose first $e_m = 0$. Then, at most e of the $\{s_k\}$ are nonzero, hence at least r-e are zero. Thus if $e \leqslant [r/2]$, at least $[(r+1)/2]$ of the $\{s_k\}$ are zero, whence $\mathrm{Maj}(\{s_k\},0) = 0$. Suppose now $e_m = a$, $a \neq 0$. Then, at most e-1 of the $\{s_k - a\}$ are nonzero, whence at least r-(e-1) of the $\{s_k\}$ are equal to a. But, if $e \leqslant [r/2]$, we have $r-(e-1) > [(r+1)/2]$, and thus $\mathrm{Maj}(\{s_k\},0) = a$. Hence, in all cases, provided that $e \leqslant [r/2]$, we have $e_m = \mathrm{Maj}(\{s_k\}, 0)$, and the theorem is proved.

If $r = 2t$ is an even number, it follows from theorem 1 that e_m can be correctly determined when t or fewer of the received symbols are in error. Similarly, if $r = 2t + 1$ is odd, then again e_m can be correctly determined whenever t or fewer errors have occurred, and in addition $t + 1$ errors may be detected when the value assumed by the greatest fraction of the $\{s_k\}$ is nonzero and has exactly $t+1$

occurrences. It follows from these considerations that any nonzero codeword with $a_m \neq 0$ has at least $r + 1$ nonzero components. We further remark that, when r is an even number, the zero is superfluous, and e_m is, under the assumption of theorem 1, correctly given by $e_m = \text{Maj}(\{s_k\})$. Whenever a set of r parity checks orthogonal on e_m can be found for each digit e_m, $m = 1, 2, \ldots, n$, then decoding can be performed according to the algorithm of theorem 1 for each of them, and we shall refer to this method as **one-step majority decoding** with orthogonal parity checks.

In the binary case ($q = 2$), the algorithm of theorem 1 can be simplified as follows. Since, in that case, each s_k is 0 or 1, we have

$$\text{Maj}(\{s_k\},0) = \begin{cases} 1, \text{ if } \Sigma\, s_k > [(r+1)/2], \\ \\ 0, \text{ otherwise,} \end{cases}$$

where $\Sigma\, s_k$ denotes the sum of the s_k (as real numbers). Hence, the **decoding rule** of theorem 1 simply is: **choose $e_m = 1$ if, and only if, the real sum $\Sigma\, s_k$ exceeds the threshold value $[(r + 1)/2]$.** This method can easily be instrumented by means of a threshold logical element.

1.3 Multistep Majority Decoding with Orthogonal Parity Checks.

The maximal value of r for which a set of r parity checks orthogonal on e_m can be found for each digit e_m , is usually much smaller than $2t+1$ for an arbitrary t-error-correcting code. In that case, on-step majority decoding is rather inefficient as compared to the error-correcting capability of the code. It is then often possible to increase the efficiency of majority decoding by proceeding in several steps of majority decisions, where at each step a new set of parity check estimates is obtained. This method, called **multistep majority decoding**, uses at each step sets of parity checks that are orthogonal on selected sums of error symbols, which we define as follows.

Definition 2. A set of r parity checks (1.8) is said to be orthogonal on the sum

$$s = b_{m_1} e_{m_1} + b_{m_2} e_{m_2} + \ldots + b_{m_p} e_{m_p} \tag{1.9}$$

if each parity check s_k of the set is of the form $s_k = s + s_k'$ where no symbol of the subset $\{e_{m_1}, e_{m_2}, \ldots, e_{m_p}\}$ is checked by s'_k, and each of the remaining n-p symbols is checked by at most one s'_k in the set. It follows from the definition that any error symbol not involved in the sum can affect at most one parity check of the set. This observation leads to theorem 2, the proof of which, being very similar to the one of theorem 1, will be omitted.

Theorem 2. Provided that $[r/2]$ or fewer of the $\{e_i\}$ that are checked by a set of r parity checks s_k orthogonal on the sum s are nonzero, then s is given correctly as that value of GF(q) which is assumed by the absolute majority of the set $\{s_k\} \cup \{0\}$, that is

$$s = \mathrm{Maj}(\{s_k\}, 0). \tag{1.10}$$

The value of s, as given by (1.10) (*), can then be viewed as a new parity check (1.9), which in turn, combined with other known parity checks, may produce new sets that are orthogonal on a subsum of s, etc... One might hope to obtain, after a number of steps, a set of parity checks orthogonal on one single symbol e_m, from which an estimate of e_m is finally obtained by theorem 1. One easily sees that this final estimate is obtained through a tree of majority decisions, where at each step at least 2t parity checks are needed to garantee t-error-correction. Roughly speaking, with a tree consisting of L levels, a total of at least

$$1+2t+(2t)^2+\ldots+(2t)^{L-1} = ((2t)^L-1)/(2t-1) \tag{1.11}$$

majority elements are required, per decoded symbol, for t-error-correction. Fortunately, however, this value can be drastically reduced in many cases by improvements of the method.

We conclude this paragraph with a simple example, to illustrate the concepts introduced.

Example 1. A (7,4) binary code has the following set of 3 linearly independent parity checks:

$$s_1 = e_1+e_2+e_4+e_7, \quad s_2 = e_1+e_2+e_3+e_5,$$
$$s_3 = e_2+e_3+e_4+e_6.$$

(*) Set $s = 0$, in case there is no majority, to avoid ambiguity.

We observe that s_1 and s_2 are orthogonal on e_1+e_2, while s_2 and s_1+s_3 are orthogonal on e_1+e_3. Hence, provided that no more than one error has occurred, the following sequence of majority decisions correctly determines e_1 :

$$s_4 = e_1+e_2 = \mathrm{Maj}(s_1, s_2),$$

$$s_5 = e_1+e_3 = \mathrm{Maj}(s_2, s_1+s_3),$$

$$s_6 = e_1 = \mathrm{Maj}(s_4, s_5).$$

In short, $e_1 = \mathrm{Maj}(\mathrm{Maj}(s_1, s_2), \mathrm{Maj}(s_2, s_1 + s_3))$. Similarly, $s_7 = e_2 + e_4 =$ $= \mathrm{Maj}(s_1, s_3)$, and we have now enough information to obtain all of the $\{e_i\}$ since

$$e_1=s_6,\ e_2=s_4+e_1,\ e_3=s_5+e_1,\ e_4=s_6+e_2,$$

$$e_5=s_2+e_1+e_2+e_3,\ e_6=s_3+e_2+e_3+e_4,$$

$$e_7=s_1+e_1+e_2+e_4.$$

Hence, four majority elements are enough to determine all of the $\{e_i\}$. This is because the four parity checks s_4, s_5, s_6 and s_7 obtained by majority decisions together with the original set $\{s_1, s_2, s_3\}$ for a nonsingular set of 7 linear equations in the 7 unknown e_i. In general, for an (n,k) code, it would be sufficient to determine k parity checks in addition to, and linearly independent from, the original set of n-k.

1.4 Threshold Decoding with Nonorthogonal Parity Checks.

The basic idea of majority decoding can be applied to more general configurations of parity checks, as we shall see.

Definition 3. A set of r parity checks is said to form an **(r,λ) configuration on e_m** if e_m is checked with coefficient 1 by each parity check of the set and if no other error digit is checked by more than λ parity checks of the set. Comparing with definition 1, one easily verifies that a set of parity checks orthogonal on e_m forms an (r,1) configuration on e_m. From the definition, it follows that e_m is able to affect all of the r parity checks in the set, but no other error digit can affect more than λ parity checks. This observation leads to the following theorem.

Theorem 3. Provided that $[(r + \lambda - 1)/2\lambda]$ or fewer of the $\{e_i\}$ that are checked by a set of r parity checks forming an (r,λ) configuration on e_m are nonzero, then e_m is given correctly as that value of GF(q) which is assumed by the absolute majority of the elements of the set S consisting of the r parity checks and λ zeros.

Proof. Assume that e among the $\{e_i\}$ that are checked are nonzero, and suppose first $e_m = 0$. Then, at most $e\lambda$ of the r parity checks are nonzero, whence at least $r-e\lambda$ are zero. Thus, the number of zeros in the set S is at least equal to $r-e\lambda+\lambda$. Suppose now $e_m = a$, $a \neq 0$. Then, at most $(e-1)\lambda$, of the r parity checks are not equal to a, whence at least $r-(e-1)\lambda$ elements in the set S are equal to a. It suffices now to observe that, for $e \leqslant [(r+\lambda-1)/2\lambda]$, we have

$$r-e\lambda+\lambda > [(r+\lambda)/2],$$

which shows that, in each case, the correct value of e_m is assumed by the absolute majority of the elements of S. This proves the theorem.

The algorithm of theorem 3 can be easily implemented in the binary case by means of a threshold logical element, since e_m is then simply given by the following rule: Choose $e_m = 1$, if, and only if, the real sum of the r parity checks exceeds the threshold value $[(r+\lambda)/2]$.

A well known class of combinatorial designs can be used to construct (r,λ) configurations on each error digit. These are the balanced incomplete block designs which are defined as follows. An incidence structure between a set A of n elements, called points, and a set B of b elements, called blocks, is said to form a **balanced incomplete block design** with parameters (n,b,r,d,λ), if the following conditions are satisfied: every point is incident to exactly r blocks, every block is incident to exactly d points, and any two distinct points are simultaneously incident to exactly λ blocks. These conditions imply the following relations among the parameters:

$$nr = bd, \quad \lambda(n-1) = r(d-1).$$

The point block (0,1)-incidence matrix N of this configuration satisfies

$$NN^T = (r-\lambda)I + \lambda J,$$

from which it follows that, for $r > \lambda$, NN^T is nonsingular, whence $b \geqslant n$. If the column vectors of N are used to define parity check rules for a code of length n, it is easily verified that the parity checks corresponding to the r blocks incident to any given point do form an (r,λ) configuration on the error digit corresponding to that point. Hence, decoding can be performed according to the algorithm of theorem 3 We shall examine later on how this idea has been exploited by use of the block designs arising in finite geometires.

Let us finally point out that this method can be extended to multi-step decoding by use of (r, λ) configurations of parity checks on selected sums of error digits. This generalization is straightforward and will not be futher discussed.

We conclude this section by showing how the (7,4) binary code of example 1 can be decoded by use of (4,2) configurations of parity checks on each error digit. Returning to example 1, we observe that the four parity checks

$$s_1 = e_1+e_2+e_4+e_7, \quad s_2 = e_1+e_2+e_3+e_5,$$

$$s_1+s_3 = e_1+e_3+e_6+e_7, \quad s_2+s_3 = e_1+e_4+e_5+e_6,$$

form a (4,2) configuration on e_1. Similar configurations can be constructed on each digit. In fact, the seven parity checks s_1, s_2, s_3, s_1+s_2, s_1+s_3, s_2+s_3, $s_1+s_2+s_3$ define the incidence structure of a block design with parameters $n = b = 7$, $r = d = 4$, $\lambda = 2$. Hence, provided that no more than one error has occurred, decoding can be performed correctly by the algorithm of theorem 3.

1.5 Notes

Most of the material of this section is based on the work of Massey [1]. Rudolph [2] was the first to apply majority decoding with (r, λ) configurations of parity checks and proved that the error-correcting capability of the algorithm described in theorem 3 was at least equal to $[r/2\lambda]$. Bastin [3] and Ng [4] independently improved Rudolph's result by showing that the error-correcting capability was at least equal to $[(r+\lambda-1)/2\lambda]$. .

1.6 References

[1] **Massey, J.L., *Threshold Decoding,* M.I.T. Press, Cambridge, mass. (1963).**

[2] **Rudolph, L.D., A class of majority logic decodable codes, *IEEE Trans. on Information Theory, IT-13* (1967), 305-307.**

[3] **Bastin, G., A propos du décodage par décision majoritaire, *Revue H.F.*, 8, (1970), 85-95.**

[4] **NG, S.W., On Rudolph's majority logic decoding algorithm, *IEEE Trans. on Information Theory, IT-16,* (1970), 651-652.**

2. Geometric Codes

We use the name geometric codes for the class of codes which use the incidence structure of points and flats in Euclidean or projective geometries over finite fields to define parity check rules, Since these geometric structures are essential to the decoding algorithm, we shall first briefly recall their main properties. For more details, we refer to [1].

2.1 Finite Geometries

Definition 4. A Euclidean Geometry EG(m,q) of dimension m over the field GF(q) consists of q^m **points** and sets of **t-flats**, for t = 0,1,...,m, which are recursively defined as follows. The 0-flats are the points of the geometry, which can be identified with the q^m distinct m-tuples of elements from GF(q). For any t-flat E_t and any point α not belonging to E_t in EG(m,q), there is a unique (t+1)-flat E_{t+1} in EG(m, q) which contains both E_t and α. This (t+1)-flat consists of the set of points

$$E_{t+1} = \{\beta + \omega(\alpha-\gamma) \mid \beta \in E_t,\ \omega \in GF(q)\},$$

where γ is any fixed point of E_t.

In the following, we shall make frequent use of the **q-ary Gaussian coefficients** $\left[{m \atop i}\right]$ defined as follows, for i = 0,1,...,m.

$$\left[{m \atop 0}\right] = 1,\ \left[{m \atop i}\right] = \prod_{j=0}^{i-1}(q^m - q^j)/(q^i - q^j),\ i=1,2,\ldots,m. \tag{2.1}$$

Note that $\left[{m \atop i}\right]$ is the number of distinct i-dimensional subspaces of an m-dimensional vector space over GF(q). We further quote the following properties of these Gaussian coefficients:

$$\left[{m \atop m-i}\right] = \left[{m \atop i}\right],$$

$$\left[{m \atop i}\right] = \left[{m-1 \atop i}\right] + q^{m-i}\left[{m-1 \atop i-1}\right]. \tag{2.2}$$

For more details, we refer to [4].

2.1.1 From definition 4, the following properties are easily derived:

(i) A t-flat contains exactly q^t points and has the structure of an Euclidean geometry EG(t,q).

(ii) An EG(m,q) contains $q^{m-t}[{m \atop t}]$ distinct t-flats, for t = 0,1,...m.

(iii) For any s,t, with $0 \leqslant s \leqslant t \leqslant m$ each s-flat is properly contained in exactly $[{m-s \atop t-s}]$ distinct t-flats in EG(m,q).

From these properties, it follows that the incidence structure, defined by inclusion between points (or 0-flats) and t-flats in EG(m,q), is a balanced incomplete block design with the parameters $n = q^m$, $b = q^{m-t}[{m \atop t}]$, $r = [{m \atop t}]$, $d = q^t$ and $\lambda = [{m-1 \atop t-1}]$, for t = 1,2,...,m-1.

Definition 5. A **projective geometry** (PG(m,q) of dimension m over the field GF(q) consists of $[{m+1 \atop 1}]$ points and sets of t-flats, for t= 0,1,...,m, which are recursively defined as follows. The 0-flats are the points of the geometry, which can be identified with the $[{m+1 \atop 1}]$ distinct one-dimensional spaces of the space of (m+1)-tuples over GF(q). In other words, PG(m,q) can be identified with a maximal set of nonzero (m+1)-tuples no two of which are dependent over GF(q). Then, for any t-flat P_t and any point α, not belonging to P_t in PG(m,q), there is a unique (t+1)-flat P_{t+1} in PG(m,q) which contains both P_t and α. This (t+1)-flat consits of the set of points

$$P_{t+1} = \{\alpha\} \cup \{\beta + \omega\alpha \mid \beta \in P_t,\ \omega \in GF(q)\}.$$

2.1.2 From this definition, the following properties are easily derived:

(i) A t-flat contains exactly $[{t+1 \atop 1}]$ points and has the structure of a projective geometry PG(t,q).

(ii) A PG(m,q) contains $[{m+1 \atop t+1}]$ distinct t-flats, for t = 0,1,...,m.

(iii) For any s,t, with $0 \leqslant s \leqslant t \leqslant m$, each s-flat is properly contained in exactly $[{m-s \atop t-s}]$ distinct t-flats in PG(m,q),

From these properties, it follows that the incidence structure, defined by inclusion between points (or 0-flats) and t-flats, is a balanced incomplete block design with the parameters $n = [{m+1 \atop 1}]$, $b = [{m+1 \atop t+1}]$, $r = [{m \atop t}]$, $d = [{t+1 \atop 1}]$ and $\lambda = [{m-1 \atop t-1}]$, for t =1,2, ...,m-1.

2.2 Basic Majority Decoding Algorithms

In this section, we shall discuss the two basic decoding algorithms that can be applied to any linear code having in its parity check space the incidence vectors of all of the t-flats of a finite geometry, EG(m,q) or PG(m,q). No assumption is made on the field on which these codes are defined, since the above incidence vectors, with all of their components equal to 0 or 1, can be defined over any field. The first algorithm makes use of the block design structure of these incidence vectors, and was first obtained by Rudolph [8]. The second algorithm proceeds in several steps and makes use of the geometric structure of this block design. It was first obtained by Goethals and Delsarte [3], and independently by Smith [9], and Weldon [12]. As we shall see, this second algorithm has, in general, a much greater error-correcting capability than the first. It has, however, the disadvantage of a greater complexity.

2.2.1 Rudolph One-Step Majority Decoding.

At the end of section 1.4, it was observed that a block design with parameters (n,b,r,d,λ) provide (r,λ)-configurations of parity checks on each error digit of a code of length n. Hence, by use of the block design structure of the incidence vectors of t-flats in a finite geometry, decoding can be performed according to the algorithm of theorem 3 for any error digit. The error-correcting capability of this algorithm is given by $[(r+\lambda-1)/2\lambda]$, where $r = [{m \atop t}]$ and $\lambda = [{m-1 \atop t-1}]$, for both EG(m,q) and PG(m,q). Hence, this capabiltiy is at least equal to the integer part of

$$\frac{1}{2}\, r/\lambda = \frac{1}{2}(q^m - 1)/(q^t - 1), \qquad (2.3)$$

but differ from this value by at most 1.

2.2.2 Multistep Majority Decoding

From the definition of a t-flat in a finite geometry, it follows that the incidence vectors of all of the t-flats which contain a given (t-1)-flat define a set of $r = [{m-(t-1) \atop t-(t-1)}]$ parity checks that are orthogonal on the sum of the error digits associated to the points of that (t-1)-flat. For, given a (t-1)-flat E, any point α not belonging to E is contained in precisely one t-flat that contains E. Hence, by the algorithm of theorem 2, a new parity check can be obtained that corresponds to the incidence vector of the (t-1)-flat E. This can be done for all of the (t-1)-flats that contain a given (t-2)-flat, which in turn define a set of parity checks orthogonal on that (t-2)-flat. Hence, by induction, we obtain after t steps a set of parity checks orthogonal on a 0-flat, that is on a single error symbol. At the i-th step, the number of orthogonal parity checks that can be used is $\left[{m-t+i \atop 1}\right]$. Hence, there are at least $\left[{m-t+1 \atop 1}\right]$, at each step, and the error-correcting capability of the algorithm is given by

$$\frac{1}{2}\left[{m-t+1 \atop 1}\right] = \frac{1}{2}\,(q^{m-t+1}-1)/(q-1). \tag{2.4}$$

Comparing this with (2.3), we conclude that, for $t > 1$, this algorithm has a better error-correcting capability than the first. However, while the first algorithm requires only one majority decision, the second requires a number growing exponentially with t, per decoded digit. On the other hand, the total number of distinct parity checks used is the same, namely $\left[{m \atop t}\right]$, but they are used quite differently in the two methods, when $t > 1$. All are used for one single majority decision in the first method, while only a small portion is used for each majority decision, in the second. Note that, for $t = 1$, the two methods are identical.

By combining the two methods, it is possible to reduce the number of steps to at most two, in all cases, while retaining the error-correcting capability of the multistep method. This method, due to Weldon [11], will be briefly examined hereunder.

2.2.3 Weldon's hybrid method.

Assuming that the incidence vectors of the $(t+1)$-flats are used as parity check rules, we may obtain, as in a first step of the second method, the parity checks corresponding to all of the t-flats, provided the number of errors does not exceed $1/2(q^{m-t}-1)/q-1)$. Now, we may apply the first method to obtain directly the estimate of a single error digit, provided the number of errors does not exceed (2.3). And, since

$$(q^m-1)/(q^t-1) > (q^{m-t}-1)/(q-1),$$

this 2-step algorithm has the same error-correcting capability $1/2(q^{m-t}-1)/(q-1)$ as the $(t+1)$-step method.

2.3 Geometric Codes - Definitions and Main Properties.

As we have pointed out in the preceding section, any linear code having in its parity check space the incidence vectors of all of the t-flats of a finite geometry over GF(q) can be decoded according to one of the algorithms described hereabove. It is therefore quite natural to define a geometric code to be the largest linear code having that property. But now the question arises: on what particular field alphabet GF(p) should the code be defined? It turns out that the code is trivial (that is of dimension 0 to 1) unless GF(q) and GF(p) are fields of the same characteristic. In the following we shall assume that p is a prime number, and q some power p^ν of that prime. In that case, the dimension of the code can be found by a method we shall briefly discuss in the next section. In the present section, we shall merely give a definition and state the main properties of these codes, which we call Euclidean geometry codes or projective geometry codes according to the type of the geometry.

2.3.1 Euclidean Geometry Codes (in short: EG codes)

Definition 6: For $q = p^\nu$, the (t, ν)-EG code of length q^m is the largest linear code over GF(p) having in its parity check space the incidence vectors of all of the $(t+1)$-flats in EG(m,q).

Let us denote by $R_p(t,m,q)$ the p-rank of the incidence matrix of all of the t-flats in EG(m,q). Then, obviously, the (t,ν)-EG code of length q^m over GF(p) has dimension

$$k = q^m - R_p(t+1, m, p^\nu). \tag{2.5}$$

For a positive integer u, we define $w_q(u)$, the **q-weight** of u, to be the (real) sum of the digits appearing in its q-ary expansion. That is, for $u = u_0 + u_1 q + u_2 q^2 + \ldots$, with $0 \leqslant u_i \leqslant q-1$, we have

$$w_q(u) = u_0 + u_1 + u_2 + \ldots \tag{2.6}$$

With this definition, $R_p(t,m,q)$, where $q = p^\nu$, can be shown to be equal to the number of distinct positive integers u less than q^m satisfying

$$\max_{0 \leqslant i < \nu} w_q(p^i u) \leqslant (m-t)(q-1). \tag{2.7}$$

For example, the binary (0,2)-EG code of length 16 has dimension $k = 16 - R_2(1,2,4)$, where $R_2(1,2,4) = 9$, since the numbers 0,1,2,3,4,6,8,9,12 all satisfy the condition (2.7).

In the particular case when t= m-1, a closed form can be given for $R_p(t,m,p^\nu)$, namely

$$R_p(m-1,m,p^\nu) = \binom{m+p-1}{p-1}^\nu, \tag{2.8}$$

cf. MacWilliams and Mann [7]. For q = p, the number of integers u, $0 \leqslant u < q^m$ with $w_q(u) = i$, is given by the coefficient of Z^i in

$$(1+Z+Z^2+...+Z^{p-1})^m = \left(\frac{1-Z^p}{1-Z}\right)^m,$$

from which it follows that $R_p(t,m,p)$ is the coefficient of $Z^{(m-t)(p-1)}$ in

$$(1-Z^p)^m\ (1-Z)^{-(m+1)}. \tag{2.9}$$

In particular, for p =,2, we have

$$R_2(t,m,2) = \sum_{i=0}^{m-t} \binom{m}{i} \tag{2.10}$$

In fact, it can be shown that the (t,1)-EG binary code of length 2^m is identical to the t-th order Reed-Muller code of length $n = 2^m$, dimension $K = \binom{m}{0} + \binom{m}{t} + ... + \binom{m}{1}$, and minimum distance 2^{m-t} .

Since the basic multistep majority decoding algorithm for the (t, ν)-EG code has the error-correcting capability $1/2(q^{m-t}-1)/(q-1)$, it follows that its minimum distance is at least equal to $(q^{m-t}-1)/(q-1)+1$. However, it can be shown that this minimum distance is at least equal to $(q+p)\,q^{m-t-2}$, by use of the BCH bound, cf. Chow [2] and Lin [6]. Note that these two bounds agree and give the correct minimum distance in the case when q=p=2, that is for the Reed-Muller codes.

2.3.2 Projective Geometry Codes (in short **PG codes**).

Definition 7. For $q = p^\nu$, the (t, ν)-**PG code** of length $(q^{m+1}-1)/(q-1)$ is the largest linear code over GF(p) having in its parity check space the incidence vectors of all of the t-flats in PG(m,q).

We shall denote by $Q_p(t,m,q)$ the rank over GF(p) of the incidence matrix of all of the t-flats in PG(m,q). Then, $Q_p(t,m,p^\nu)$ is equal to the number of distinct positive integers u less than $(q^{m+1}-1)/(q-1)$ satisfying

$$\max_{0 \leqslant i < \nu} w_q[p^i u(q-1)] \leqslant (m-t)(q-1). \tag{2.11}$$

The dimension k of the p-ary (t, ν)-PG code of length $(q^{m+1}-1)/(q-1)$ is then simply given by

$$k = (q^{m+1}-1)/(q-1) - Q_p(t,m,p^\nu). \tag{2.12}$$

Observing that there exists a one-to-one correspondence between the numbers u', with

$$(q^m-1)/(q-1) \leqslant u' < (q^{m+1}-1)/(q-1),$$

satisfying (2.11), and the numbers u, with $0 \leqslant u < q^m$, satisfying (2.7), we conclude that the numbers $Q_p(t,m,q)$ satisfy the following recurrence relation,

$$Q_q(t,m,q) - Q_p(t,m-1,q) = R_p(t,m,q),$$

from which we obtain

$$Q_p(t,m,q) = \sum_{i=0}^{m-t} R_p(t,m-i,q). \tag{2.13}$$

In particular, from (2.8), we obtain

$$Q_p(m-1,m,p^\nu) = 1 + \binom{m+p-1}{p-1}^\nu, \tag{2.14}$$

cf., for instance, Goethals and Delsarte [3].

The class of q-ary $(1,\nu)$-PG codes of length $n = (q^3-1)/q-1) = q^2+q+1$, was originally introduced by Weldon [10] under the name **Different set codes**, for which Graham and MacWilliams [5] obtained the dimension

$$k = q^2 + q - \binom{p+1}{p-1}^\nu;$$

in agreement with (2.12) and 2.14).

2.4 References

[1] Carmichael, R.D., *Introduction to the theory of groups of finite order*, Dover, N.Y. (1956) Chapter 11.

[2] Chow, D.K., On threshold decoding of cyclic codes, *Information and Control, 13* (1968), 471-483,

[3] Goethals J.M., and P. Delsarte, On a class of majority logic decodable cyclic codes, *IEEE Trans. on Information Theory, IT-14* (1968), 182-188.

[4] Goldman J. and G.C. Rota, On the foundation of combinatorial theory IV: Finite vector spaces and Eulerian genrating functions, *Studies in Appl. Math. 49* (1970), 239-258.

[5] Graham R.L. and F.J. MacWilliams, On the number of information symbols in difference set cyclic codes, *Bell System Techn. J., 45* (1966), 1057-1070.

[6] Lin, S., On a class of cylcic codes, *Error Correcting Codes* (H.B. Mann, ed.), J. Wiley N.Y. (1968), 131-148.

[7] Mac Williams, F.J., and H.B. Mann, On the p-rank of the design matrix of a difference-set, *Information and Control, 12* (1968), 477-488.

[8] Rudolph, L.D., A class of majority logic decodable cyclic codes, *IEEE Trans. on Information Theory, IT-13* (1967), 305-307.

[9] Smith, K.J.C., Majority decodable codes derived from finite geometries, *Univ. North Carolina, Inst. of Statistics Mimeo Series* No. 587 (1968).

[10] Weldon, E.J., Difference-set cyclic codes, *Bell System Techn. J., 45*, (1966), 1045-1056.

[11] Weldon, E.J., Some results on majority-logic decoding, *Error-Correcting Codes* (H.B. Mann, ed.), J. Wiley, N.Y. (1968), 149-162.

[12] Weldon, E.J., Euclidean geometry codes, *Combinatorial Mathematics and its Applications* (R.C. Bose and T.A. Dowling eds.) Univ. North Carolina Press, Chapell Hill, N.C. (1969), 377-387.

3. Generalized Reed-Muller Codes

Geometric codes appear to be strongly related to the class of codes introduced by Kasami, Lin and Peterson [2], and Weldon [7], under the name of generalized Reed-Muller codes. This latter class appears as a natural generalization of the codes discovered by Muller [4] and studied by Reed [6], who first obtained a multistep majority logic decoding algorithm for these codes, similar although not identical to the one described in § 2.2.2. These codes, called Reed-Muller codes, will first be described. We shall then proceed to the generalization and briefly indicate how the geometric codes described hereabove appear in this setting.

3.1 Reed-Muller codes

3.1.1 Let $V(m,2)$ be the m-dimensional vector space of m-tuples over the binary field $GF(2)$. Then, any function f from $V(m,2)$ into $GF(2)$ can be described by a polynomial in the m indeterminates $x_1, x_2, \ldots, x_m$ over $GF(2)$,

$$f(x_1, x_2, \ldots, x_m) = \Sigma A(i_1, i_2 \ldots, i_m)\, x_1^{i_1} x_2^{i_2} \ldots x_m^{i_m},$$

with degree at most equal to one with respect to each variable x_i. For convenience, we shall use the following conventions: we shall write

$$\mathbf{x} \text{ for } (x_1, x_2, \ldots, x_m),$$

$$\mathbf{x}^i \text{ for } x_1^{i_1} x_2^{i_2} \ldots x_m^{i_m}, \tag{3.1}$$

where $i = i_1 + 2i_2 + 2^2 i_3 + \ldots + 2^{m-1} i_m$.

With these conventions, the above polynomial is simply written as follows:

$$f(\mathbf{x}) = \sum_{i=0}^{2^m-1} A_i \mathbf{x}^i. \tag{3.2}$$

The set $F(m,2)$ of all functions $f\colon V(m,2) \rightarrow GF(2)$ is a 2^m-dimensional space, with basis $\{\mathbf{x}^i \mid 0 \leq i \leq 2^m - 1\}$. Clearly, the set of monomials $\mathbf{x}^i$ with degree $(i_1 + i_2 + \ldots + i_m)$ less than or equal to r, for some integer number r, $0 \leq r < m$, generates a subspace of $F(m,2)$, whose dimension is given by $(\binom{m}{0} + \binom{m}{1} + \ldots + \binom{m}{r}$.

The map ϕ: f $\rightarrow$ v(f), where v(f) is the 2^m-tuple $(v_0, v_1, \dots, v_{2^m-1})$ of values of the function f in V(m,2), clearly induces an isomorphism between F(m,2) and the vector space $V(2^m,2)$ of 2^m-tuples over GF(2). Hence, the image under ϕ of the set of all functions of degree less than or equal to r is a linear binary code of length $n=2^m$, and dimension $k=\binom{m}{0}+\binom{m}{1}+\dots+\binom{m}{r}$. This is the **r-th order Reed-Muller code of length 2^m**, in short **RM(m,r)**. In other words, any code word in RM(m,r) is described by the 2^m-tuple $v(f) = (v_0, v_1, \dots, v_{2^m-1})$ of values of a function f of degree less than or equal to r. The function $x_1 x_2 \dots x_r$ has degree r and has exactly 2^{m-r} nonzero values in V(m,2). It can be shown that any function of degree r or less has at least 2^{m-r} nonzero values, from which it follows that the minimum distance of the code RM(m,r) is exactly 2^{m-r}. In fact, any codeword of weight 2^{m-r} in RM(m,r) is described by a function of the form $f = y_1 y_2 \dots y_r$, where the y_i's are linearly independent linear functions $y_i = b_i + \sum_j a_{ij} x_j$. It follows that any such codeword is the incidence vector of an (m-r)-flat in EG(m,2).

The dual code of RM(m,r) can be shown to be the Reed-Muller code RM(m,m-1-r), by use of the following easily verified arguments:

(i) The componentwise product of two vectors v(f) and v(g) is the vector v(fg).

(ii) the sum of the coordinates of any v(f) is nonzero modulo 2 if, and only if, the function f has degree m.

(iii) dim.RM(m,r) + dim.RM(m,m-1-r) = 2^m.

From the above discussion, we conclude that the Reed-Muller code RM(m,t) has in its parity check space the incidence vectors of all of the (t+1)-flats in EG(m,2), and since it is the largest code having this property, it follows that a (t,1)-EG code of length 2^m is nothing but a Reed-Muller code RM(m,t).

3.1.2. Any Reed-Muller code RM(m,r) of degree $r<m$ has the property that the sum of the coordinates of every codeword is zero modulo 2. Hence, for any $v(f) \in$ RM(m,r), with

$$v(f) = (v_0, v_1, \dots, v_{2^m-1}), \quad v_0 = f(0,0,\dots,0),$$

we may write

$$v_0 = v_1 + v_2 + \dots + v_{2^m-1} \tag{3.2}$$

For the remaining coordinates $v_1, v_2, \dots v_{2^m-1}$, we shall adopt an ordering which will provide RM(m,r) with the structure of an extended cyclic code, cf. [5]. This ordering is obtained as follows. The field $GF(2^m)$ has the structure of an m-dimensional vector space over GF(2). Given a primitive element α, any nonzero element α^i in the field can be expressed as a linear combination of the first m powers of α, that is

$$\alpha^i = \sum_{j=1}^{m} a_{i,j}\alpha^{j-1}, \quad \text{for } i = 1,2,\ldots,2^m-1. \tag{3.3}$$

We shall identify the nonzero m-tuple $(a_{i,1}, a_{i,2}, \ldots, a_{i,m})$ with the field element α^i. With this convention, we may write $f(\alpha^i)$ for $f(a_{i,1}, a_{i,2}, \ldots, a_{i,m})$, and define v_i in $\mathbf{v}(f)$ to be

$$v_i = f(\alpha^i), \quad \text{for } i = 1,2,\ldots,2^m-1. \tag{3.4}$$

Now, let $\mathbf{v}(f) = (v_0, v_1, v_2, \ldots, v_{2^m-1})$ be any codeword in RM(m,r), where the v_i's are defined as in (3.2) and (3.4), and let us show that the vector $(v_0, v_2, v_3, \ldots, v_{2^m-1}, v_1)$ belongs to the same code. To that end, we define A to be the companion matrix of the minimal polynomial of α. Then, postmultiplying by A the m-tuple $(a_{i,1}, a_{i,2}, \ldots, a_{i,m})$ corresponding to α^i, cf. (3.3), merely produces the m-tuple $(a_{i+1,1}, a_{i+1,2}, \ldots, a_{i+1,m})$ corresponding to α^{i+1}. Hence, defining $g(\mathbf{x})$ to be $g(\mathbf{x}) = f(\mathbf{x}A)$, we have $g(\alpha^i) = f(\alpha^{i+1})$, whence, with v_i defined by (3.4),

$$\mathbf{v}(g) = (v_0, v_2, v_3, \ldots, v_{2^m-1}, v_1) \tag{3.5}$$

and since $g(\mathbf{x})$ clearly has degree at most equal to that of $f(\mathbf{x})$, the vector (3.5) belongs to RM(m,r), which was to be shown. This suffices to show that RM(m,r) has the structure of an extended cyclic code.

Any cyclic code of length 2^m-1 is uniquely defined by a polynomial g(z), called **generator polynomial**, which divides $(z^{2^m-1}-1)$ over GF(2). To any codeword $(v_1, v_2, \ldots, v_{2^m-1})$ is associated a polynomial $v(z) = \sum v_i z^{i-1}$ divisible by g(z). The code may as well be defined by the set of nonzero field elements α^i that are zeros of g(z), since the zeros of $(z^{2^m-1}-1)$ are the 2^m-1 nonzero elements of $GF(2^m)$. More generally, for any integer $n \mid q^m-1$, a cyclic code of length n over GF(q) is defined by a generator polynomial g(z) which divides z^n-1 over GF(q). Since

$$(z^n-1) \mid (z^{q^m-1}-1), \quad \text{for } n \mid q^m-1,$$

the code may as well be defined by the set of elements of $GF(q^m)$ which are zeros of g(z). The dimension k of the code is related to the degree of g(z) by n-k = deg.g(z). If z-1 does not divide g(z), the (n+1,k) code over GF(q) obtained by adding an extra coordinate $v_0 = -(v_1+v_2+...+v_n)$ to all codewords $(v_1,v_2,...,v_n)$ of the cyclic code generated by (g(z)), is called an **extended cyclic code.** The dual of the cyclic code generated by g(z) is the cyclic code generated by $h^*(z)$, where $h^*(z)$ is the reciprocal polynomial of

$$h(z) = (z^n-1)/(g(z)) \tag{3.6}$$

and if $(z-1) \nmid g(z)$, the dual of the extended cyclic code generated by g(z) is the extended cyclic code generated by $h^*(z)/(z-1)$.

We shall now give, without proof, a characterization of the Reed-Muller codes as extended cyclic codes.

Theorem. The Reed-Muller code RM(m,t) is the extended cyclic code of length 2^m generated by the polynomial g(z) having as zeros the elements α^i of $GF(2^m)$ with exponents i satisfying

$$0<i<2^m-1, \quad w_2(i) \leqslant m-t-1 \tag{3.7}$$

Remark. Compare (2.7) and (3.7).

3.2 Generalization of the Reed-Muller Codes

3.2.1 Primitive Codes

Let V(m,q) be the m-dimensional vector space of m-tuples over GF(q). Then, any function f from V(m,q) into GF(q) can be described by a polynomial in the m indeterminates $x_1,x_2,...,x_m$ over GF(q), with degree at most equal to (q-1) with respect to each variable x_i. For convenience, we shall denote by $\mathbf{x}^i$ a typical monomial

$$x_1^{i_1}x_2^{i_2}...x_m^{i_m}, \quad 0\leqslant i_k\leqslant q-1, \tag{3.8}$$

appearing in the expansion of $f(\mathbf{x}) = f(x_1,x_2,...,x_m)$, by using the one-to-one correspondence between the integer numbers $i, 0\leqslant i\leqslant q^m-1$, and the m digits i_k appearing in their q-ary expansion

$$i = i_1 + i_2 q + i_3 q^2 + ... + i_m q^{m-1}. \tag{3.9}$$

The degree of the monomial (3.8) is then simply given by the q-weight $w_q(i)$ of the number (3.9). The set of monomials $\mathbf{x}^i$ of degree $w_q(i) \leqslant R$, for some integer number R, $0 \leqslant R \leqslant m(q-1)$, clearly generates a subspace of the q^m-dimensional space F(m,q) of all functions f: V(m,q) → GF(q).

The map ϕ: f → $\mathbf{v}$(f), where $\mathbf{v}$(f) is the q^m-tuple of values of the function f in V(m,q), clearly induces an isomorphism between F(m,q) and V(q^m,q). Hence, the image under ϕ of the set of all functions of degree less than or equal to R is a linear q-ary code of length q^m. This is the **R-th order** (primitive) **generalized Reed-Muller code of length** q^m, in short **RM(m,R,q)**. Any codeword of this code is described by the q^m-tuple $\mathbf{v}(f) = (v_0, v_1, ..., v_{q^m-1})$ of values of a function of degree R or less. The dimension of the code is equal to the number of monomials (3.8) of degree less than or equal to R, that is to the number of distinct integers i satisfying

$$0 \leqslant i \leqslant q^m - 1, \quad w_q(i) \leqslant R. \tag{3.10}$$

Let r and s be, respectively, the quotient and the remainder obtained upon dividing R by (q-1), that is

$$R = r(q-1) + s. \tag{3.11}$$

Then, it can be shown that any function f ∈ F(m,q) of degree R or less has at least

$$(q-s)q^{m-1-r} \tag{3.12}$$

nonzero values in V(m,q). In particular, the function f of degree R, defined by

$$f(\mathbf{x}) = \prod_{i=1}^{r} (1 - x_i^{q-1}) \prod_{j=1}^{s} (\omega_j - x_m), \tag{3.13}$$

where $\omega_1, \omega_2, ..., \omega_s$ are distinct elements of GF(q), has exactly $(q-s)q^{m-1-r}$ nonzero values. Hence, the minimum distance of the code RM(m,R,q) is exactly given by (3.12). By use of the following easily verified arguments:

(i) the componentwise product of the two vectors $\mathbf{v}$(f) and $\mathbf{v}$(g) is the vector $\mathbf{v}$(fg),

(ii) the sum of the coordinates of any $\mathbf{v}(f)$ is nonzero in GF(q) if, and only if, the function f has degree m(q-1),

(iii) $\dim.RM(m,R,q) + \dim.RM(m,m(q-1)-R-1,q) = q^m$,

it can be shown that the dual code of RM(m,R,q) is the code RM(m,m(q-1)-R-1,q).

Now, let $y_1, y_2, \ldots, y_{m-t}$, be (m-1) linarly independent linear functions $y_i = b_i + \Sigma\, a_{ij} x_j$, , and let us consider the function

$$f(\mathbf{x}) = \pi_{i=1}^{m-t} (1 - y_i^{q-1}), \tag{3.14}$$

of degree (m-t) (q-1). This function f has values:

1, if and only if, $y_1 = y_2 = \ldots = y_{m-t} = 0$,

0, otherwise.

Hence, $\mathbf{v}(f)$ is the (0,1)-incidence vector of a t-flat in EG(m,q) = V(m,q). It follows that the incidence vectors of all of the t-flats in EG(m,q) belong to any RM(m,R,q) code with $R \geq (m-t)(q-1)$.

Let $q = p^\nu$. Then, GF(p) is a subfield of GF(q). Given a linear code C over GF(q), the **subfield subcode** of C over GF(p) consists of those codewords of C all of whose components lie in GF(p). Now, let us consider the code RM(m,(m-t) (q-1),q), where $q = p^\nu$ and p is a prime number. From the above, it clearly appears that its subfield subcode over GF(p) contains the incidence vectors of all of the t-flats of EG(m,q). In fact, it can be shown that these incidence vectors generate that subfield subcode. In other words the p-rank of the incidence matrix of these t-flats, that is $R_p(t,m,p^\nu)$, is equal to the dimension of the subfield subcode of RM(m,(m-t)(q-1)) over GF(p). Now, let $f(\mathbf{x})$ be any function in F(m,q). Its values in V(m,q), that is the components v_i of $\mathbf{v}(f)$, all lie in GF(p) if, and only if,

$$f^p(\mathbf{x}) \equiv f(\mathbf{x}) \text{ modd}(x_i^q - x_i), \quad i = 1,2,\ldots,m. \tag{3.15}$$

This latter condition implies that, for any monomial (3.8) appearing in the expansion of $f(\mathbf{x})$, the monomial $x_1^{pi_1} x_2^{pi_2} \ldots x_m^{pi_m}$, , reduced modd $(x_i^q - x_i)$, also appear in $f(\mathbf{x})$ and has degree $w_q(pi)$. It follows that, for any $\mathbf{v}(f)$ belonging to the subfield subcode over GF(p) of RM(m,(m-t) (q-1)) only those monomials $\mathbf{x}^u$ whose exponent u satisfies

$$w_q(up^j) \leqslant (m\text{-}t)(q\text{-}1), \quad j = 0,1,2,\ldots,\nu\text{-}1, \tag{3.16}$$

can appear in f(**x**). Hence, the dimension $R_p(t,m,q)$ of this subfield subcode is equal to the number of distinct integers u less than q^m satisfying (3.16), or equivalently (2.7).

The generalized Reed-Muller codes and their subfield subcodes can be given the structure of extended cyclic codes, by use of a suitable ordering of the coordinates. The reasoning is similar to the one used in §3.1.2 and will not be reproduced here. We merely give, without proof, a characterization of these codes as extended cyclic codes.

Theorem The generalized Reed-Muller code RM(m,R,q) is the extended cyclic code of length q^m over GF(q) generated by the polynomial g(z) having as zeros those elements α^u of GF(q^m) whose exponents u with respect to the primitive element α satisfy

$$0 < u < q^m\text{-}1, \quad w_q(u) \leqslant m(q\text{-}1)\text{-}R\text{-}1. \tag{3.17}$$

Its subfield subcode over GF(p), where $q = p^\nu$, is the extended cyclic code over GF(p) generated by the polynomial having as zeros those α^u for which u satisfies

$$0 < u < q^m\text{-}1, \quad \min_{0 \leqslant i < \nu} w_q(up^i) \leqslant m(q\text{-}1)\text{-}R\text{-}1. \tag{3.18}$$

Now, since the (t,ν)-EG code of length q^m over GF(p) is the largest code orthogonal to the incidence vectors of all of the (t + 1)-flats in EG(m,q), and since these generate the subfield subcode of RM(m,(m-t-1)(q-1),q) over GF(p), it follows that we have the following corollary.

Corollary. The (t,ν)-EG code of length q^m over GF(p) is the extended cyclic code generated by the polynomial having as zeros those α^u for which u satisfies

$$0 < u < q^m\text{-}1, \quad \max_{0 \leqslant i < \nu} w_q(up^i) \leqslant (m\text{-}t\text{-}1)(q\text{-}1). \tag{3.19}$$

3.2.2 Nonprimitive Codes

Let us consider two nonzero m-tuples **a**, **b** in V(m,q) that are multiples of each other over GF(q), that is $\mathbf{b} = \omega.\mathbf{a}$, for some $\omega \neq 0$ in GF(q). Then, for any function f(**x**) in F(m,q) whose polynomial expansion only contains monomials (3.8) of degree $w_q(i)$ divisible by (q-1), we have the property that f(**b**) = f(**a**). We shall denote by $F_{q-1}(m,q)$ the subspace of F(m,q) generated by the set of monomials $\mathbf{x}^i$ of degree $w_q(i)$ divisible by (q-1) and less than m(q-1). By using the congruence $i \equiv w_q(i) \bmod (q-1)$, we may identify this set of monomials with

$$\{\mathbf{x}^i \mid 0 \leq i < q^m-1, \; i \equiv 0 \bmod (q-1)\} \tag{3.20}$$

Hence, dim. $F_{q-1}(m,q) = (q^m-1)/(q-1)$.

For any two dependent m-tuples **a**, **b**, over GF(q), and for any f(**x**) in $F_{q-1}(m,q)$, we have f(**a**) = f(**b**). Thus, we may restrict our attention to a maximal set of $(q^m-1)/(q-1)$ m-tuples in V(m,q), no two of which are dependent over GF(q), that is to the set of points of the projective geometry PG(m-1,q). Then, clearly, the map $\psi : f \rightarrow \mathbf{u}(f)$, where **u**(f) is the $[(q^m-1)/(q-1)]$-tuple of values of f in PG(m-1,q), sets up an isomorphism between $F_{q-1}(m,q)$ and the vector space $V(N_m,q)$, where for convenience we write N_m for $(q^m-1)/(q-1)$. Thus, the image under ψ of any subspace of $F_{q-1}(m,q)$ is a linear code of length N_m over GF(q). In particular, for some integer number r, $0 \leq r < N_m$, the image of the set of polynomials of degree less than or equal to r(q-1) in $F_{q-1}(m,q)$ is the **r-th order nonprimitive generalized Reed-Muller code of length** $(q^m-1)/(q-1)$, in short **NPRM(m,r,q)**. Any codeword of this code is described by the N_m-tuple of values in PG(m-1,q) of a function f of degree r(q-1) or less belonging to $F_{q-1}(m,q)$. The dimension of this code is equal to the number of integers u satisfying

$$0 \leq u < (q^m-1)/(q-1), \quad w_q[u(q-1)] \leq r(q-1). \tag{3.21}$$

The minimum distance is easily shown to be equal to $(q^{m-r}-1)/(q-1)$, and the codewords of minimum weight are multiples of incidence vectors of (m-1-r)-flats in PG(m-1,q).

We shall now examine briefly how the projective geometry codes can be described in this setting. Given any set of (m-t) linearly independent linear homogeneous functions $y_i = \Sigma a_{ij}x_j$ of the (m+1) variables $x_0, x_1, \dots, x_m$, the set of points $\mathbf{x} = (x_0, x_1, \dots, x_m)$ in PG(m,q) satisfying $y_1 = y_2 = \dots = y_{m-t} = 0$, is a t-flat in PG(m,q). It follows that the vector **u**(f) defined by the function

$$f(\mathbf{x}) = \prod_{i=1}^{m-t} (1-y_i^{\,q-1}),$$

belonging to $F_{q-1}(m+1,q)$ is the incidence vector of the above r-flat. Since $\deg f(\mathbf{x}) = (m-t)(q-1)$, we conclude that the incidence vectors of all the t-flats in PG(mq,) belong to the subfield subcode of any NPRM $(m+1,r,q)$ with $r \geqslant m-t$. In fact, it can be shown that the subfield subcode of NPRM $(m+1,m-t,p^\nu)$ over GF(p) is generated by these incidence vectors of t-flats, from which we conclude that the (t,ν)-PG code of length $(q^{m+1}-1)/(q-1)$ **over GF(p) is the dual of the subfield subcode of NPRM** $(m+1,m-t,p^\nu)$.

By use of a suitable ordering of the coordinates, the nonprimitive generalized Reed-Muller codes and their subfield subcodes can be shown to be cyclic. We conclude this section by giving a characterization of these codes as cyclic codes. For a primitive element α in $GF(q^{m+1})$, we denote by β the primitive N_{m+1}-th root of unity $\beta = \alpha^{q-1}$, where $N_{m+1} = (q^{m+1}-1)/(q-1)$.

Theorem. The nonprimitive generalized Reed-Muller code NPRM$(m+1,r,q)$ is the cyclic code of length N_{m+1} over GF(q) generated by the polynomial having as zeros the elements β^u of $GF(q^{m+1})$ for which u satisfies

$$0 < u < N_{m+1}, \quad w_q[u(q-1)] \leqslant (m-r)(q-1). \tag{3.22}$$

Its subfield subcode over GF(p), where $q = p^\nu$, is the cyclic code generated by the polynomial having as zeros the elements β^u for which u satisfies

$$0 < u < N_{m+1}, \quad \min_{0 \leqslant i < \nu} w_q[up^i(q-1)] \leqslant (m-r)(q-1). \tag{3.23}$$

Corollary. The (t,ν)-PG code of length N_{m+1} over GF(p) is the cyclic code generated by the polynomial having as zeros the elements β^u for which u satisfies

$$0 \leqslant u < N_{m+1}, \quad \max_{0 \leqslant i < \nu} w_q[up^i(q-1)] \leqslant (m-t)(q-1). \tag{3.24}$$

Remark. Compare (3.24) and (2.11).

3.3 Notes

The material of this section is based primarily on the work of Kasami, Lin, Peterson and Weldon, cf. [2], [3], [7], although the approach used here follows more closely [1]. Our main goal was to describe the interrelation between geometric codes and generalized Reed-Muller codes, and in particular to obtain a characterization of these codes as cyclic (or extended cyclic) codes. The cyclic nature of these codes permits, indeed, to reduce the complexity of a decoding circuit to the one needed for estimating a single error digit.

For more details, we refer the reader to [5].

3.4 References

[1] Delsarte, P., J.M. Goethals and F.J. MacWilliams, On generalized Reed-Muller codes and their relatives, *Information and Control, 16* (1970), 403-442.

[2] Kasami, T., S. Lin and W.W. Peterson, New generalizations of the Reed-Muller codes — Part I: primitive codes, *IEEE Trans. on Information Theory, IT-14* (1968), 189-199.

[3] Kasami, T., S. Lin and W.W. Peterson, Polynomial codes, *IEEE Trans. on Information Theory, IT-14* (1968), 807-814.

[4] Muller, D.E., Application of Boolean algebra to switching circuit design and to error detection, *IRE Trans. Electronic Comput., EC-3* (1954), 6-12.

[5] Peterson, W.W., and E.J. Weldon, *Error-Correcting Codes* (2nd Ed.), MIT Press, Cambridge, Mass. (1972).

[6] Reed, I.S., A. Class of multiple-error-correcting codes and the decoding scheme, *IRE Trans. on Information Theory, PGIT-4* (1954), 38-49.

[7] Weldon, E.J., New generalizations of the Reed-Muller codes — Part II: nonprimitive codes, *IEEE Trans. on Information Theory, IT-14* (1968) 199-205.

4. Threshold Decoding with t-Design

4.1. Definitions and Main Properties

A **t-design** $D_\lambda(t,d,n)$ is a collection of d-subsets of an n-set, called **blocks**, having the property that any t-subset is contained in a constant number λ of blocks, cf. [1]. A 2-design is also called a **balanced** incomplete block **design**, and a t-design with $\lambda = 1$ is usually called a **Steiner system**. For any $i \leq t$, a t-design $D_\lambda(t,d,n)$ yields an i-design $D_{\lambda_i}(i,d,n)$, where

$$\lambda_i = \lambda\binom{n-i}{t-i}/\binom{d-i}{t-i} \tag{4.1}$$

is the number of blocks containing any given i-subset. In particular, we have $\lambda_t = \lambda$, and by convention, $\lambda_0 = \lambda\binom{n}{t} / \binom{d}{t}$ denotes the number of blocks. Similarly, it can be shown that for any i,j, with $i + j \leq t$, the number of blocks containing any i-subset, but containing no element of any given j-subset disjoint from that i-subset, is a constant, denoted by λ_{ij}, where $\lambda_{i,0} = \lambda_i$ by definition. From the following, quite obvious, recurrence

$$\lambda_{i+1,j} = \lambda_{i,j} - \lambda_{i,j+1}, \tag{4.2}$$

these numbers are easily obtained from the λ_i's . For example, the well-known Steiner system $D_1(5,8,24)$ has $\lambda_5=1$, $\lambda_4=5$, $\lambda_3=21$, $\lambda_2=77$, $\lambda_1=253$, $\lambda_0=759$, from which the $\lambda_{i,j}$ with $i+j \leq 5$ are easily obtained, by use of (4.2); they are given in Table I.

i \ j	0	1	2	3	4	5
0	759	506	330	210	130	78
1	253	176	120	80	52	
2	77	56	40	28		
3	21	16	12			
4	5	4				
5	1					

Table I

4.2 Threshold Decoding of Binary Codes for which the Dual Code Contains t-Designs

A binary vector of length n and weight d can be viewed as the incidence vector of a d-subset of an n-set. Some binary codes have the property that the collection of codevectors of a given weight d yields a t-design. The most celebrated example is the **extended Golay code** (24,12) which contains 759 codevectors of weight 8, 2576 codevectors of weight 16, where each of these collections yields a 5-design, namely the Steiner system $D_1(5,8,24)$ and two other designs, $D_{48}(5,12,24)$ and $D_{78}(5,16,24)$, respectively. Since this code is self-dual, its codevectors can be used as defining parity check rules. In particular, we may take the 253 codevectors of weight 8 containing a "1" in any given position, say the first. These define a (77,21) configuration of parity checks on the first error digit, and we may apply Rudolph's method with an error-correcting capability of $[(77+20)/42] = 2$. The code, however, is capable of correcting any combination of 3 errors, since the minimum weight is 8. A more careful analysis shows that this is indeed possible, by making use of the 5-design property. More precisely, we shall show that, according to the number of errors, and according as the first error digit e_1 is equal to 1 or to 0 the number of parity checks equal to 1 in the above set of 253 is given by the following table:

No. of errors	$e_1 = 1$	$e_1 = 0$
1	253	77
2	176	112
3	141	125
4	128	128(*)

Table II

We shall treat the case when the number of errors is equal to 3. The other cases are treated similarly. Only the case marked with a (*) requires further arguments.
Assume first $e_1 = 1$ and $e_i = e_j = 1$, all other $e_k = 0$.
From Table I, we deduce that among the 253 parity checks that check e_1, there are exactly:

21 that check e_i and e_j,
56 " " e_i but not e_j,
56 " " e_j but not e_i,
120 " " neither e_i nor e_j.

Hence, there are exactly 21 + 120 = 141 parity checks equal to 1.
Similarly, if $e_1 = 0$, $e_i = e_j = e_k = 1$, all other $e_m = 0$, we have, among the 253 parity checks that check e_1,

5 that check e_i, e_j and e_k,
16 " " e_i, e_j, but not e_k,

16	that check	e_i, e_k, but not e_j,
16	" "	e_j, e_k, " " e_i
40	" "	e_i, but not e_j nor e_k,
40	" "	e_j, " " e_k " e_i,
40	" "	e_k, " " e_i " e_j,
80	" "	neither e_i, nor e_j, nor e_k.

Hence, there are exactly $5 + 3 \times 40 = 125$ parity checks equal to 1 in this case.
From the above discussion, it follows that e_1 is correctly determined by the following rule:

$e_1 = 1$, if the (real) sum of the parity checks is > 141,

$e_1 = 0$, if it is < 125,

and 4 errors have occurred if it is equal to 128.

4.3 In conclusion, we have shown that the guaranteed error-correcting capability of Rudolph's method can be improved by making use of the t-design property of a set of parity checks. Let us mention, for example, that the (48,24) extended quadratic residue code can be decoded in a similar way. Some refinements of the method are also possible, by proceeding in several steps. For example, a two-step method has been devised for the Golay (23,12) code, which makes use of 56 parity checks forming an (16,4) configuration on the sum of two error digits, in a first step, and then proceeds trivially in a second step.

4.4 References

[1] Assmus, E.F., Jr., and H.F. Mattson, Jr., On tactical configurations and error-correcting codes, *J. Combin. Theory*, *2* (1967), 243-257.

[2] Goethals, J.M., On the Golay perfect binary code, *J. Combin. Theory*, *11* (1971), 178-186.

[3] Goethals, J.M., On t-designs and threshold decoding, *Univ. North Carolina, Inst. Statist. Mimeo Ser.* No. 600.29 (1970).

SOME CURRENT RESEARCH IN DECODING THEORY

L.D. Rudolph

Systems and Information Science
Syracuse University
Syracuse, New York

1. Introduction

Two central problems of coding for noisy channels are :

1.) Find high-performance codes

2.) Devise efficient but practical decoding methods.

These two goals are incompatible of course, and the major practical problem is to find compromise solutions which yield the best performance for a given cost.

It is intuitively clear that the complexity of decoding increases ever more rapidly as the upper limit in performance dictated by the channel capacity is approached. Because of the steep slope of the performance-complexity curve as it approaches the performance limit, we are quite willing to suffer a small reduction in performance in order to obtain a large reduction in complexity. The problem is to make the trade in the most advantageous way possible.

One method of trading performance for complexity is to impose restrictions on the code which force the code to be suboptimal but which allow simplifications in decoding. Examples of such restrictions are: quantization of time and/or amplitude (as in digital communication); fixing the memory of the code (as in trellis codes and block codes) : imposing algebraic structure (linearity, cyclicity, geometric properties, etc.). In order to obtain the reduction in complexity paid for by the loss of performance, the decoder must fully utilize the restrictions placed on the code. In section 2, we describe a digital decoding procedure for cyclic codes which achieves very low complexity for cyclic codes with a rich subcode structure by taking full advantage of the cyclic property.

A second method of trading performance for complexity is to use a suboptimal decoding procedure. Examples of suboptimal decoding methods are: information-lossy quantization followed by digital decoding ; t-error-correcting and bounded-distance decoding ; definite decoding. Here, one must be careful that the loss in performance is justified by the reduction in complexity. This is not always so, as in the following case. The usual decoding procedure proposed for high-rate linear binary codes is 0-1 quantization followed by digital decoding. In our view this usually constitutes a poor trade because, regardless of the reduction in complexity, the performance improvement over no coding is virtually nil. In Section 3, we consider an approach to decoding high-rate codes which avoides the information loss inherent in the initial 0-1 quantization.

2. Sequential Code Reduction

In this section we consider a digital decoding method for cyclic codes which we call sequential code reduction [1]. We first formulate a general t-error-correcting decoding algorithm for cyclic codes and then introduce the concept of sequential code reduction by means of an example. For simplicity, only binary cyclic codes will be considered.

2.1 General t-error-correcting Algorithm

A t-error-correcting decoding algorithm for an (n,k) code is guaranteed to correct all errors of weight t or less. Some t-error-correcting algorithms (e.g. threshold decoding) correct some patterns of more than t errors, but this excess correction capability is due to accident rather than design and can be considered a bonus obtained at no extra cost.

Let $c = (c_0,\ldots,c_{n-1})$ be the transmitted code word, $e = (e_0,\ldots,e_{n-1})$ the error vector, and $c \oplus e = \tau = (\tau_0,\ldots,\tau_{n-1})$ the received word, where '$\oplus$' denotes mod 2 vector addition. Let H be a reduced parity check matrix for the code and $\hat{e} = (\hat{e}_0,\ldots,\hat{e}_{n-1})$ and $\hat{c} = (\hat{c}_0,\ldots,\hat{c}_{n-1})$ the decoder's estimate of e and c respectively. $W_H(x)$ denotes the Hamming weight of x. Since we are considering cyclic codes only, the decoding algorithm need only be capable of correctly determining c_0 whenever t or fewer errors have occurred. A general algorithm to accomplish this is :

1.) Calculate the syndrome $s = H\tau$.

2.) Solve for $\hat{e}_0$ in

$$\begin{cases} H\hat{e} = s \\ W_H(\hat{e}) \leqslant t. \end{cases}$$

3.) Set $\hat{c}_0 = \tau_0 \oplus \hat{e}_0$.

The second step of this algorithm (the only nontrivial one) may be viewed in the following way. There are 2^k solutions to the linear matrix equation $H\hat{e} = s$. The effect of the nonlinear constraint $W_H(\hat{e}) \leqslant t$ is to reduce the number of solutions from 2^k to exactly one (under the assumption that $W_H(e) \leqslant t$). This reduction is usually accomplished in one step using a combinational logic circuit. However – and this is the basis for sequential code reduction – there is no reason why this cannot be accomplished sequentially in stages.

The basic idea is a simple one and can be illustrated quite easily by means of an example. We will consider the decoding of the (7,4) single-error-correcting Hamming code, first by a conventional two-step majority decoding algorithm and then by two-stage sequential code reduction.

2.2 Decoding Algorithms for the (7,4) code

The first step of the general t-error-correcting decoding algorithm just given is to calculate the syndrome

$$\overset{s}{\begin{bmatrix} s_1 \\ s_2 \\ s_3 \end{bmatrix}} = \overset{H}{\begin{bmatrix} 1\,1\,1\,0\,1\,0\,0 \\ 1\,1\,0\,1\,0\,0\,1 \\ 1\,0\,1\,0\,0\,1\,1 \end{bmatrix}} \overset{r}{\begin{bmatrix} r_0 \\ r_1 \\ r_2 \\ r_3 \\ r_4 \\ r_5 \\ r_6 \end{bmatrix}}$$

where H is a reduced parity check matrix for the (7,4) code.

The second step is to solve for $\hat{e}_0$ in the equations

$$\begin{cases} \overset{H}{\begin{bmatrix} 1\,1\,1\,0\,1\,0\,0 \\ 1\,1\,0\,1\,0\,0\,1 \\ 1\,0\,1\,0\,0\,1\,1 \end{bmatrix}} \overset{\hat{e}}{\begin{bmatrix} \hat{e}_0 \\ \hat{e}_1 \\ \hat{e}_2 \\ \hat{e}_3 \\ \hat{e}_4 \\ \hat{e}_5 \\ \hat{e}_6 \end{bmatrix}} = \begin{bmatrix} s_1 \\ s_2 \\ s_3 \end{bmatrix} \\ W_H(\hat{e}) \leqslant 1. \end{cases}$$

The number of solutions for ê in the unconstrained linear equation is $2^k = 2^4 = 16$. The problem is to find the one solution for ê that also satisfies $W_H(\hat{e}) \leqslant 1$. In majority decoding, this is accomplished by deriving new parity checks from the old (the syndrome) using the nonlinear majority function. These new parity checks are valid only under the assumption that $W_H(e) \leqslant 1$. The effect of adding these parity checks to the syndrome, and the corresponding rows to the parity check matrix H, is to increase the rank of H and thereby decrease the number of solutions for $\hat{e}$. New parity checks are added until all of the solutions for $\hat{e}$ have the same value for $\hat{e}_0$. In our example, the addition of three new parity checks

$$s_4 = \text{maj}\ \{0, s_1, s_2\} = \widehat{e_0 \oplus e_1}$$
$$s_5 = \text{maj}\ \{0, s_1, s_3\} = \widehat{e_0 \oplus e_2}$$
$$s_6 = \text{maj}\ \{0, s_4, s_5\} = \hat{e}_0$$

yields the extended decoding equation

$$\overset{H'}{\begin{bmatrix} 1\,1\,1\,0\,1\,0\,0 \\ 1\,1\,0\,1\,0\,0\,1 \\ 1\,0\,1\,0\,0\,1\,1 \\ 1\,1\,0\,0\,0\,0\,0 \\ 1\,0\,1\,0\,0\,0\,0 \\ 1\,0\,0\,0\,0\,0\,0 \end{bmatrix}} \overset{\hat{e}}{\begin{bmatrix} \hat{e}_0 \\ \hat{e}_1 \\ \hat{e}_2 \\ \hat{e}_3 \\ \hat{e}_4 \\ \hat{e}_5 \\ \hat{e}_6 \end{bmatrix}} = \overset{s'}{\begin{bmatrix} s_1 \\ s_2 \\ s_3 \\ s_4 \\ s_5 \\ s_6 \end{bmatrix}}$$

The rank of H' is 6, so there are two solutions for $\hat{e}$. However, both solutions have $\hat{e}_0 = s_6$, so the process of reducing the solution space by the addition of new parity checks is complete.

The final step of the algorithm is to obtain $\hat{c}_0$ from

$$\hat{c}_0 = r_0 \oplus \hat{e}_0 = r_0 \oplus s_6 .$$

The corresponding conventional two-step majority decoder is shown in Fig. 1.

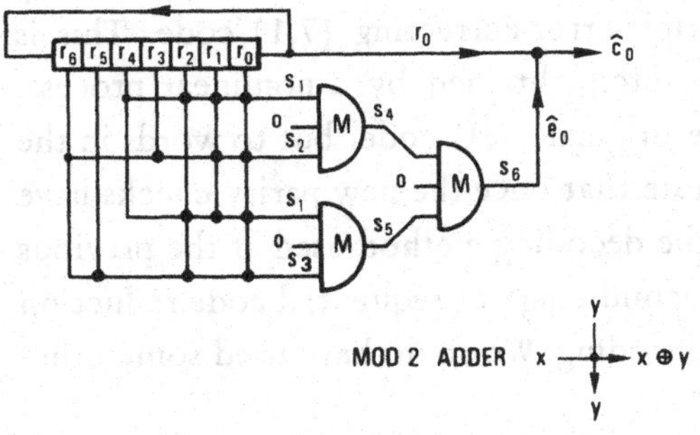

Fig. 1 Conventional two-step majority decoding circuit.

Let s_4, s_4', s_4'', etc. denote the sequence of outputs from the upper left majority gate that results when the received word r is ring-shifted in the buffer. Since $s_4 = \widehat{e_0 \oplus e_1}$, and the code is cyclic, we have

$$s_4 = \widehat{e_0 \oplus e_1}$$
$$s_4' = \widehat{e_1 \oplus e_2}$$
$$s_4'' = \widehat{e_2 \oplus e_3}$$
$$\vdots$$
$$s_4^{vi} = \widehat{e_6 \oplus e_0} .$$

But note that

$$s_4 \oplus s_4' = \widehat{e_0 \oplus e_1} \oplus \widehat{e_1 \oplus e_2} = \widehat{e_0 \oplus e_2} = s_5 .$$

It is therefore not necessary to have a separate majority gate to calculate s_5 if we are willing to implement the equation $s_5 = s_4 \oplus s_4'$. The resulting two-stage sequential code reduction decoding circuit is shown in Fig. 2.

2.3 Sequential Code Reduction: Interpretation and Open Questions

Essentially what we have done in the circuit of Fig. 2 is to use only the uppermost path of the combinational majority-logic decoding tree of the decoder in Fig. 1. This is made possible by inserting a one-word delay between the levels of majority logic. The trick is to use the cyclic property at each level, not just the last. In general, this allows us to reduce the logical complexity of the decoder from a bounded exponential function of the number of levels to a linear function at the cost of a linear increase in buffer storage and decoding time (or delay if additional buffering is used). We note here that the idea of obtaining new parity checks as linear combinations of parity checks already computed was first used by Massey [2] in his demonstration that L-step orthogonalization never requires more than k majority decisions.

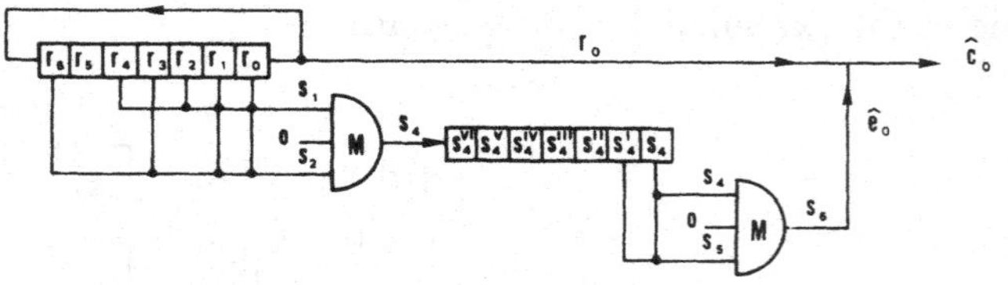

Fig. 2 **Two-stage sequential code reduction circuit.**

The term 'sequential code reduction' stems from the observation that we have a different decoding problem for a different code at each stage of the decoding process. Thus, once we obtain s_4, s_4', s_4'', etc. (Fig. 2), we have a new decoding problem: single-error-correction for the triple-error-correcting (7,1) code. This is because the new parity checks s_4, s_4', s_4'', etc., obtained by a nonlinear process, correspond not to words in the dual of the original (7,4) code, but to words in the dual of the (7,1) code. It is important to note that once the new parity checks have been found, we are not tied in any way to the decoding method used at the previous stage. In fact, error-trap decoding at the second stage of sequential code reduction of the (7,4) code is simpler than majority decoding. We could have used some other decoding method at the first stage as well.

The example used here does not show much if any reduction in complexity through the use of sequential code reduction. This is because the (7,4) code is very small. To get a more realistic idea of the savings involved, consider the three-step orthogonalizable, 31-error-correcting (8191,5812) Euclidean-geometry code. The conventional three-level majority decoding circuit requires 3907 majority decisions and 238,328 parity check calculations, whereas majority decoding using sequential code reduction requires 3 majority decisions and 186 parity check calculations with an increase in buffer storage and decoding delay of three.

When sequential code reduction can be applied to an L-step orthogonalizable code, the majority decoder takes the form of a string of L subcode decoders, each consisting of 2t orthogonal parity checks and one 2t-input majority gate. We conjecture that such a decoding circuit is applicable to any L-step orthogonalizable code. We have verified this for all finite-geometry codes of length up to $n = 10{,}000$ and for all cyclic Reed-Muller codes. To give a general proof of the conjecture for all finite-geometry codes, it would be sufficient to show the existence of a "generator flat" in every finite projective and Euclidean geometry. The interested reader is referred to [1] for details.

Unfortunately, many of the best cyclic codes are not L-step orthogonalizable. However, majority sequential code reduction with one majority gate at each stage is still applicable; the difference is that at least one of the subcode decoders must use $M > 2t$ nonorthogonal parity checks and an M-input majority gate. The complexity of the decoder in this case depends heavily on the values of M at stages where nonorthonal parity checks are required. The values of M depend in turn on what functions we choose to compute at these stages. We now arrive at an impasse: unless we can estimate the complexity of a decoding function (i.e. the minimum number of parity checks required to compute it), we cannot optimize the design. The question of the existence of a finite M for any Boolean function has been settled [3], but the complexity question is completely open. About the only result known for decoding functions is that the complexity of two decoding functions in the same coset with respect to the dual code have the same complexity. A first try at developing a theory of the complexity of Boolean functions along these lines is reported in [4].

Another open problem is to find classes of cyclic codes which lend themselves naturally to decoding by sequential code reduction but which are more efficient than the classical L-step orthogonalizable codes. The best such codes found to date are the generalized finite-geometry codes [5 - 7]. Some of these are L-step orthogonalizable, some are not. A related open question is whether or not BCH codes lend themselves easily to decoding by sequential code reduction.

3. Analog Threshold Decoding

In this section we consider an approach to decoding linear binary codes which does not require 0-1 quantization prior to decoding. Although the approach is general, in practice it would probably only be used for high-rate codes because of the rapid increase of decoder complexity with the error-correcting capability of the code. We first discuss the intuitive grounds on which we would expect such a decoding method to exist and then illustrate the idea by an example. We would like to emphasize at the outset that the material in this section is highly speculative.

3.1 A Decoding-Complexity Folk Theorem

I think that most coding theorists who have wrestled with the problem of designing practical decoders for linear block codes would agree with the following:

Folk Theorem:
A code and its dual have about the same decoding complexity.

An intuitive argument might go something like this. In order to decode a low-rate (n,k) code (i.e. a code for which $k < n/2$) we store the 2^k code words and decode the received word by matching it against the stored code words and selecting the best match. To decode the (n,n-k) dual code, we store the 2^k coset leaders and decode the received word by subtracting from it the most likely error pattern (coset leader) using syndrome decoding. In each case, it was necessary to store 2^k words. Following this line of reasoning, we would expect rate 1/2 codes to be the most difficult to decode. This certainly seems to be the case in practice.

Now it is well known that correlation decoding (where a matched filter is provided for each code word and decoding is performed by noting which filter shows the best match with the received waveform) is a highly efficient analog decoding method for very-low-rate codes, e.g. the m-sequence codes. The folk theorem says that there should exist an analog decoding method of comparable complexity for very-high-rate codes, e.g. the Hamming codes. And following the intuitive argument above, we might expect such a decoding procedure to be somehow analogous to syndrome decoding, i.e. a decoding procedure which makes strong use of the group property (which correlation decoding does not).

Our approach is to start with a one-step majority logic decoding function and then attempt to extend this to allow threshold decoding of the received word in its unquantized analog form. We again proceed by example using the (7,4) code. A conventional one-step majority decoding function using nonorthogonal parity checks [8] will be derived, and its extension to analog threshold decoding will be considered.

3.2 Decoding algorithm for the (7,4) code

Let H* be a complete parity check matrix for the (7,4) code, i.e. a parity check matrix whose rows are all of the $2^{n-k}=8$ words of the dual code. Then a vector c is a code word if and only if

$$\overset{H^*}{\begin{bmatrix} 0&0&0&0&0&0&0 \\ 1&1&1&0&1&0&0 \\ 1&1&0&1&0&0&1 \\ 1&0&1&0&0&1&1 \\ 0&1&0&0&1&1&1 \\ 1&0&0&1&1&1&0 \\ 0&0&1&1&1&0&1 \\ 0&1&1&1&0&1&0 \end{bmatrix}} \begin{bmatrix} c_0 \\ c_1 \\ c_2 \\ c_3 \\ c_4 \\ c_5 \\ c_6 \end{bmatrix} = 0.$$

Solving for c_0 (where we add c_0 to both sides where necessary) and substituting τ for c on the right hand side, we obtain the complete set of estimators for c_0 :

$$\begin{bmatrix} 1 \\ 1 \\ 1 \\ 1 \\ 1 \\ 1 \\ 1 \\ 1 \end{bmatrix} \begin{bmatrix} c_0 \end{bmatrix} = \begin{bmatrix} 1&0&0&0&0&0&0 \\ 0&1&1&0&1&0&0 \\ 0&1&0&1&0&0&1 \\ 0&0&1&0&0&1&1 \\ 1&1&0&0&1&1&1 \\ 0&0&0&1&1&1&0 \\ 1&0&1&1&1&0&1 \\ 1&1&1&1&0&1&0 \end{bmatrix} \begin{bmatrix} \tau_0 \\ \tau_1 \\ \tau_2 \\ \tau_3 \\ \tau_4 \\ \tau_5 \\ \tau_6 \end{bmatrix}$$

The decoding function $\hat{c}_0$ is now computed by summing over a subset of the best estimators (those of lowest weight), weighted according to their reliability, and thresholding on the result. There are a number of subsets that work ; for our purposes we choose the following subset :

$$
\begin{aligned}
\hat{c}_0^{(0)} &= r_0 \\
\hat{c}_0^{(1)} &= r_1 \oplus r_2 \oplus r_4 \\
\hat{c}_0^{(2)} &= r_1 \oplus r_3 \oplus r_6 \\
\hat{c}_0^{(3)} &= r_2 \oplus r_5 \oplus r_6 \\
\hat{c}_0^{(4)} &= r_3 \oplus r_4 \oplus r_5
\end{aligned}
$$

Note that each received bit except r_0 appears in exactly two of the estimators. This suggests that we include the estimate $\hat{c}_0^{(0)}$ twice (which is equivalent to assigning it a weight of 2). Since a single error can cause only two of the six estimators to give an incorrect estimate, the following function is a 1-error-correcting decoding function for the (7,4) code :

$$
\hat{c}_0 = \begin{cases} 1 \text{ if } \hat{c}_0^{(0)} + \sum_{i=0}^{4} \hat{c}_0^{(i)} > 3 \\ 0 \text{ otherwise} \end{cases}
$$

where '+' denotes addition over the real numbers. Note that we are using one more estimator than is necessary; we could discard any one of the six estimators and threshold on 2 1/2 instead of 3. The threshold decoding circuit is shown in Fig. 3.

We would now like to extend this type of decoding function to allow decoding of the received word r in its analog form, i.e. we wish to treat r as a vector of real numbers. The first task is to extend modulo 2 addition to the real numbers. There are many ways to do this of course, but the natural way (for reasons having to do with group characters and Fourier transforms) is the following: For binary variables x and y,

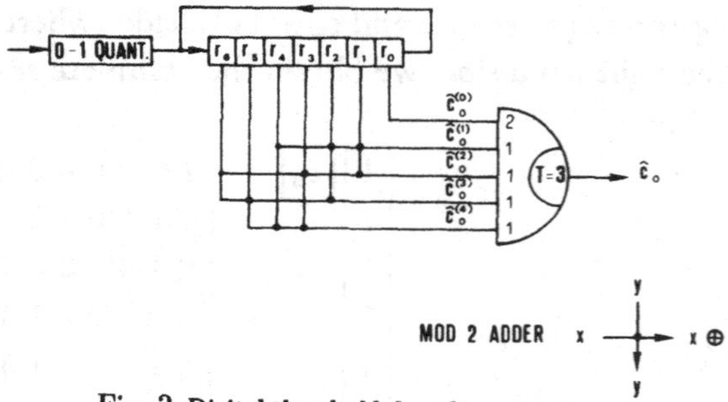

Fig. 3 Digital threshold decoding circuit

$$
x \oplus y = \frac{1 - \cos \pi (x+y)}{2} .
$$

If we ignore the translation and scal'ng, we see that a modulo 2 sum can be replaced

by cos π of the corresponding real sum. In the case of the (7,4) code, this gives the following set of real valued estimators:

$$
\begin{aligned}
\hat{\gamma}_0^{(0)} &= \cos\pi\,(\tau_0) \\
\hat{\gamma}_0^{(1)} &= \cos\pi\,(\tau_1+\tau_2+\tau_4) \\
\hat{\gamma}_0^{(2)} &= \cos\pi\,(\tau_1+\tau_3+\tau_6) \\
\hat{\gamma}_0^{(3)} &= \cos\pi\,(\tau_2+\tau_5+\tau_6) \\
\hat{\gamma}_0^{(4)} &= \cos\pi\,(\tau_3+\tau_4+\tau_5).
\end{aligned}
$$

Direct substitution would yield the analog threshold decoding function

$$
\hat{c}_0 = \begin{cases} 1 \text{ if } 2\,\hat{\gamma}_0^{(0)} + \sum_{i=1}^{4} \hat{\gamma}_0^{(i)} < 0 \\ 0 \text{ otherwise} \end{cases}
$$

While this function would correct all single digital errors (as it must, since it is an extension of the digital threshold decoding function), it is not suitable for analog decoding. Our criterion for an analog decoder for a code with minimum Hamming distance d (and thus minimum Euclidean distance $\sqrt{d}$) is that every error pattern of Euclidean weight less than $\sqrt{d}/2$ be corrected. The analog decoding function obtained above by direct substitution fails this test. To see why, we go back and reconsider the specification of the (7,4) code in 7-dimensional real space.

The natural first attempt is the following: a real vector c is a word in the (7,4) code if and only if

$$
\cos\pi\left(\begin{bmatrix} 1&1&1&0&1&0&0 \\ 1&1&0&1&0&0&1 \\ 1&0&1&0&0&1&1 \end{bmatrix}\begin{bmatrix} c_0\\ c_1\\ c_2\\ c_3\\ c_4\\ c_5\\ c_6 \end{bmatrix}\right) = \begin{bmatrix} 1\\ 1\\ 1 \end{bmatrix}
$$

where the matrix multiplication is over the reals. The words of the (7,4) code satisfy this of course, but so do other vectors. For example, an even integer may be added to any position of any word in the (7,4) code and the above defining relation will still be satisfied. This is not objectionable, however, since the minimum distance of the full lattice extension of the (7,4) code is the same as that of the (7,4) code alone. But the vector whose components are all equal to 1/2 also satisfies the defining relation and this is bothersome because the distance of this vector from the all-0 code word is less than the minimum Euclidean distance of the (7,4) code. We can eliminate such unwanted 'code words' by forcing each component of a code vector to be an integer. This is easily accomplished by requiring that $\cos \pi(2c_i) = 1$ for $i = 0, 1, \ldots, 6$,. The corresponding defining relation for the lattice extension of the (7,4) code is

$$\cos \pi \left(\begin{bmatrix} 1&1&1&0&1&0&0 \\ 1&1&0&1&0&0&1 \\ 1&0&1&0&0&1&1 \\ 2&0&0&0&0&0&0 \\ 0&2&0&0&0&0&0 \\ 0&0&2&0&0&0&0 \\ 0&0&0&2&0&0&0 \\ 0&0&0&0&2&0&0 \\ 0&0&0&0&0&2&0 \\ 0&0&0&0&0&0&2 \end{bmatrix} \begin{bmatrix} c_0 \\ c_1 \\ c_2 \\ c_3 \\ c_4 \\ c_5 \\ c_6 \end{bmatrix} \right) = \begin{bmatrix} 1 \\ 1 \\ 1 \\ 1 \\ 1 \\ 1 \\ 1 \\ 1 \\ 1 \\ 1 \end{bmatrix}.$$

The dual of the lattice extension of the (7,4) code is simply the row space, over the integers, of this extended parity check matrix. We may now proceed in much the same way we did in the case of digital threshold decoding.

We wish to choose a subset of the estimators corresponding to words in the dual code. We have an infinite number to choose from now, so we generate a list of estimators, starting with the most reliable, and stop when we have a sufficient number of estimators to construct a decoding function that will correct all errors of Euclidean weight less than $\sqrt{d}/2$ (in our case less than $\sqrt{3}/2$). The first 17 estimators in order of their reliability are:

$$\begin{array}{lll}
\hat{\gamma}_0^{(0)} & = \cos \pi\,(\tau_0 &) \\
\hat{\gamma}_0^{(1)} & = \cos \pi\,(\tau_1 + \tau_2 + \tau_4 &) \\
\hat{\gamma}_0^{(2)} & = \cos \pi\,(-\tau_1 + \tau_2 + \tau_4 &) \\
\hat{\gamma}_0^{(3)} & = \cos \pi\,(\tau_1 - \tau_2 + \tau_4 &) \\
\hat{\gamma}_0^{(4)} & = \cos \pi\,(\tau_1 + \tau_2 - \tau_4 &) \\
\hat{\gamma}_0^{(5)} & = \cos \pi\,(\tau_1 + \tau_3 + \tau_4 &) \\
\hat{\gamma}_0^{(6)} & = \cos \pi\,(-\tau_1 + \tau_3 + \tau_6 &) \\
\hat{\gamma}_0^{(7)} & = \cos \pi\,(\tau_1 - \tau_3 + \tau_6 &) \\
\hat{\gamma}_0^{(8)} & = \cos \pi\,(\tau_1 + \tau_3 - \tau_6 &)
\end{array}$$

$$
\begin{aligned}
\hat{\gamma}_0^{(9)} &= \cos\pi\,(\tau_2 + \tau_5 + \tau_6)\\
\hat{\gamma}_0^{(10)} &= \cos\pi\,(-\tau_2 + \tau_5 + \tau_6)\\
\hat{\gamma}_0^{(11)} &= \cos\pi\,(\tau_2 - \tau_5 + \tau_6)\\
\hat{\gamma}_0^{(12)} &= \cos\pi\,(\tau_2 + \tau_5 - \tau_6)\\
\hat{\gamma}_0^{(13)} &= \cos\pi\,(\tau_3 + \tau_4 + \tau_5)\\
\hat{\gamma}_0^{(14)} &= \cos\pi\,(-\tau_3 + \tau_4 + \tau_5)\\
\hat{\gamma}_0^{(15)} &= \cos\pi\,(\tau_3 - \tau_4 + \tau_5)\\
\hat{\gamma}_0^{(16)} &= \cos\pi\,(\tau_3 + \tau_4 + \tau_5)
\end{aligned}
$$

If we include $\hat{\gamma}_0^{(0)}$ eight times and each of the other estimates once, we obtain the analog threshold decoding function

$$
\hat{c}_0 = \begin{cases} 1 \text{ if } 8\,\hat{\gamma}_0^{(0)} + \sum_{i=1}^{16} \hat{\gamma}_0^{(i)} < 0 \\ 0 \text{ otherwise.} \end{cases}
$$

But now an interesting thing happens. If we espand the estimators using trigonometric identities, collect terms and scale down the coefficients, we get

$$
\begin{aligned}
\hat{c}_0 = 2\ &\cos\pi\,(\tau_0)\\
&+ \cos\pi\,(\tau_1)\cos\pi\,(\tau_2)\cos\pi\,(\tau_4)\\
&+ \cos\pi\,(\tau_1)\cos\pi\,(\tau_3)\cos\pi\,(\tau_6)\\
&+ \cos\pi\,(\tau_2)\cos\pi\,(\tau_5)\cos\pi\,(\tau_6)\\
&+ \cos\pi\,(\tau_3)\cos\pi\,(\tau_4)\cos\pi\,(\tau_5).
\end{aligned}
$$

In other words, we get the same estimators that we used for digital threshold decoding except that 0-1 quantization has been replaced by the cos π operator and mod 2 addition has been replaced by real product. The corresponding analog threshold decoding circuit is shown in Fig. 4.

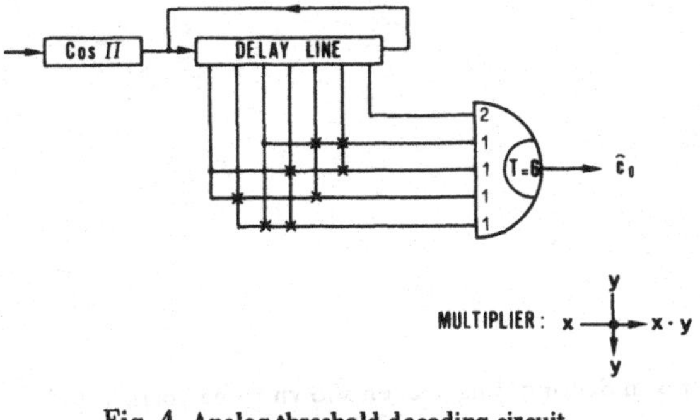

Fig. 4 Analog threshold decoding circuit.

The tale ends here. A proof that the decoder of Fig. 4 will correct all errors of Euclidean weight less than $\sqrt{3}/2$ is not yet complete, but the rough preliminary analyses carried out by the author and C.R.P. Hartmann, his collaborator in this research, strongly imply that it does. If so, then it would appear that this decoder should be asymptotically equivalent to correlation decoding. Whether this particular approach to analog decoding of high-rate codes will turn out to be the answer to the decoding-complexity folk theorem - or indeed whether the theorem is even true – is anybody's guess at this point*.

* Since these lectures were given, the conjectures in Section 3 have been shown to be correct [9], and much more powerful and general results have been obtained [10, 11].

References

[1] Rudolph L.D. and C.R.P. Hartmann, "Decoding by Sequential Code Reduction," *IEEE Trans. on Inform. Theory*, vol. IT-19, pp. 549-555, July 1973.

[2] Massey, J.L., *Threshold Decoding*, Cambridge Mass.: M.I.T. Press, 1963.

[3] Rudolph, L.D., and W.E. Robbins, "One-Step Weighted-Majority Decoding," *IEEE Trans. on Inform. Theory*, vol. IT-18, pp. 446-448, May 1972.

[4] Robbins, W.E., and L.D. Rudolph, "On Two-Level Exclusive-OR Majority Networks," *IEEE Trans. on Computers*, vol. C-23, pp. 34-40, January 1974

[5] Delsarte, P. "A Geometric Approach to a Class of Cyclic Codes," *J. Combinatorial Theory*, vol. 6, pp. 340-358, May 1969.

[6] Lin, S. and E.J. Weldon, Jr., "New Efficient Majority-Logic Decodable Cyclic Codes," presented at the 1972 IEEE International Symposium on Information Theory, Asilomas, Calif.

[7] Hartmann, C.R.P., J.B. Ducey and L.D. Rudolph, "On the Structure of Generalized Finite Geometry Codes," *IEEE Trans. on Information Theory*, IT-20, pp. 240-252,, March 1974.

[8] Rudolph, L.D. "A Class of Majority Logic Decodable Codes," *IEEE Trans. on Inform. Theory*, vol. IT013, pp. 305-307, April 1967.

[9] Rudolph, L.D. and C.R.P. Hartmann, "Maximum-Radius Analog Threshold Decoding", presented at the 1975 IEEE International Symposium on Information Theory, Notre Dame, Ind.

[10] Rudolph, L.D. and C.R.P. Hartmann, "Algebraic Analog Decoding", presented at the IEEE Information Theory Workshop, Lenox Mass-, June 1975. To be submitted to the *IEEE Trans. on Inform. Theory*.

[11] Hartmann, C.R.P., and L.D. Rudolph, "On Optimum Symbol-by-Symbol Decoding Rule for Linear Codes", Submitted to the *IEEE Trans. on Inform. Theory*.

PROCEDURES OF SEQUENTIAL DECODING

K. Sh. Zigangirov

This course is devoted to the foundations of sequential decoding. Sequential decoding was first suggested by J. Wozencraft [1]. Then it has been essentially improved by American and Soviet scientists.

The present course is based on papers of the author [2-8, 11]. (See also [10]). However, a familiarity with Gallager's [9] may also be helpful.

1. Information Transmission

The general model for a communication system consists of an **information source**, an **encoder** (or **coder**), a **modulator**, a **communication line**, a **demodulator**, a **decoder** and a **recipient** (Fig. 1). The source generated the information sequence $u_1, u_2, \ldots, u_i, \ldots$ of binary symbols. Any symbol u_i may take, with equal probabilities, either the value zero or one

$$P(u_i = 0) = P(u_i = 1) = \frac{1}{2}, \quad i = 1,2,\ldots \tag{1.1}$$

In this case each symbol conveys **1 bit** (i.e. $\ell n\, 2$ nat) of information. The problem is to transmit with negligible distortion.

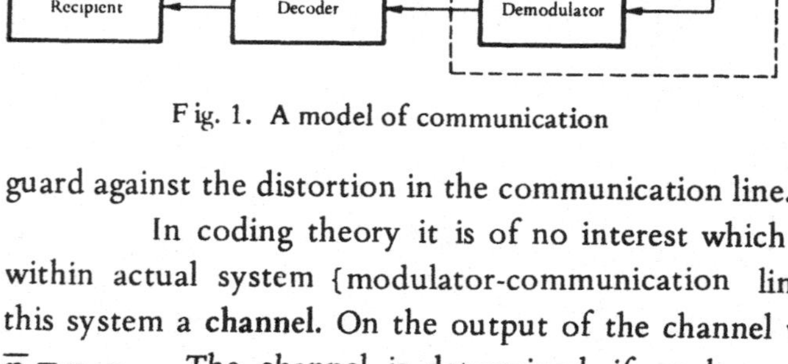

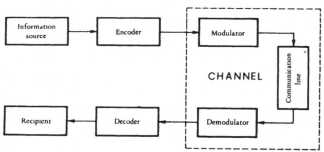

Fig. 1. A model of communication

The information sequence is going into the coder, which transforms this sequences into encoded sequence $\bar{\xi} = \xi_1, \xi_2, \ldots$. We suppose that the symbol ξ_i may take values in a discrete input alphabet $\{a_1, a_2, \ldots a_K\}$. The purpose of this transformation is to guard against the distortion in the communication line.

In coding theory it is of no interest which physical processes take place within actual system {modulator-communication line-demodulator}. We will call this system a **channel**. On the output of the channel we have the **received sequence** $\bar{\eta} = \eta_1, \eta_2, \ldots$. The channel is determined, if we know the conditional probabilities $p(\bar{\eta}/\bar{\xi})$, that $\bar{\eta}$ is received, given the encodes sequence $\bar{\xi} = \xi_1, \xi_2, \ldots$. The symbols of the received sequence may take values in an output alphabet $\{b_1, b_2, \ldots, b_J\}$.

The channel is discrete and memoryless if

$$P(\bar{\eta}/\bar{\xi}) = P(\eta_1/\xi_1)\, P(\eta_2/\xi_2)... \tag{1.2}$$

This channel is determined if we know the transitional probabilities

$$P(b_j/a_k) = P_{kj},\ k = 1,2,...K,\ j = 1,2,...J \tag{1.3}$$

The simplest channel is the **binary symmetric channel** (BSC, see Fig. 2). For this channel

$$\begin{aligned} P_{11} &= P_{22} = q \\ P_{12} &= P_{21} = p \end{aligned} \tag{1.4}$$

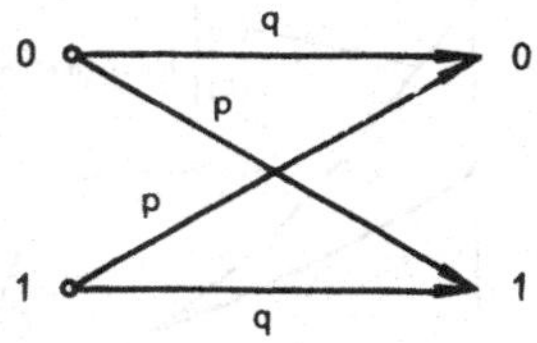

Fig. 2. The binary symmetric channel

We will discuss sequential decoding by means of the B S C example.

The decoder has to restore the information sequence as far as possible.

Historically the first of coding techniques was the **block method.** In this case coder divides the information sequence into blocks of N symbols. The encoder transforms each block into a code word of ν code symblos. Code symbols are transmitted through the channel. The value

$$R = N/\nu - \ln 2 \ \frac{\text{nat}}{\text{symbol}} \tag{1.5}$$

is called the **transmission rate,** ν is called the **block length.** The block length ν characterizes the complexity of coding.

A block of N information symbols is decoded at the output of the decoder correctly, if all decoded symbols coincide with the corresponding symbols of the information sequence at the input of the encoder. Otherwise we speak of **erroneous decoding.** We denote the probability of this event by $P(\&)$.

An encoding-decoding method is efficient if the probability of error decays exponentially with the block length. This means that for each $\epsilon > 0$ there is a value ν_0 such that for any $\nu > \nu_0$ we have

$$e^{-(E+\epsilon)\nu} < P(\&) < e^{-(E-\epsilon)\nu} \tag{1.6}$$

The coefficient $E = E(R)$ is called the **reliability function.**

It has been proved that there exists a block encoding-decoding method, for which

$$E(R) \geqslant \underline{E}(R). \tag{1.7}$$

I.e. the value $\underline{E}(R)$ is a lower bound of the reliability function.

On the contrary, there is no encoding-decoding method for which the reliability function exceeds $\bar{E}(R)$. For a discrete memoryless channel the bounds $\underline{E}(R)$ and $\bar{E}(R)$ are shown in Fig. 3. (See the continuous curves). C is called the **capacity** of the channel, R_{cr} is called **critical rate.**

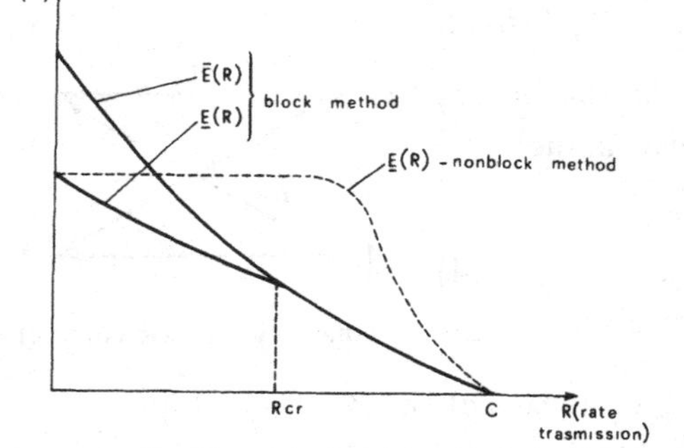

Fig. 3. Typical bounds for the reliability function /—— block method; ------ non-block method of encoding-decoding

The dashed curves in Fig. 3 are lower bounds for nonblock encoding--decoding, one way of which is convolutional encoding and sequential decoding. Obviously, for given complexity the non-block methods have essentially better reliability functions than block methods.

2. Convolutional Coding

In the case of sequential decoding one adopts for coding the information sequence the method of convolutional coding. The convolutional encoder (or coder) for BSC consists of a **shift register**, some stages of which are connected by **summators modulo 2**, and **commutator**. We will denote the length of the register by N and the number of commutators by m. The simplest kind of convolutional coder for BSC is shown in Fig. 4.

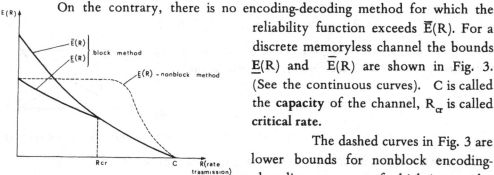

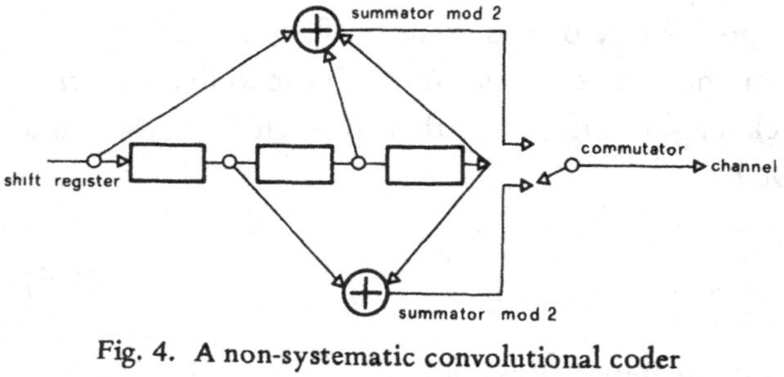

Fig. 4. A non-systematic convolutional coder

Subblocks of ℓ information symbols are fed into the coder. As a new subblock appears at the coder input, present in the right ℓ stages, are removed from the register. Each of the symbols is shifted by ℓ stages to the right and the new subblock is introduced into the firtst ℓ left stages.

In the summators the symbols are added to modulo 2 and then fed to the commutator. The transmission rate is

$$R = \ell/m \, \ell n2 \left(\frac{\text{nat}}{\text{symbol}}\right) \tag{2.1}$$

One may define a code by specifying which stages of the register are connected to a given summator. The connections to summator i are described by the **generating sequence** g_i, Viz.,

$$g_i = (g_{i1}, g_{i2}, \dots g_{iN}), \tag{2.2}$$

Here

$$g_{ij} = \begin{cases} 1, & \text{if the j-th stage of the register is connected to i-th summator} \\ 0, & \text{otherwise} \end{cases}$$

The matrix $G = |g_{ij}|$, $i = 1,2,\dots,m$, $j = 1,2,\dots N$ is called the **generating matrix.** The coder is **systematic** if

$$\begin{aligned} \bar{g}_1 &= (1,0,0, \dots \quad 0) \\ \bar{g}_2 &= (0,1,0, \dots \quad 0) \\ &\quad \vdots \\ \bar{g}_\ell &= (0,0,\dots, 1, \dots 0) \quad \leftarrow \ell\text{-th position} \end{aligned} \tag{2.4}$$

and other generating sequences are arbitary. In Fig. 5 the simplest systematic convolutional coder for BSC is shown.

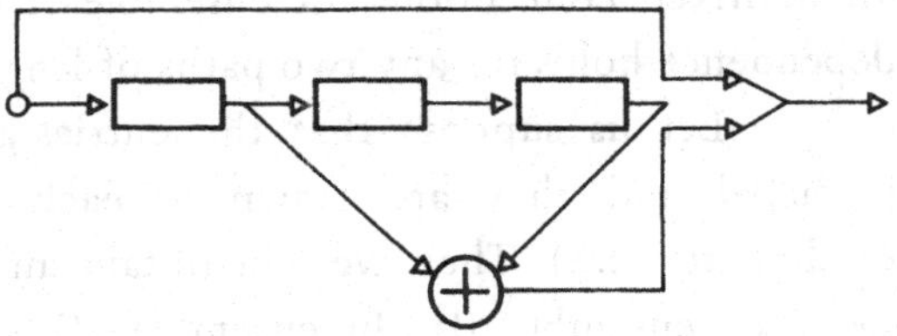

Fig. 5. Systematic convolutional coder

The **constraint length** ν is the number of code symbols that come out of the coder between when a given information symbol enters the coder and when it passes out of the coder.

We call the number of such elements as registers, adders (modulo 2), etc., **coder complexity**, and the number of operations associated with a subblock coding **complexity**. Obviously the coder complexity as well as the coding complexity are proportional to ν .

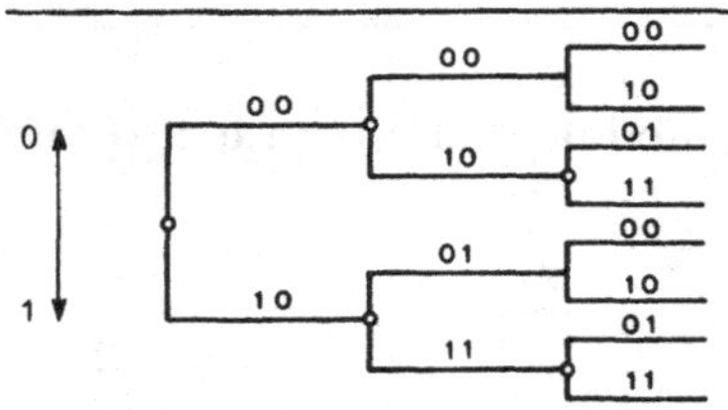

Fig. 6. Code tree for the convolutional encoder shown in Fig. 5

It is convenient to illustrate convolutional coding by a code tree. Fig. 6 shows the code tree for the convolutional coder depicted in Fig. 5.

The points denote the **nodes** of the code tree. The node on the far left is the **initial node.** The lines between the nodes are called **branches.** Any set of some consecutive branches is a path.

The **length of a path** is the number of code symbols along the path. The depth of the node is the number of code symbols along the path from the initial node to considered node.

Assume that each entry g_{ij} of the generating matrix is drawn randomly and independently with the probabilities

(2.5) $$P(g_{ij} = 0) = P(g_{ij} = 1) = \frac{1}{2}, \quad i = 1,2,...m, j = 1,2,...N.$$

In this way one obtains some member of the ensemble of $2^{Nm} = L^{\nu}$ convolutional coders. Here $L = 2^{\ell}$ is the number of branches left to the considered node. We call this ensemble the **ensemble of** (nonsystematic) **convolutional coders** with **fixed connections,** or **ensemble I.**

Pairwise independence can be proved for code paths, the length of which is equal to the **constraint length** ν. (Observe that pairwise independence is a looser property than total independece [9]).

Let us assume, that as a new information subblock is introduced the components g_{ij} are changed, viz. drawn randomly and independently according to (2.5). We obtain in this way an ensemble of (nonsystematic) convolutional coders with nonfixed connections, or ensemble II. In this case the property of pairwise independence holds for any two paths of length τ (where τ is arbitrary).

Let us suppose that the entries g_{ij} are changed as a new subblock is introduced and they are drawn **to each information sequence** independently, according to (2.5). Then we will obtain an ensemble of (nonsystematic) random coders, or ensemble III. In ensemble III all different code symbols are totally independent.

The corresponding ensemble of systematic coders is defined in a similar way. In this case the first ℓ generating sequences are specified according to (2.4), the last $(m-\ell)$ sequences are drawn randomly according to (2.5), as we did for ensembles of nonsystematic coders.

The Hamming distance d between two code sequences $\bar{\xi}$ and $\bar{\xi}'$, is the number of symbols which do not coincide. We present some results concerning the minimum distance d_{min} between code sequences.

The **characteristic parameter** ρ with respect to the rate is the root of the following equation

$$R = \rho \ln 2\rho + (1-\rho)\ln 2(1-\rho), \tag{2.6}$$

Theorem 2.1. For each $\epsilon > 0$ there exists some ν_0 such that for each $\nu > \nu_0$ there is a coder within the ensemble I, for which the minimum distance between any code sequences of length ν satisfies

$$d_{min}/\nu \geqslant \rho - \epsilon. \tag{2.7}$$

(Here ν denotes the constraint length and ρ is the characteristic parameter with respect to the rate R).

Remark: Theorem 2.1 holds also for ensembles II and III and for systematic codes; (2.7) is the **Varshamov-Gilbert bound.**

Let us suppose we are interested in distance between code sequences, the length τ of which exceeds the constraint length ν. Let $k = \tau/\nu$, and $R^{(k)}$ denote the rate for which the characteristic parameter satisfies

$$k[\rho \ln 2\rho + (1-\rho)\ln 2(1-\rho)] = (k-1)\ln 2(1-\rho) \tag{2.8}$$

Theorem 2.2. For each $\epsilon > 0$ there exists some ν_0 and τ_0, such that for each $\nu > \nu_0$ and $\tau > \tau_0$ there is a coder with a constraint length ν for which the minimum distance between code sequences of length τ satisfies inequality

$$d_{min}/\nu \geqslant \begin{cases} \dfrac{-R\dfrac{k}{k-1}}{\ln\left(2e^{-R\frac{k}{k-1}} - 1\right)} - \epsilon, & 0 < R < R^{(k)} \\ k(\rho - \epsilon) & R^{(k)} \leqslant R \end{cases} \tag{2.9}$$

Here $k = \tau/\nu$, $R^{(k)}$ according to (2.8), ρ is the characteristic parameter with respect to the rate R.

When $k \to \infty$ the bound (2.9) tends to Costello's bound. I.e.,

$$d_{min}/\nu \geq \frac{-R}{\ln(2e^{-R}-1)} - \epsilon, \quad 0 < R < 1. \tag{2.10}$$

Bounds (2.9) and (2.10) hold also for ensemble III, but it had not been proved for ensemble I yet. Bounds (2.9) and (2.10) do not hold for systematic codes.

For systematic codes one has to use instead of Costello's bound the Varshamov-Gilbert bound (2.7) or the following looser bound

$$d_{min}/\nu \geq \frac{m-\ell}{m} \cdot \frac{-R}{\ln(2e^{-R}-1)} - \epsilon . \tag{2.11}$$

(Here m and ℓ denote the number of the code and information symbols within any subblock, respectively).

3. The Stack Algorithm

The stack algorithm (i.e., the maximum likelihood algorithm for sequential decoding) has been introduced in the author's Paper [2] as well as in F. Jelinek's paper [10]. We are going to study this algorithm in detail for the BSC.

Let us consider some node 0 of depth t within the coding tree. Let the number of the symbol, different from the corresponding symbols of the received sequence, along the path from this node to the initial node, be d (d is the Hamming distance). We define the likelihood function of this node as

$$\overline{\overline{Z}} = d \ln 2p + (t-d) \ln 2q - tB \tag{3.1}$$

where p is cross-over probability of the BSC, q = 1-p. The bias B has to meet

$$R \leq B < C \tag{3.2}$$

(Here $C = q\ln 2q + p\ln 2p$ stands for the capacity of the BSC).

If B = R the likelihood function is identical to Fano's likelihood function.

If

$$B = \begin{cases} R_{comp}, & \text{when } R < R_{comp} \\ R, & \text{when } R \geq R_{comp} \end{cases}, \tag{3.3}$$

the likelihood function is identical to the computational likelihood function. Here $R_{comp} = \ell n2 - \ell n\,(1 + \sqrt{4pq})$ denotes the **computational cutoff rate** for BSC. Let the likelihood function of the initial node be equal to 0.

The stack algorithm of sequential decoding makes a comparison (by computing the likelihood function) between the received sequence and the code path for which the likelihood function takes its maximal value among all the likelihood functions stored in the decoder memory.

Next we describe the procedure in more detail.

Step 1. Compute the values of the likelihood functions for 2^{ℓ} branches leaving the initial node and place them into the operative memory of the decoder.

Step 2. Pick the node corresponding to the largest likelihood function and compute the values of the likelihood function for the branches, leaving this node, and place these into the decoder memory, deleting the original node.

Step 3. Arrange the values of the likelihood function already placed within the memory, in decreasing order.

Step 4. If the top path is of depth $\tau + (i\text{-}1)m$, $i = 1,2,\ldots,$for the first time, go to step 5, otherwise return to step 2. The value τ is the **back-search limit.**

Step 5. Make the decision that the i-th information subblock belongs to the path. Discard from the memory the nodes belonging to paths having an i-th information subblock different from the decoded subblock.

Thus the sequential stack decoder consists of a **buffer, device for computation of the likelihood function, copy of the convolutional coder,** which is actually used for encoding, an **operative memory** and a **computing device**. They are all connected.

The buffer is a conventional shift register with a great length. It is an essential part of the decoder as the number of computations for sequential decoding is a random variable. In the buffer the symbols of the received sequence wait until the decoder proceeds to these symbols.

The block "copy of the convolutional coder" can generate the coding sequence for each path of the code tree.

The device for computing the likelihood function can compute the likelihood function for the received sequence and the examined coding sequence generated by the copy of the coder. The control device is operating on all blocks.

The operative memory contains the likelihood functions of the examined nodes.

There are two ways for realizing operative memory. First, one may have a list of all examined nodes, arranged in order of decreasing likelihoods. After each calculation of the likelihood function the decoder has to rearrange the order of the nodes in the memory, which is a quite difficult task.

In the second case all the examined nodes are arranged in the order of entering the memory. For each examined node is also indicated the preceding and following node (in the sense of decreasing likelihoods). In this case it is not necessary to rearrange nodes in the memory at each step, but the search for the maximum of the likelihood function is still an involved operation.

Let us rewrite the likelihood function (3.1) as

$$\bar{\bar{Z}} = \alpha'(t-d) + \alpha'' d \tag{3.4}$$

Here t is the depth of the examined node, d is the Hamming distance between path leading from the initial to the examined node and received sequence,

$$\begin{aligned} \alpha' &= \ln 2q - B \\ \alpha'' &= \ln 2p - B \end{aligned} \tag{3.5}$$

We can choose α' and α'' in a way that

$$\alpha' = i\Delta, \quad \alpha'' = -j\Delta, \tag{3.6}$$

where i and j are integers, and Δ a positive real number.

In this case $\bar{\bar{Z}}/\Delta$ is an integer and we can store this in the memory rather than the likelihood itself. The memory consists of bins; each bin corresponds to some value of the likelihood function. In the bin are contained all the nodes which have the corresponding value of the likelihood function. One can pick for examination any arbitrary node from the bins with a maximal likelihood function. Practically the number of bins is not large.

F. Jelinek has proposed a similar algorithm not assuming limitation (3.6). Also the memory consists of bins. The i-th bin ($i = 0, \pm 1, \pm 2, \ldots$) corresponds to nodes, the likelihood function $\bar{\bar{Z}}$ of which is

$$i\Delta \leqslant \bar{\bar{Z}} \leqslant (i+1)\Delta \tag{3.7}$$

At each instant the decoder can pick any node from the bin, which corresponds to the maximal likelihood function. This is the modification of the stack algorithm, **due to Jelinek.**

Next we discuss another modification of the stack algorithm. Let us suppose that the nodes following any given node are in some way ordered. For example, the first node may correspond to the information subblock 0,0,...0, the second to the information subblock 1,0,0...0,...., the 2^{ℓ}-th node to the information subblock 1,1...1. We then define a stack algorithm in the following way:

Step 1. Compute the value of the likelihood function for the first node, next to the initial node and place it into decoder memory.

Step 2. Take the node corresponding to the largest likelihood function. Compute the value of the likelihood function for the first nonexamined node following it, and place it into the decoder memory. If all branches leaving a certain node have been examined, delete this node from the memory.

The definition of steps 3-5 is the same as that for the original stack algorithm. This **ordered modification** of the stack algorithm has less decoding complexity than that of the original algorithm, but the implementation of this algorithm is slightly more complicated.

Another modification of the stack algorithm is a **threshold modification.** Steps 1 through 3 and 5 of this modification are the same as steps 1 through 3 and 5 of the original stack algorithm. However step 4 is defined as follows

Step 4. If the likelihood function along the top takes, for the first time, a value greater than or equal to $z(i-1)+a$, go to step 5. (Here (i-1) is the number of the last decoded subblock nodes, $z(i-1)$ is the likelihood function at the (i-1) the node of the decoded path, $a > 0$. Otherwise return to step 2.

Observe that this modification has a special threshold $a > 0$. This threshold is of the same relevance as the threshold in sequential analysis.

4. Characteristics of the Algorithm

The main characteristics of sequential decoders are **reliability** and **decoding complexity.**

Information subblocks are decoded by decoder in the same order as they enter to encoder. The decoder makes, as a rule, not a single error, but a set of errors. This phenomenon is called **"error propagation"**. If the decoder decodes, e.g., the first information subblock incorrectly, the second, the third and further blocks will be erroneously decoded with large probability.

Let us denote by $\mathcal{E}_i$ the event that i-th information has been incorrectly decoded, C_i stands for the event that the 1-th, 2-th,...i-th subblocks have been correctly decoded. C_0 is the trustworthy event. Let $P(\mathcal{E}_i/C_{i-1})$ denote the conditional probability of error in i-th subblock provided the preceding subblocks are decoded correctly. $P(\mathcal{E}_i/C_{i-1})$ is called the probability of undetectable error, because the decoder cannot detect this error.

Another characteristic is decoder complexity. We call the number of such elements in the coder as adder, or a stage of the shift register as **decoder complexity.** From this point of view the most complicated parts of the decoder are the buffer and the operative memory. (The complexity is about 10^5).

The decoding complexity is the number of computer operations necessary to decode an information subblock. Unfortunately we cannot define the complexity in terms of real computer operations, because this does not only depend on the decoding algorithm, but also on the programme, the computer, etc.

One decoding operation is the cycle of computations necessary to examine one node (computing its likelihood function).

Some additional definitions are still necessary. The path of a coding tree, which corresponds to the transmitted sequence is called a **correct path.** If the path from some node of coding tree to the initial node crosses no node of the correct path except the initial, we say that this node belongs to the **first incorrect** subtree. If the path from some node of the coding tree crosses no other node of the correct path except initial, first, second, . . . (i - 1)-th we say that this node belongs to **i-th incorrect subtree.**

As far as decoding of the i-th information subblock is concerned we count : (1) all operations used for the examination of i-th node of the correct path, and (2) all operations used for the examination of i-th incorrect subtree. The number of operations, ω_i used for examination of i-th subblock is a random variable. Let us consider the conditional expectation $M(\omega_i/C_i)$ of the number of operations on subblock, provided that the 1-th, 2th,...,i-th subblock has been correctly decoded. We call this conditional expectation (which does not depend on i) **decoding complexity.**

Let us define Gallager's function for the BSC

$$G(x) = x\ln 2 - (x+1)\ln\left(q^{1/1+x} + p^{1/1+x}\right) \tag{4.1}$$

for x >0, the computational cutoff rate

$$R_{comp} = G(1) = \ln 2 - \ln(1 + \sqrt{4pq}),$$

γ -parameter of the rate R – the positive root of

$$\left.\frac{dG(x)}{dx}\right|_{x=\gamma} = R \qquad (4.3)$$

β -parameter of the rate R, – the positive root of

$$G(\beta)/\beta = R, \qquad (4.3)$$

as the critical rate R_{cr}, the rate for which $\gamma = 1$, and as $R^{(k)}(k \geqslant 1)$ the rate for which

$$\beta(R\frac{k}{k-1}) = \gamma(R). \qquad (4.5)$$

The probability of error $P(\&_i/C_{i-1})$ depends on the transitional probability p of the BSC, the transmission rate R, the constraint length ν and the back search limit τ. Let $k = \tau/\nu$. Let us confine ourselves to $\tau \geqslant \nu$ (The case when $\nu \geqslant \tau = \tau_0$ is similar to $\tau_0 = \tau = \nu$).

Theorem 4.1 Assume that (a) a (nonsystematic) coder from ensemble II (with constraint length ν) is used in BSC for coding the information and (b) a stack decoder with a back search limit $\tau, \tau \geqslant \nu$, is used for decoding.

Then for the conditional expectation $M_{II}P(\&_i/C_{i-1})$ (over the ensemble II) of the random variable $P(\&_i/C_{i-1})$ $i = 1,2,...$ we have

$$M_{II}P(\&_i/C_{i-1}) \leqslant \nu \text{ const } e^{-R_{comp}\nu} \qquad (4.6)$$

when

$$R \leqslant R_{comp}\frac{k-1}{k}, \quad B = R_{comp}\,; \qquad (4.7)$$

if

$$R_{comp}\frac{k-1}{k} \leqslant R_{cr} \qquad (4.8)$$

$$M_{II}P(\&_i/C_{i-1}) < \nu \text{ const } e^{-(R_{comp}-R)k\nu}, \qquad (4.9)$$

when

$$R_{comp}\frac{k-1}{k} < R \leqslant R_{cr},\quad B = R_{comp}\,; \tag{4.10}$$

if

$$R_{comp}\frac{k-1}{k} > R_{cr}\,, \tag{4.11}$$

$$M_{II}P(\mathcal{E}_i/C_{i-1}) \leqslant \text{const}\; e^{-G[\beta(R\frac{k}{k-1})]\nu} \tag{4.12}$$

when

$$R_{cr} < R \leqslant R_{comp}\frac{k-1}{k},\quad B = R\frac{k}{k-1}\,; \tag{4.13}$$

$$M_{II}P(\mathcal{E}_i/C_{i-1}) \leqslant \text{const}\; e^{-[G(\gamma)-\gamma R]k\nu}, \tag{4.14}$$

when

$$\max[R_{cr}, R^{(k)}] < R \leqslant C,\quad B = G(\gamma)/\gamma\,; \tag{4.15}$$

all the constants do not depend on τ and ν (The notation is a previously).

For the proof of theorem 4.1 see [11].

Corollary There exists (a) a coder within ensemble *II* with constraint length ν, (b) a stack decoder with a back search limit ν, $\tau \geqslant \nu$, such that the conditional probability of error $P(\mathcal{E}_i/C_{i-1})$ is upper bounded: (I) by the left side of (4.6) under condition (4.7), or (II) (4.9) under conditions (4.8) and (4.10), or (III) (4.12) under conditions (4.11), (4.13), or (IV) (4.14) under condition (4.15).

The corollary immediately follows from Theorem 4.1, as all coders and decoders with the ensemble cannot have characteristics worse than the average characteristics of the ensemble.

Similarly to the block coding case, coefficient $E(R)$ within the exponent in the probability of error is called reliability function.

From Theorem 4.2 one may obtain the following upper bound for the reliability function: $E(R) \leqslant \bar{E}(R)$. The upper bound $E(R)$ as a function of R is shown in Fig. 7.

From a practical point of view the most favorable case is when the decoder operates within the first region. In this case it is necessary to have a back search limit according to

$$\tau \geqslant \frac{R_{comp}}{R_{comp}-R}\,\nu \qquad (4.16)$$

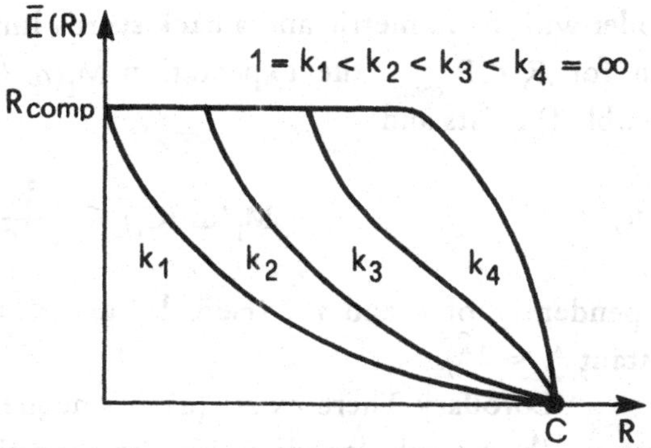

Fig. 7. Typical bounds for the reliability function of sequential decoding

Theorem 4.1 is correct also for coders from ensemble III. For $\tau = \nu$ the theorem holds also for ensemble I. Unfortunately, for $\tau > \nu$ we have no result concerning the reliability function of coders from ensemble I. But bounds for ensemble II are also used for ensemble I in engineering.

The bounds for systematic coders are, generally speaking, worse than those for nonsystematic coders. From this point of view, nonsystematic coders are better than systematic ones.

Theorem 4.1 is correct also for the modification of the stack algorithm, i.e. for ordered modification. For the threshold modification and Fano metric (which is optimal in this case), we have

$$M_{II}(\mathcal{E}_i/C_{i-1}) \leqslant e^{-a} \qquad (4.17)$$

where a is the threshold.

Next we consider the complexity of decoding. Let us denote by h_1 the minimal root of

$$qe^{h_1\alpha'} + pe^{h_1\alpha''} = 1. \qquad (4.18)$$

Here p is the crossover probability of the BSC, $q = 1 - p$, α' and α'' are (3.5). For $R < C$ equation (4.18) has one negative root. Let us also denote by ϵ an upper bound of the conditional probability of error $P(\mathcal{E}_i/C_{i-1})$.

Theorem 4.2 Assume that: (a) an encoder from ensemble I (or ensemble II) with constraint length ν is used for coding information for a BSC (b) and a stack decoder with Fano metric and a back search limit $\tau = \nu$ is used for decoding. Then for $R < R_{comp}$ the expectation $M_I(\omega_i/C_i)$ ($M_I(.)$ means expectation over ensemble I) exists and

(4.19)
$$M_I(\omega_i/C_i) \leqslant \frac{1}{1-\epsilon} A \frac{h_1}{1+2h_1},$$

independently of ν and τ. (Here h_1 and ϵ are described above, A stands for a constant $A \approx 2^{\ell}$)

Corollary There exist: (a) an encoder within ensemble I with a constraint length ν, (b) a stack decoder with Fano metric and a back search limit equal to ν, such that the expectation $M(\omega_i/C_i)$ (here $M(.)$ means mathematical expectation over the ensemble of information sequences and noise in channel) exists and is upperbounded by the right side of (4.19).

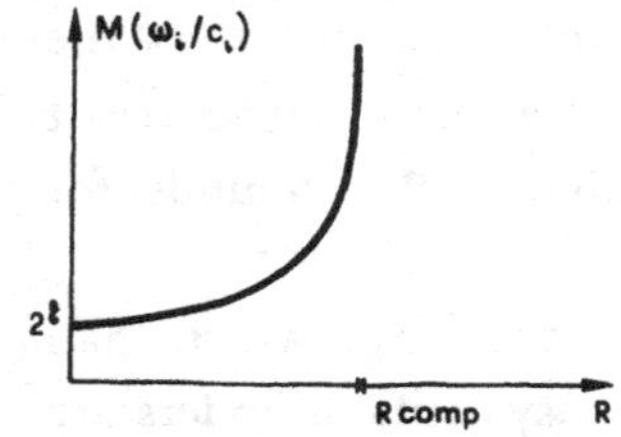

Fig. 8. Typical bounds $M(\omega_i/C_i)$ for sequential decoding

The upper bound of $M(\omega_i/C_i)$ as a function of R, is shown in Fig. 8.

Theorem 4.2 holds for ensemble II as well as III also for $\tau \geqslant \nu$ and systematic coders.

Let us prove theorem 4.2 in the simplest case : $\tau = \nu = \infty$, and $i = 1$. In this case C_1 is a trustworthy event and $M_I(\omega_1/C_1) = M_I(\omega_1)$.

The stack decoder examines at each instant the node for which the likelihood function is maximal. Let us suppose that there is a minimum value of the likelihood function at a node of some path. Obviously, the nodes of the code tree for which the values of the likelihood function are less than this minimum will not be examined.

We will prove that, provided $R < C$, the minimum value of the likelihood function μ for nodes of the correct path exists with probability 1, and

(4.20)
$$P(\mu \leqslant x) \leqslant e^{-h_1 x}, \quad x \leqslant 0;$$

here h_1 is root of (4.18).

In the case of BSC we consider the value of the likelihood function for nodes along the correct path as a random process with independent increments. Let us denote by $z(t)$ the value of the likelihood function at the nodes of depth t.

In this case

(4.21)
$$\bar{\bar{Z}}(t) = \mathcal{H}_0 + \mathcal{H}_m + \ldots \mathcal{H}_{t-m}, \quad t = m \ldots.$$

here $\mathcal{H}_i$, $i = 0, m, \ldots,$ stands for the increment of the likelihood function at the branch of the correct path, which is going from node of depth i.

Let us define

$$\mu(t) = \min_{t' \geqslant t} [\bar{\bar{Z}}(t') - \bar{\bar{Z}}(t)], \quad t = 0, m, 2m, .. \tag{4.22}$$

We are interested in the distribution of $\mu(0)$. By definition

$$P[\mu(t) < x] = \overline{P[\mu(t + m) < x - \mathcal{H}_t]}\,, \quad t = 0, m, \ldots \tag{4.23}$$

where the bar means the mathematical expectation over all values of $\mathcal{H}_t$. But

$$\mathcal{H}_t = \begin{cases} m\ \alpha' \text{ with probability } q^m \\ (m - 1)\alpha' + \alpha'' \text{ with probability } C_m^1 \cdot q^{m-1} p \\ \vdots \\ m\ \alpha'' \text{ with probability } p^m \end{cases}$$

From (4.22) and (4.23) we have

$$P[\mu(t) < x] = \sum_{k=0}^{m} C_m^k\, q^k\, p^{m-k}\, P[\mu(t + m) < x - k\alpha' - (m - k)\alpha''] \tag{4.24}$$

Let us denote the distribution function of the $\mu(t)$ by $\Phi(t)$ (Observe that it does not depend on t). From (4.24) we have

$$\Phi(x) = \sum_{k=0}^{m} C_m^k\, q^k\, p^{m-k}\, \Phi[x - k\alpha' - (m - k)\alpha''] \qquad x \leqslant 0 \tag{4.25}$$

Obviously

$$\Phi(x) = 1\,, \quad x > 0 \tag{4.26}$$

One can prove that each function for which

$$\Phi^*(x) \geqslant \sum_{k=0}^{m} C_m^k\, q^k\, p^k\, \Phi^*[x - k\alpha' - (m - k)\alpha''] \quad x \leqslant 0 \tag{4.27}$$

$$\Phi^*(x) > 1, \quad x > 0 \tag{4.28}$$

upperbounds the solution of (4.25)-(4.26), viz.,

$$\Phi^{*}(x) \geqslant \Phi(x). \tag{4.29}$$

Observe that

$$\Phi^{*}(x) = e^{-h_1 x} \quad , \quad -\infty < x < \infty, \tag{4.30}$$

is according to (4.27) - (4.28), h_1 is as previously defined. This proves (4.20).

Let us now suppose, that the minimal value μ of the likelihood function at nodes of correct path is equal to x. We are interested in the expectation of the computation number, which the decoder has to make within first incorrect subtree, under conditions $\mu = x$.

Let z denote the value of the likelihood function at some node of the first incorrect tree. Let W(z,x) be the number of computations at the nodes following this particular node. Following the previously outlined method one can show that

$$W(\bar{\bar{Z}},x) = 2^{\ell} \sum_{k=0}^{m} C_m^k \left(\frac{1}{2}\right)^m W\left[\bar{\bar{Z}}+k\alpha' + (m-k)\alpha'', x\right]+1, \quad Z \leqslant x \tag{4.31}$$

where ℓ and m are numbers of information and coding symbols within the subblock respectively.

Here we take into account the condition that the probabilities of increment of the likelihood function in ensemble I are

$$\mathcal{H}_t = \begin{cases} m\alpha', \text{ with probability } \left(\frac{1}{2}\right)^m \\ (m-1)\alpha' + \alpha'', \text{ with probability } C_m^1 \left(\frac{1}{2}\right)^m \\ \vdots \\ m\alpha'' \text{ with probability } \left(\frac{1}{2}\right)^m \end{cases} \tag{4.32}$$

Obviously

$$W(z,x) = 0, \quad z < x \tag{4.33}$$

One may prove, that for such function satisfying the inequalities

$$W^{*}(z,x) \geqslant 2^{\ell} \sum_{k=0}^{m} C_m^k \left(\frac{1}{2}\right)^m W^{*}[z + k\alpha' + (m-k)\alpha'',x] + 1, \quad z \geqslant x, \tag{4.34}$$

$$W^{*}(z,x) \geqslant 0, \quad z < x, \tag{4.35}$$

the solution of (4.31)-(4.33) satisfies the following upper bounds

$$W^*(z,x) \geqslant W(z,x) \tag{4.36}$$

Let us try to express $W^*(z,x)$ in the form

$$W^*(z,x) = \begin{cases} A\, e^{\lambda(z-x)}, & z \geqslant x \\ 0, & z < x \end{cases} \tag{4.37}$$

where A and λ denote some constants. If we put (4.37) into (4.34), through (4.35), we obtain that

$$A \simeq 2^{\ell}, \tag{4.37}$$

λ is the minimal root of the equality

$$\frac{1}{2}\, e^{\lambda\alpha'} + \frac{1}{2}\, e^{\lambda\alpha''} = e^{-R}, \tag{4.38}$$

(R denotes the transmission rate).

But at the initial node z = 0. Then the expectation of the number of computations for the first incorrect subtree is

$$M_I(\omega) = \int M_I(\omega_1/\mu = x)\, dP(\acute{\mu} < x) = \int W(0,x)\, d\Phi(x). \tag{4.39}$$

One can also prove, that we can operate with upper bounds of $W(0,x)$ and $\Phi(x)$ as with $W(0,x)$ and $\Phi(x)$. In this way we obtain upper bounds for $M_I(\omega_1)$.

In case of Fano metric, $\lambda_1 = 1+h_1$. If the transmission rate $R < R_{comp}$, then $h_1 < -1/2$ and the mathematical expectation $M_I(\omega_i/C_i)$ exists and satisfies (4.19).

For computational metrics the expectation of the number of computations is nonessentially larger than that for Fano metric.

Theorem 4.2 is also correct for a coder from ensemble III.

In addition it is possible to prove the next theorems.

Theorem 4.3: Let us suppose that the conditional probability $P(\&_i/C_{i-1})$ is upperbounded by inequality $P(\&_i/C_{i-1}) < \epsilon$. Then there exists (a) a coder within ensemble I (or ensemble II) and (b) a corresponding stack decoder with computational metric for which distribution function of the computation number is lowerbounded by

$$P(\omega_i < x) \geqslant 1 - \text{const } x^{-\beta} - \epsilon \tag{4.40}$$

where

$$\beta = \begin{cases} R/R_{comp}, \text{ if } R < R_{comp} \\ \beta\text{-parameter of rate, if } R \geqslant R_{comp} \end{cases} \tag{4.41}$$

and the constant does not depend on x.

Theorem 4.4: Let us suppose, that the conditional probability $P(\mathcal{E}_i/C_{i-1})$ is upperbounded by inequality $P(\mathcal{E}_i/C_{i-1}) < \epsilon$. Then there exists (a) an encoder within ensemble III and (b) a corresponding stack decoder with Fano metric, for which the distribution function of the computation number is lowerbounded by

$$P(\omega_i < x) \geqslant 1 - \text{const } x^{-\beta} - \epsilon \tag{4.42}$$

where β is β-parameter of rate R and the constant does not depend on x.

The distributions (4.40) and (4.42) are Pareto distributions.

From theorems 4.3 and 4.4 it follows that the distribution of the operative memory overflow and distribution of buffer overflow are algebraic functions of memory or buffer volume.

5. The Algorithm with Recurrences

The main shortcoming of stack algorithm consits in the need of large opertive memory. The well-known Fano algorithm of sequential decoding is free from this shortcoming. Here we will describe another algorithm of sequential decoding – **the algorithm with recurrences**, which was introduced in paper [2].

Operative memory of decoder with recurrences consists of τ/m bins, each of which can store $(2^{\ell}+1)$ numbers. Here m is the number of a code symbol in the branch and 2^{ℓ} is the number of the branches which are going from each node. Each bin stores information concerning some node of the examined path. This information consists of the value of the likelihood function in this node and the values of 2^{ℓ} threshold $T_1, T_2, \ldots, T_{2^{\ell}}$, one of them corresponds to this node and the $2^{\ell}-1$ others correspond to the adjacent nodes are ordered in some way.

Now we can define an algorithm with recurrences. Notation t_0 means the depth of the examined node, $z_i(t)$ means the value of the likelihood function of i-th node, whose depth is equal to t, $T_i(t)$ means the value of the threshold of i-th node, whose depth is equal to t.

Step 1. Compute the values of the likelihood functions for 2^ℓ nodes, following the initial node: $z_1(m), z_2(m) \ldots z_{2^\ell}(m)$. Take (*)

$$T_1(m) = [\frac{z_1(m)}{\Delta}]\,\Delta; \quad T_2(m) = [\frac{z_2(m)}{\Delta}]\,\Delta, \ldots, T_{2^\ell}(m) = [\frac{z_{2^\ell}(m)}{\Delta}]\,\Delta$$

Step 2. Pick the node which corresponds to the largest threshold. If the depth of this node t_0 is not less than the depth of each other node, go to step 3. Otherwise discard from the memory the information about nodes whose depth is larger than t_0.

Step 3. Compute the values of the likelihood functions for 2^ℓ nodes following to the examined node: $z_1(t_0+m), z_2(t_0+m), \ldots, z_{2^\ell}(t_0+m)$. Take

$$T_1(t_0+m) = [\frac{z_1(t_0+m)}{\Delta}]\,\Delta,$$

$$T_2(t_0+m) = [\frac{z_2(t_0+m)}{\Delta}]\,\Delta, \ldots, T_{2^\ell}(t_0+m) = [\frac{z_{2^\ell}(t_0+m)}{\Delta}]\,\Delta.$$

Change the values of the thresholds at nodes of the examined path, taking in this case max $T_i(t_0+m)$.(Thresholds in adjacent nodes do not change).

Steps 4 and 5 are analogous to steps 4 and 5 of the stack algorithm.

It is possible to prove that for a decoder with recurrences the complexity of the operative memory is a linear function of τ, the upper bound of the probability of error is as in theorem 4.1, for the complexity of decoding theorem 4.2 is correct, but the constant A is larger than for the stack decoder.

For comparison, recall that for Fano decoder the complexity of the operative memory is constant,for the error probability theorem 4.1 is correct, for the complexity of decoding theorem 4.2 is correct with constant A which is essentially larger than for decoder with recurrences.

In the author's papers [3–8] these results were generalised to the continuous channel, to the channel with memory, channel with synchronisation errors and so on.

(*) Symbol [x] means whole part of x

References

[1] Wozencraft, J.M.: "Sequential Decoding for Reliable Communication", Sc.D. Thesis, M.I.T., 1957.

[2] Zigangirov, K.Sh.: "Some Sequential Decoding Procedures", Probl. Peredachi Inform. 4, 1966.

[3] Zigangirov, K.Sh., M.S. Pinsker and B.S. Tsybakov: "Sequential Decoding in Continuous Channels", Probl. Peredachi Inform., 4, 1967.

[4] Zigangirov, K.Sh. "Algorithm of Sequential Decoding in which the Error Probability Increases in Agreement with Random Coding Bound", Probl. Peredachi Inform. 2, 1968.

[5] Zigangirov, K.Sh. "Sequential Decoding in a Channel with Deletion and Insertion" Probl. Peredachi Inform., 2, 1969.

[6] Vvdenskaja N.D. and K. Sh. Zigangirov: "On the Computation Time Distribution of the Sequential Decoding", Probl. Peredachi Inform., 4, 1969.

[7] Zigangirov, K.Sh. and V.V. Ovchinnikov: "Sequential Decoding in Channel with Error Burst", Probl. Peredachi Inform., 1, 1971.

[8] Zigangirov, K.Sh.: "Sequential Transmission from Source with Variable Rate", Probl. Peredachi Inform., 2, 1971.

[9] Gallager, R.G.: "Information Theory and Reliable Communication" John Wiley and Sons, N.Y., 1968.

[10] Jelinek, F.: "Fast Sequential Decoding Algorithm Using a Stack", IBM Journ. Research Develop. 11, 1969.

[11] Zigangirov, K.Sh.: "On the Error Probability of Sequential Decoding on the BSC", IEEE Trans., IT-18, 1, 1972.

DECODING COMPLEXITY AND CONCATENATED CODES

V.V. Ziablov

1. Encoding and Decoding Complexity
2. Decoding Complexity for Low-Density Parity-Check Codes
3. Generalized Concatenated Codes
4. Decoding of Generalized Concatenated Codes.

1.Encoding and Decoding Complexity.

1.1 Introduction

Let $\Psi:X\rightarrow Y$ be some Boolean function, where X and Y are sets of binary words of length n_1 and n_2, respectively. It is obvious that encoding and decoding can be viewed as such a Boolean function.

Assume that the Boolean function is constructed by means of logic elements from the following fixed set:

– AND element (2 inputs - complexity 1),
– OR element (2 inputs - complexity 1),
– NOT element (1 input - complexity 1),
– ADDER (2 inputs - complexity 1),
– THRESHOLD element (ℓ inputs - complexity ℓ).

A suitable combination of these elements can be represented as a directed graph which will be referred to as a logic circuit. In such a directed graph there are n_1 points from which lines start only (such points will be referred to as input points); and n_2 points where lines end only (such points will be referred to as output points).

The points which are neither input nor output points will be referred to as functional points. In each functional point there is a logic element.

The complexity $\kappa(S)$ of the circuit S is defined as the sum of the complexities of the single elements in all the functional points.

The circuit S is said to realize the function $\Psi:X\rightarrow Y$ if for every word $x\epsilon X$ fed to the input points of the circuit, the word $y=\Psi(x)\epsilon Y$ is obtained at the output points. Let $G(\Psi)$ be the set of all circuits which realize the function Ψ.

Definition. The complexity $\kappa(\Psi)$ of a function Ψ ("computational work" in Savage's terminology) is defined by:

$$\kappa(\Psi) = \min_{S \,\epsilon\, G(\Psi)} \kappa(S). \tag{1}$$

Hence, the complexity of a function is the complexity of the simplest circuit which realizes this function.

Assume that there exists a path from some input point to some output point of a circuit. The length of this path is defined as the number of functional points lying on the path. Let L_s be the longest path in a circuit S.

Definition. The computational delay $L(\Psi)$ of a function Ψ is defined as

$$L(\Psi) = \min_{S \in G(\Psi)} L_s. \tag{2}$$

Suppose now that the function $\Psi:X \to Y$ can also be realized by L interacting sequential machines $\{M_i, i=\overline{1,L}\}$, the i-th containing $\kappa(M_i)$ logic elements and binary cells and working for T_i clock cycles, $i=\overline{1,L}$. The computational work (Savage) is defined as

$$\kappa'(\Psi) = \sum_{i=1}^{L} \kappa(M_i)\, T_i\,. \tag{3}$$

It should be noted that $\kappa(\Psi) \leqslant \kappa'(\Psi)$.

In this case the complexity of a machine (e.g. the complexity of an encoder or decoder) is defined as

$$\kappa'(M) = \sum_{i=1}^{L} \kappa(M_i)\,. \tag{4}$$

Notice that when the decoder is a machine which performs the decoding in $T > 1$ clock cycles the complexity of decoding is not the same as the complexity of the decoder.

In this case the realization time (i.e. the encoding time or the decoding time) is defined as

$$L'(\Psi) = \sum_{i=1}^{L} T_i. \tag{5}$$

1.2 Statement of the Problem and Main Results

Let P_e be the error probability and take $Q = -\log_2 P_e$ as the quality of coding. Further, let κ stand for the encoding and decoding complexity and L stand for the encoding and decoding delay. Both κ and L are increasing functions of Q:

$$\kappa = f(Q), \quad L = \varphi(Q) \tag{6}$$

and we are obviously interested in functions which grow as slow as possible when Q increases. Then the problem is to get an estimate of such slowly increasing functions.

In terms of coding theory this problem may lead to several questions, such as:

1. What are the estimated functions

$$\kappa(\Psi_E) = f_1(Q), \quad L_E = \varphi_1(Q)$$
$$\kappa(\Psi_D) = f_2(Q), \quad L_D = \varphi_2(Q), \tag{7}$$

where E stands for encoding and D stands for decoding, κ and L are the computational work and the computational delay, respectively?

2. What are the estimated functions

$$\kappa(\Psi_E) = f_3(n), \quad L_E = \varphi_3(n)$$
$$\kappa(\Psi_D) = f_4(n), \quad L_D = \varphi_4(n), \tag{8}$$

if it is required that the number of correctable errors and erasures grows as $\alpha n (\alpha > 0)$ in terms of the code length n?

3. What are the estimated functions

$$\kappa'(E) = f_5(n) \quad \text{if } L'_E \leqslant cn$$
$$\kappa'(D) = f_6(n) \quad \text{if } L'_D \leqslant cn, \tag{9}$$

if it is required that the number of correctable errors grows as $\alpha n (\alpha > 0)$ in terms of the code length n? ($\kappa'(E)$ and $\kappa'(D)$ are the complexities of the encoder and of the decoder, respectively, c is a constant.)

Some of these problems were investigated at the Institute for the Problems of Information Transmission (USSR) and the following solutions were obtained:

1. For problem (7) it was shown that for all transmission rates less than channel capacity, $R < C$, there exists a sequence of codes such that

$$\kappa(\Psi_E) \leqslant C_1 Q, \quad L_E \leqslant C_1^* \log Q$$
$$\kappa(\Psi_D) \leqslant C_2\, Q \log Q, \quad L_D \leqslant C_2^* \log Q, \tag{10}$$

where C_i and C_i^* $(i=1,2)$ are constants. Further, it was shown that it is possible to construct a sequence of codes such that:

$$\kappa(\Psi_E) \leqslant K_1 Q^{1+\epsilon}, \quad L_E \leqslant K_1^* Q^{\epsilon}$$
$$\kappa(\Psi_D) \leqslant K_2 Q^{1+\epsilon}, \quad L_D \leqslant K_2^* Q^{\epsilon}, \tag{11}$$

where K_i and $K_i^*(i=1,2)$ are constants, and

$$\epsilon \leqslant \sqrt{\frac{\log\log Q}{\log Q}}$$

approaches zero as Q increases.

2. For problem (8) similar results were obtained: there exists a sequence of codes such that

$$\kappa(\Psi_E) \leqslant C_3 n \quad L_E \leqslant C_3^* \log n$$
$$\kappa(\Psi_D) \leqslant C_4\, n \log n, \quad L_D \leqslant C_4^* \log n\,. \tag{12}$$

Further, it is possible to construct a sequence of codes such that

$$\kappa(\Psi_E) \leqslant K_3 n^{1+\epsilon}, \quad L_E \leqslant K_3^* n^{\epsilon}$$
$$\kappa(\Psi_D) \leqslant K_4\, n^{1+\epsilon}, \quad L_D \leqslant K_4^* n^{\epsilon} \tag{13}$$

where ϵ is an arbitrary positive constant. Notice that the number of correctable errors depends on ϵ.

3. As to problem (9), for all $R < C$ it is possible to construct a sequence of codes

such that

$$\kappa(E) = C_5\, Q, \quad \kappa(E) = K_5\, n, \quad L_E = K_5^{*}\, n\,, \tag{14}$$

$$\kappa(D) = C_6\, Q, \quad \kappa(D) = K_6\, n, \quad L_D = K_6^{*} n\,.$$

A solution for this problem is provided e.g. by a sequence of Forney concatenated codes. Thus it is of great interest to solve another problem, namely to find a sequence of codes with similar encoding and decoding complexities, but with a better error correction capability.

All the results sketched above were obtained in the papers [1-5].

Some of the above problems which can be solved by using low-density parity-check codes and concatenated codes will be considered in the sequel.

2. Decoding Complexity for Low-Density Parity-Check Codes.

2.1. Ensembles of Low-Density Parity-Check Codes

Consider two ensembles of binary low-density parity-check codes: (A) Gallager's ensemble of low-density parity-check codes; and (B) the ensemble of low-density parity-check codes introduced by Pinsker and Ziablov.

The ensemble (A), which will be referred to as $\mathscr{A}(n,h,\ell)$ can be defined as follows.

Let A be an m x n binary matrix of the form

$$A = \|I_1 I_2 \ldots I_h\|. \tag{1}$$

where I_j is the m x m unit matrix, and $n = m\,.\,h$. Consider now all possible permutations on the columns of matrix A in (1), and assume that each of the m x n matrices thus obtained is given the same probability. Let $\mathscr{A}(n,h,1)$ be the corresponding matrix ensemble. Then the code ensemble (A) can be defined by means of the ensemble of parity check matrices

$$H = \left\| \begin{matrix} A_1 \\ A_2 \\ \vdots \\ A_\ell \end{matrix} \right\| \tag{2}$$

where $A_i \in \mathscr{A}(n,h,1)$ and each of the A_i, $i = 1,2,\ldots,\ell$ is chosen independently in the ensemble.

It is obvious that the rate of every code in the ensemble $\mathscr{A}(n,h,\ell)$ satisfies

$$R > 1 - \frac{\ell m}{n} = 1 - \frac{\ell}{h} . \tag{3}$$

Notice that there are ℓ 1's in each column and h 1's in each row in the parity-check matrix.

The ensemble (B),which will be referred to as $\mathscr{M}(n,m,\ell)$, can be specified as follows.

Consider the ensemble $\mathscr{M}(n,m,1)$ of all m x n binary matrices having only one symbol 1 in each column. Every matrix in this ensemble is assigned the same probability. Then the code ensemble (B) can be given through the ensemble of parity-check matrices

$$H = \left\| \begin{matrix} M_1 \\ M_2 \\ \cdot \\ \cdot \\ \cdot \\ M_\ell \end{matrix} \right\| \tag{4}$$

where $M_i \in \mathscr{M}(n,m,1)$ and each of the matrices m_i , $i = 1,2,\ldots,\ell$, is chosen independently in the ensemble.

It is obvious that the rate of every code in the ensemble $\mathscr{M}(n,m,\ell)$ satisfies

$$R > 1 - \frac{\ell m}{n} = 1 - \ell\mu, \quad \mu = m/n . \tag{5}$$

Notice that in each of the parity-check matrices (4) there are ℓ 1's in each column, and a random number of 1's in each row.

The codes in the ensemble (A) will be used to correct errors, and the codes in the ensemble (B) will be used to correct erasures.

2.2 An Algorithm for Correcting Erasures

To correct erasures we shall use those checks in the parity-check matrix which contain the erasures. Such checks will be referred to as correction checks. If there is a correction check for a single erasure it is very simple to correct it.

From the received vector containing erasures we construct the following binary vector of length 2n:

(6) $$u=(u_1 v_1 \quad u_2 v_2 \quad \dots \quad u_n v_n)$$

where

$$u_i = \begin{cases} \text{received symbol, if it is not an erasure ,} \\ 0, \text{ if it is an erasure,} \end{cases}$$

$$v_i = \begin{cases} 1, \text{ if the i-th received symbol is an erasure ,} \\ 0, \text{ if the i-th received symbol is not an erasure .} \end{cases}$$

The decoding algorithm can be viewed as the successive use for T times of the same function f

(7) $$u^T = f(f(\dots f(f(u^\circ))\dots)), \quad u^\circ = u,$$

where u^T is the result of decoding. If $v_i = 0,\ i=\overline{1,n}$ then all erasures are corrected. A single use of the function f

(8) $$u^{j+1} = f(u^j)$$

will be referred to as a decoding step.

The computation of the function f consists of the following operations:

– correction checks are selected,

– erasures, for which there exist correction checks, are corrected,

– all values of v_i corresponding to corrected erasures are changed to 0.

The function f can be realized by a circuit F with complexity

(9) $$\kappa(F) = (7\ell + 2)n \ .$$

Hence the overall algorithm for erasure correction can be implemented by a circuit with complexity.

(10) $$\kappa(\Psi) = \kappa(F)T = (7\ell + 2)nT \ .$$

There are, however, two open questions here:

1. Is there any code which can be decoded by means of this procedure?
2. What number T of decoding steps is needed to correct erasures?

2.3 Existence of Low Complexity Codes for Erasure Correction

Assume that for every set of τ erasures $(\tau < \alpha_\gamma n, \alpha_\gamma > 0)$ more than $\gamma\tau$ of them $(0 < \gamma < 1)$ possess a correction check each. It is obvious that after the first decoding step the number of erasures τ_1 is upper bounded by

$$\tau_1 \leqslant (1-\gamma)\,\tau_0 \tag{11}$$

where τ_0 is the number of erasures before the first step. If τ_i is the number of erasures after the i-th decoding step, then:

$$\tau_i \leqslant (1-\gamma)\tau_{i-1} \leqslant (1-\gamma)^i \tau_0 \,. \tag{12}$$

If after T decoding steps we have

$$\tau_T \leqslant (1-\gamma)^T \tau_0 < 1 \tag{13}$$

this means that all erasures have been corrected in no more than T decoding steps. Hence the number of decoding steps satisfies:

$$T \leqslant \frac{\ln \tau_0}{-\ln(1-\gamma)} \leqslant \frac{\ln n}{-\ln(1-\gamma)} \,. \tag{14}$$

Let P_γ be the probability that for some set of τ erasures $(\tau \leqslant \alpha_\gamma n, \alpha_\gamma > 0)$ there are less than $\gamma\tau$ erasures for each of which there is a correction check. It is obvious that if $P_\gamma < 1$ then there exist codes which can correct every set of τ erasures $(\tau \leqslant \alpha_\gamma n)$ with a decoding complexity

$$\kappa(\Psi) \leqslant (7\ell + 2)nT < c\,\ln n, \tag{15}$$

c being a constant.

The existence of such codes is exhibited by the following theorem:

Theorem 2.1 If $\gamma < (\ell - 2)/\ell$,, in the ensemble $\mathcal{M}(n,m,\ell)$ the probability P_γ is asymptotically upperbounded by

$$P_\gamma \leqslant C_\gamma n^{-\ell(1-\gamma)+2}(1 + 0(1)), \; n \to \infty, \tag{16}$$

where C_γ is a constant and α_γ ($\alpha_\gamma > 0$) is any number smaller than the smallest positive root of the equation

$$H(\alpha) + \ell\varphi(\alpha,\mu) = 0, \tag{17}$$

where

$$H(\alpha) = -\alpha \ln \alpha - (1-\alpha) \ln (1-\alpha),$$

$$(18)$$

$$\varphi(\alpha,\mu) = \alpha \ln \frac{\alpha}{e\mu} + \inf_{0<s} \{\mu \ln (e^s - s) - \alpha \ln s\}.$$

The main idea of the proof is the following. Let $P_\gamma(\tau)$ be the probability that for a fixed set of τ erasures there are less than $\gamma\tau$ erasures for each of which there is a correction check. Then it is obvious that

$$P_\gamma \leqslant \sum_{\tau=1}^{\alpha_\gamma n} \binom{n}{\tau} P_\gamma(\tau). \tag{19}$$

Let $P(\tau, j)$ be the probability that for a fixed set of τ erasures there are j correction checks. Then it is obvious that

$$P_\gamma(\tau) \leqslant \sum_{j=0}^{\gamma\tau\ell} P(\tau, j). \tag{20}$$

Consider now the ensemble of matrices $\mathcal{M}(\tau,m,\ell)$. Each of these matrices consists only of those columns of the parity-check matrix which correspond to erasures. Let $N_j(\tau,m,\ell)$ be the number of matrices in this ensemble containing j rows with all 1's, and let $N(\tau,m,\ell)$ be the overall number of matrices in this ensemble. Then it is obvious that

$$P(\tau,j) = \frac{N_j(\tau,m,\ell)}{N(\tau,m,\ell)}. \tag{21}$$

Estimations of these numbers can be obtained from the following enumeration formulae:

$$\sum_{\tau=0}^{\infty}\sum_{j=0}^{m\ell} N_j(\tau,m,\ell)\frac{s^{\tau\ell}}{(\tau!)^{\ell}}\cdot t^j = (e^s - s + ts)^{m\ell},$$

$$\sum_{\tau=0}^{\infty} N(\tau,m,1)\frac{s^{\tau}}{\tau!} = e^{ms}, \quad N(\tau,m,\ell) = [N(\tau,m,1)]^{\ell}. \tag{22}$$

Substitutions in these equations easily lead to the proof.

If the number of decoding steps is estimated more accurately, then it is seen that the complexity of the erasure decoding procedure asymptotically satisfies

$$\kappa(\Psi) \leqslant \frac{7\ell + 2}{\ln \frac{\ell}{2}} n \ln n\,(1 + 0(1)), \quad n \to \infty. \tag{23}$$

In this case the number of guaranteed corrected erasures is of the order of 40-60 % of the Gilbert-Varsharmov bound (Fig. 1).

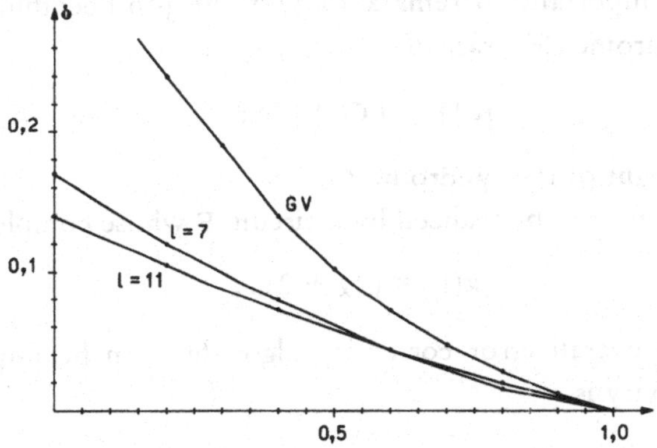

Fig. 1

2.4 An Algorithm for Error Correction

Consider a binary vector $u = (y,C)$, where y is a binary word of length n and C is the syndrome corresponding to y.

The symbol y_i is said to affect the check C_k, if there is a 1 at the intersection of the i-th column with the k-th row of the parity-check matrix. It is obvious that every symbol affects ℓ checks and that every check is affected by h symbols. Two symbols are called unaffected if in the parity-check matrix there is no check affected simultaneously by both symbols.

The check C_k, $k = \overline{1,m\ell}$ is said violated if there is an odd number of symbols with errors affecting this check.

Split the symbols of a codeword in $q = \ell\,(h-1) + 1$ or less subsets M_s such that there are only unaffected symbols in every subset M_s.

The error correcting algorithm can be viewed as the successive use of the same function f T times

(24) $$u^T = f(f(\ldots f(f(u^\circ))\ldots)),$$

where u^T is the result of the decoding. If $u^T = (y,0)$ then the result is a codeword. A single use of the function f:

(25) $$u^{j+1} = f(u^j)$$

will be referred to as a decoding step. The computation of the function f consists in the following operations. In each subset M_s, $s = \overline{1,q}$, where more than $\ell/2$ checks are violated, all 0's are changed into 1's and all 1's are changed into 0's.

It is very important to remark that, if the j-th decoding step involves t changes then the syndrome C^{j+1} satisfies

(26) $$|C^j| - |C^{j+1}| \geqslant t\,,$$

where $|C^j|$ is the weight of the syndrome C^j.

The function f can be realized by a circuit F whose complexity is

(27) $$\kappa(F) = (3\ell + 2)n\,.$$

Hence, the overall error correcting algorithm can be implemented by a circuit whose complexity is

(28) $$\kappa(\Psi_D) = \kappa(F)T = (3\ell + 2)n\,T\,.$$

In this case as well as for erasure correction, there are two open questions:

1. Is there any code which can be decoded using this procedure?
2. What number T of decoding steps is needed to correct the errors?

2.5 Existence of Low Complexity Codes for Error Correction

Assume that for every set of τ errors ($\tau \leqslant 2\alpha^* n$, $\alpha^* > 0$) there are more than $\gamma\tau\ell$ violated checks ($\gamma > 0.5$) and let τ_0 ($\tau_0 < \alpha^* n$) be the number of errors. Then for every j the number of errors satisfies

$$\tau_j < 2\alpha^* n, \tag{29}$$

where τ_j is the number of errors after j decoding steps.

Assume that at some moment $\tau_j = 2\alpha^* n$. Then the number of violated checks satisfies

$$|C^j| \geqslant \gamma\tau_j > \alpha^* n\ell .$$

At the beginning the number of violated checks satisfies

$$|C^0| < \tau_0\, \ell < \alpha^* n\ell$$

and besides it must decrease at every decoding step. The proof of inequality (29) is obtained from this contradiction.

Consider the decoding procedure. After the first step the number of violated check satisfies

$$|C^1| < (2\gamma - 1)\ |C^0| \tag{30}$$

and after the j-th step it satisfies

$$|C^j| \leqslant (2\gamma - 1)\ |C^{j-1}| \leqslant (2\gamma - 1)^j\, |C^0|\,. \tag{31}$$

If for j = T we have

$$|C^T| \leqslant (2\gamma - 1)^T\, |C^0| < 1, \tag{32}$$

then all errors have been corrected in no more than T decoding steps. Hence the number of decoding steps satisfies

$$T < \frac{\ln |C^0|}{-\ln(2\gamma - 1)} < a(\gamma) \ln n, \tag{33}$$

where $a(\gamma)$ is a constant.

Let P_γ be the probability that there are no more than $\gamma\tau\ell$ violated checks for a set of τ errors ($\tau \leqslant \alpha_\gamma n$, $\alpha_\gamma > 0$). It is obvious that, if for $\gamma > 0.5$, $P_\gamma < 1$, then there exists a code which can correct every set of τ errors ($\tau < \alpha^* n$, $2\alpha^* = \alpha_\gamma$), with decoding complexity

$$\kappa(\Psi_D) \leqslant (3\ell + 2)nT < b(\gamma) n \ \ln n, \tag{34}$$

where $b(\gamma)$ is a constant.

The existence of such codes is exhibited by the following theorem.

Theorem 2 In the ensemble $\mathscr{A}(n,h,\ell)$ the probability P_γ can be estimated as

(35) $$P_\gamma \leqslant g(\gamma) n^{-\ell(1-\gamma)+2}(1 + 0(1)), \quad n \to \infty,$$

where $g(\gamma)$ is a constant.

The main idea of the proof is the same as for Theorem 1.

If $R \to 1$ then the number $(\alpha^* n)$ of guaranteed correctable errors satisfies

(36) $$R \geqslant 1 - 22H(2\alpha^*),$$

where

$$H(\alpha) = -\alpha \log_2 \alpha - (1-\alpha) \log_2 (1-\alpha).$$

It is interesting to notice that we can construct a concatenated encoding and decoding procedure with the best binary code of length $n_1 = \log \log n$ and rate R_1 $(R < R_1 < C)$ as "inner" code and with a low-density parity-check code of rate R_2 $(R = R_1 \cdot R_2)$ as "outer" code. In this case it can be shown that for all rates $R < C$ the error probability P_e decreases with increasing n as

(37) $$P_e < 2^{-\beta n},$$

where $\beta > 0$. The decoding complexity of this system satisfies.

(38) $$\kappa(\Psi_D) \leqslant \kappa(\Psi_1) + \kappa(\Psi_2),$$

where $\kappa(\Psi_1)$ and $\kappa(\Psi_2)$ are the decoding complexities of the "inner" and the "outer" code, respectively. But the complexities $\kappa(\Psi_1)$ and $\kappa(\Psi_2)$ satisfy

(39) $$\kappa(\Psi_1) \leqslant c_1 2^{n_1} n_1 \cdot \frac{n}{n_1} = c_1 n 2^{n_1} = c_1 n \log n$$
$$\kappa(\Psi_2) \leqslant c_2 \, n \log n$$

where c_i, $i = 1,2$, are constants.

From (38) and (39) we obtain

(40) $$\kappa(\Psi_D) \leqslant cn \log n,$$

where c is a constant.

Hence, low-density parity-check codes allow us to solve a lot of existence problems concerning codes of low decoding complexity.

3. Generalized Concatenated Codes.

3.1 Definition of Generalized Concatenated Codes.

Let α be a binary word of length $n = n_1 n_2$. This word will be represented as

$$\alpha = (\alpha_1 \alpha_2 \ldots \alpha_{n_2}) \tag{1}$$

where $\alpha_j, j = \overline{1, n_2}$ is a subword of length n_1.

Generalized concatenated codes will be described by means of:

1) n_2 binary $n_1 \times n_1$ matrices $H_0^{(j)}, j = \overline{1, n_2}$;
2) $m + 1$ (n_2, b_i) group codes over $GF(2^{a_i})$, $i = \overline{1, m+1}$, $(\sum_{i=1}^{m+1} a_i = n_1)$.

In the sequel the (n_2, b_i) code over $GF(2^{a_i})$ will be referred to as the i-th code of the second level ("outer" code in Forney's terminology).

Consider the linear forms

$$\alpha_j H_0^{(j)T} = (\gamma_{1j}\, \gamma_{2j} \ldots \gamma_{(m+1)j}),\ j = \overline{1, n_2}\ , \tag{2}$$

where γ_{ij} is a binary word of length a_i . The word γ_{ij} can be interpreted as an element of $GF(2^{a_i})$.

For every binary word α of length $n = n_1 n_2$ we can construct $m + 1$ words

$$\gamma_i = (\gamma_{i1}\, \gamma_{i2} \ldots \gamma_{in_2}),\ i = \overline{1, m+1}, \tag{3}$$

where the γ_{ij}'s are the elements of $GF(2^{a_i})$ computed by means of equation (2).

Definition A binary word α of length $n = n_1 n_2$ is a codeword of the generalized concatenated code of order m, if each of the words γ_i, $i = \overline{1, m+1}$, corresponding to α is a codeword of the i-th second level code.

It can be easily shown that generalized concatenated codes are linear codes.

3.2 Investigating a Class of Generalized Concatenated Codes

In what follows, we will consider that subclass of generalized codes which is defined by means of the following specifications:

1. the same $n_1 \times n_1$ matrix H_0 is taken as matrix $H_0^{(j)}$ for all j, $(j = \overline{1,n_2})$, where H_0 has the form:

$$H_0 = \begin{Vmatrix} I_{a_1} & & & & & \\ P_{11} & I_{a_2} & & & \bigcirc & \\ P_{21} & P_{22} & I_{a_3} & & & \\ \cdot & & & & & \\ \cdot & & & & & \\ P_{m_1} & P_{m_2} & P_{m_3} & \dots & P_{mm} & I_{a_{m+1}} \end{Vmatrix} = \begin{Vmatrix} \tilde{H}_0 \\ \tilde{H}_1 \\ \tilde{H}_2 \\ \\ \\ \tilde{H}_m \end{Vmatrix} \tag{4}$$

where

$$\tilde{H}_0 = \| I_{a_1} \ 0 \dots 0 \|$$
$$\tilde{H}_i = \| P_{i1} P_{i2} \dots P_{ii} I_{a_{i+1}} \ 0 \dots 0 \| , \ i = \overline{1,m} , \tag{5}$$

P_{is} is a binary $a_{i+1} \times a_s$ matrix and I_{a_s} is a unit matrix;

2. an (n_2, b_i) linear code over $GF(2^{a_i})$ with minimum distance d_{2i} is chosen as the i-th second level code.

The Reed-Solomon codes or the lengthened Reed-Solomon codes are usually chosen as second level codes.

3.3 Some Expressions and a Geometrical Interpretation.

From (2) and (4) one can derive:

$$\gamma_{ij} = \alpha_j \tilde{H}^T_{i-1}, \ i = \overline{1,m+1}, \ j = \overline{1,n_2} . \tag{6}$$

In the case $b_{m+1} = 0$ or

$$\gamma_{m+1} = (\gamma_{m+1,1} \, \gamma_{m+1,2} \dots \gamma_{m+1,n_2}) = (00\dots0)$$

for $i = m + 1$ eq. (6) can be written as

$$\alpha_j \tilde{H}^T_m = 0, \ j = \overline{1,n_2} .$$

Using expressions (5), equation (6) can be rewritten as

$$\gamma_{ij} = (\alpha_{1j} \, \alpha_{2j} \dots \alpha_{i-1,j} \ 0 \dots 0) \, \tilde{H}^T_{i-1} + \alpha_{ij} ,$$

where α_{sj} is a binary word of length a_s.

If we let

$$\beta_{ij} = (\alpha_{1j}\ \alpha_{2j}\ \dots\ \alpha_{i-1,j}\ 0\dots0)\ \tilde{H}^T_{i-1} \qquad (8)$$

then

$$\gamma_{ij} = \beta_{ij} + \alpha_{ij} \quad \text{or} \quad \alpha_{ij} = \beta_{ij} + \gamma_{ij} \qquad (9)$$

for all $i(i = \overline{1,m+1})$ and all $j(j = \overline{1,n_2})$.

The codeword α can be interpreted as an $n_1 \times n_2$ table, where the subword α_j is the j-th column.

In this table there are $m+1$ regions B_i, $i = \overline{1,m+1}$. The region B_1 consists of the upper a_1 rows of the table; the region B_2 consists of the a_2 rows immediately below; etc. (see Figure 2 and Figure 3).

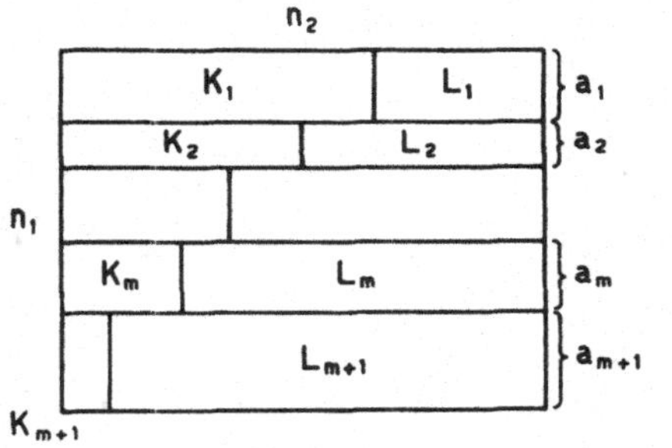

Fig. 2

In each region B_i there are two subregions K_i and L_i. The region K_i consists of the information symbols and the region L_i consists of the parity-check symbols (Fig. 2 and Fig. 3).

Notice that the structure of the generalized concatenated codes can also be specified through the values a_i and b_i.

The number k of information symbols and the number r of parity-check symbols are defined by

$$k = \sum_{i=1}^{m+1} a_i b_i, \quad r = \sum_{i=1}^{m+1} a_i(n_2 - b_i)\,. \qquad (10)$$

The rate R of a generalized concatenated code is defined by

$$R = \frac{k}{n} = \sum_{i=1}^{m+1} \frac{a_i}{n_1} \cdot R_{2i}, \qquad (11)$$

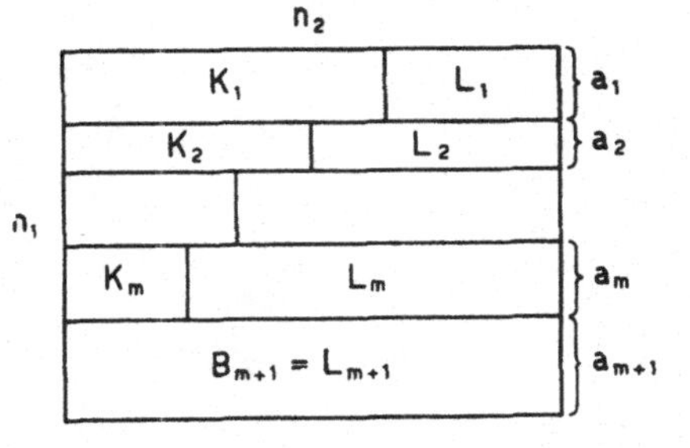

Fig. 3

where $R_{2i} = \frac{b_i}{n_2}$ is the rate of the i-th second level code.

3.4 Codes of the First Level

Consider m linear binary (n_1, k_{1i}) codes with minimum distance d_{1i}, $i = \overline{1,m}$, where the number of information symbols satisfies

Since in the i-th case

$$\gamma_s = 0 \text{ for all s } (s=\overline{i+1,m+1})$$

$$\alpha_j H_i^T = 0, \quad j=\overline{1,n_2}.$$

As a consequence, α_j is a codeword of the i-th first level code for every j, $j=\overline{1,n_2}$.

In the codeword α of the generalized concatenated code there are at least d_{2i} subwords α_j such that each of them has weight d_{1i} or more. Therefore the overall weight of the codeword α must be $d_{2i}d_{1i}$ or more. Thus inequality (13) follows.

As an example we consider the generalized concatenated code of order $m = 2$ with the following parameters

$$n_1 = 7, n_2 = 9, a_1 = a_2 = 3, a_3 = 1, b_1 = 6, b_2 = 2, b_3 = 0.$$

To play the role of matrix H_0 of equation (4) we can choose the following matrix:

$$H_0 = \left\| \begin{matrix} 1\,0\,0 & 000 & 0 \\ 0\,1\,0 & 000 & 0 \\ 0\,0\,1 & 000 & 0 \\ 1\,1\,0 & 100 & 0 \\ 0\,1\,1 & 010 & 0 \\ 1\,0\,1 & 001 & 0 \\ 1\,1\,1 & 111 & 1 \end{matrix} \right\| .$$

Therefore the first code of the first level is the (7,3) code with $d_{11} = 4$ which is specified by the parity-check matrix

$$H_0 = \left\| \begin{matrix} 1\,1\,0 & 1\,0\,0 & 0 \\ 0\,1\,1 & 0\,1\,0 & 0 \\ 1\,0\,1 & 0\,0\,1 & 0 \\ 1\,1\,1 & 1\,1\,1 & 1 \end{matrix} \right\| .$$

The second first-level code is the (7,6) code with $d_{12} = 2$ specified by the parity-check matrix

$$H_2 = \| 111\ 111\ 1 \| .$$

The lengthened (9,6) Reed-Solomon code over $GF(2^3)$ with $d_{21} = 4$ is chosen as the first code of the second level. The lengthened (9,2) Reed-Solomon code over $GF(2^3)$ with $d_{22}=8$ is chosen as the second code of the second level.

Hence, the overall generalized concatenated code has the following parameters:

$$n = 63,\ k = 24,\ d \geqslant \min\ \{4\cdot 4;\ 2\cdot 8\} = 16\ .$$

It is interesting to remark that the BCH code of the same length ($n = 63$) and with the same minimum distance ($d = 16$) has only 23 information symbols.

3.6 An Encoding Algorithm

In the sequel we describe an encoding algorithm which can be interpreted as a procedure realized in $m+1$ steps.

The i-th encoding step consists in computing the check subwords α_{ij}, $j = \overline{b_i + 1, n_2}$. This calculation can be carried out as follows:

1. Let $i = 1$.
2. For all $j(j = \overline{1,n_2})$ the word β_{ij} is computed according to (8):

$$\beta_{ij} = (\alpha_{1j}\alpha_{2j} \dots \alpha_{i-1,j}\ 0\dots0)\tilde{H}^T_{i-1}\ .$$

Remark that $\beta_{1j} = 0$ for all j.

3. For all $j(j = \overline{1,b_i})$ the word γ_{ij} is computed according to (9):

$$\gamma_{ij} = \alpha_{ij} + \beta_{ij}\ .$$

Notice that, since $\beta_{1j} = 0$, then

$$\gamma_{1j} = \alpha_{1j},\quad j = \overline{1,n_2}\ .$$

4. Let γ_{ij}, $j = \overline{1,b_i}$, be the information symbols of the i-th second level code.

Then the check symbols γ_{ij}, $j = \overline{b_{i+1}, n_2}$ can be computed using the encoding algorithm for this code. As a result of the encoding, we have the codeword

$$\gamma_i = (\gamma_{i1}\gamma_{i2} \dots \gamma_{in_2})$$

of the i-th second level code. When $i = 1$, then

$$\gamma_1 = (\alpha_{11}\alpha_{12} \dots \alpha_{1n_2})\ .$$

5. The subword α_{ij}, $j = b_i + 1, n_2$, is computed according to expression (9):

$$\alpha_{ij} = \beta_{ij} + \gamma_{ij} .$$

6. If i = m, then change $i \to i + 1$, and go back to 2. If $i = m + 1$, the encoding procedure is over.

The encoding complexity satisfies

$$\kappa(E) = \sum_{i=1}^{m+1} \kappa(E_i) , \tag{14}$$

where $\kappa(E_i)$ is the complexity of the i-th encoding step. As it is seen from the encoding algorithm, the complexity of the i-th encoding step is given by

$$\kappa(E_i) = \kappa(E_{2\,i}) + \kappa(B_i) + 0_i \tag{15}$$

where $\kappa(E_{2i})$ is the encoding complexity of the i-th code of the second level, $\kappa(B_i)$ is the complexity of computation of expression (8) and 0_i is a quantity much smaller than $\kappa(E_{2i})$ or $\kappa(B_i)$.Substituting (15) in (14), one gets

$$\kappa(E) = \sum_{i=1}^{m+1} \kappa(E_{2i}) + \sum_{i=1}^{m+1} \kappa(B_i) + \sum_{i=1}^{m+1} 0_i . \tag{16}$$

The encoding algorithm can be implemented by means of an encoder whose complexity is of the order of n, and which performs a number of clock cycles of the order of n too.

3.7 Correction Capability

The correction capability of the generalized concatenated codes can be better than that of Forney's concatenated codes, which coincide with the generalized concatenated codes of order m = 1. This is easier understood if we consider two generalized concatenated codes of the same length n = 63 and the same minimum distance d = 16, but different orders. Let one of them have order m = 1 and the other order m = 2. The number of information symbols is k = 18 in the first code and k = 24 in the second one. This difference can be explained if we consider the geometrical interpretation of the two codes (Fig. 3).

In Fig. 5 the ratio $d/n = \delta$ is represented as a function of the rate R for the generalized concatenated codes of length n = 1023 and different orders.

The rate R of the generalized concatenated codes of order m = 3 with $\delta = d/n \geqslant 0.141$ is given in Fig. 6 as a function of the length n.

Consider the ratio d/n in the asymptotic case. Let $\delta_m(R)$ be a lower

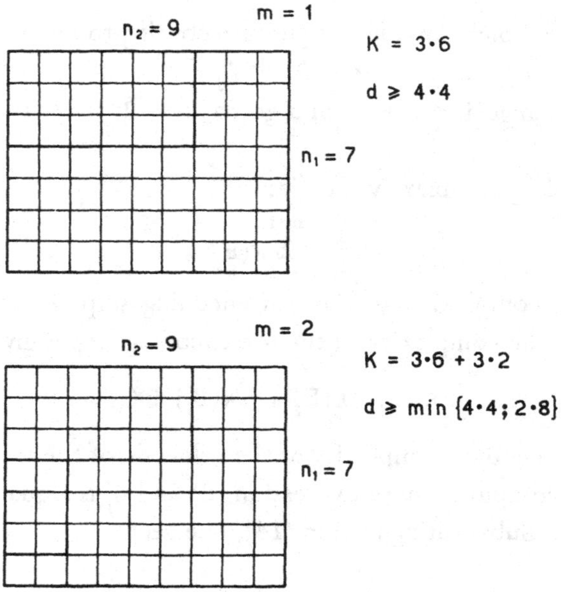

Fig. 4

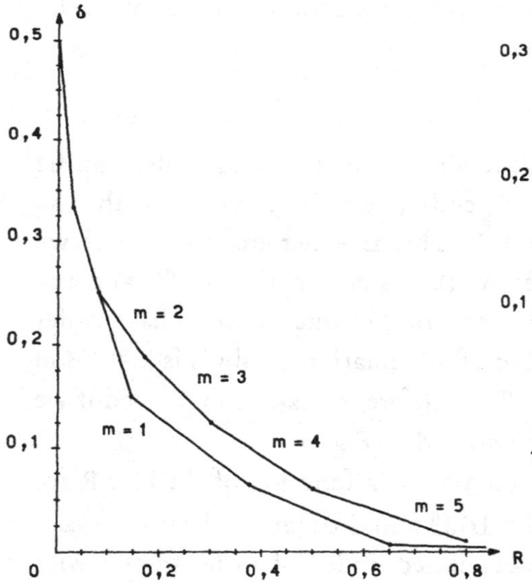

Fig. 5

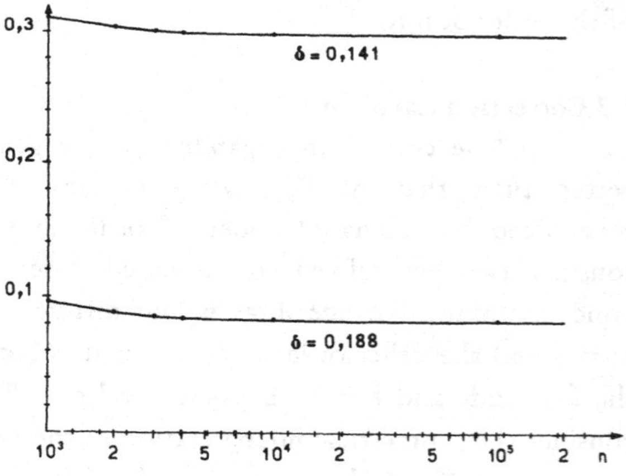

Fig. 6

bound for this ratio in the case of genralized concatenated codes of order m and rate R. This lower bound satisfies

$$\delta_m(R) \geqslant \max_{Z} \varphi_m(Z)(1 - \frac{R}{Z_m}), \tag{17}$$

where

$$Z = (Z_1 Z_2 \dots Z_m),$$

$$Z_i = \frac{1}{n_1}\sum_{s=1}^{i} a_s,$$

and the function $\varphi_m(Z)$ is specified by

$$\varphi_m(Z) = Z_m \sum_{i=1}^{i} \frac{Z_i - Z_{i-1}}{H^{-1}(1-Z_i)}, \tag{18}$$

where $H(x) = -x\log_2 x - (1-x)\log_2(1-x)$ is the binary entropy function and $H^{-1}(1-Z_i)$ is its inverse, so that $0 < H^{-1}(1-Z_i) < 0.5$.

It can be shown that

$$\delta_m(R) < \delta_{m+1}(R) < \delta_\infty(R) < \delta_{GV}(R), \tag{19}$$

where $\delta_{GV}(R)$ is Gilbert-Varshamov bound.

These bounds $\delta_m(R)$ for different values of m are sketched in Fig. 7 as functions of the rate R.

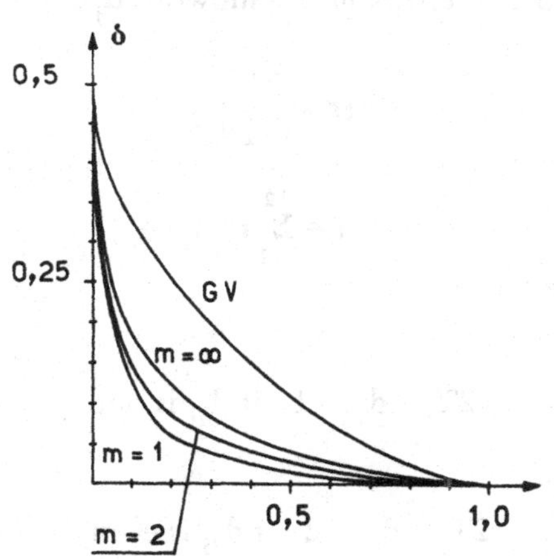

Fig. 7

4. Decoding for Generalized Concatenated Codes.

4.1 Error Syndrome

Let $\hat{\alpha}$ be a received word of length n (word with errors). Let $\hat{\alpha}_j$ be a received subwords, which can be represented as

$$\hat{\alpha}_j = \alpha_j + e_j ,$$

where e_j is the error vector corresponding to the subword α_j.

Consider the expression

$$\hat{\alpha}_j H_i^T = \alpha_j H_i^T + e_j H_i^T ,$$

which can also be written as

$$\hat{\alpha}_j H_i^T = (\gamma_{i+1,j}\ \gamma_{i+2,j} \cdots \gamma_{m+1,j}) + e_j H_i^T .$$

Let $c_{ij} = e_j H_i^T$ be the syndrome corresponding to the i-th code of the first level. Then

(1) $$c_{ij} = \hat{\alpha}_j H_i^T + (\gamma_{i+1,j}\ \gamma_{i+2,j} \cdots \gamma_{m+1,j}) .$$

It is obvious that the syndrome c_{ij} can be computed if the γ_{sj} , $s=\overline{i+1,m+1}$, are known.

Let t^*_{ij} be the number of errors corresponding to the syndrome c_{ij} and let t_{oj} be the actual number of errors in the subword $\hat{\alpha}_j$.

Let

(2)
$$t_i^* = \sum_{j=1}^{n_2} t^*_{ij} ,$$
$$t = \sum_{j=1}^{n_2} t_{oj} .$$

Let

(3)
$$2T_i = d_{1i} - 1 \quad \text{if } d_{1i} \text{ is odd} ,$$
$$2T_i = d_{2i} - 2 \quad \text{if } d_{1i} \text{ is even} .$$

In the sequel we assume that

1. Every pattern of t_{oj} errors can be corrected if $t_{oj} < T$ and can be located if $T < t_{oj} < d_{1i} - T$, for all T ($0 \leqslant T \leqslant T_i$), by means of the syndrome c_{ij}.
2. Every pattern of e errors and τ erasures in a codeword of the i-th second level code can be corrected if $2e + \tau < d_{2i}$.

4.2. The Key Algorithm

Let

$$2t = 2 \sum_{j=1}^{m} t_{oj} < d_{1i} d_{2i}, \quad i = \overline{1, m+1} \ . \qquad (4)$$

We also assume that all the words γ_{sj}, $s = \overline{i+1, m+1}$, $j = \overline{1, n_2}$, all the syndromes c_{ij}, $j = \overline{1, n_2}$ and the numbers of errors t^*_{ij}, $j = \overline{1, n_2}$, corresponding to these syndromes are known. Then the words γ_{ij}, $j = \overline{1, n_2}$, the syndromes $c_{i-1,j}$, $j = \overline{1, n_2}$ and the numbers of errors $t_{i-1,j}$, $j = \overline{1, n_2}$, can be computed by means of a key algorithm AK(i).

The algorithm AK(i) can be described as follows:

1. Let $T = 0$.
2. If $t^*_{ij} \leqslant T$ then the errors in the subword $\hat{\alpha}_j$, $j = \overline{1, n_2}$, are corrected; if $t^*_{ij} > T$, the errors are only located.
3. The words $\hat{\gamma}_{ij}$ are computed by

$$\hat{\gamma}_{ij} = \hat{\alpha}_j H^T_{i-1} \ .$$

4. The words

$$\hat{\gamma}_i = (\gamma_{i1} \ \gamma_{i2} \cdots \gamma_{in_2})$$

are decoded by means of the i-th code of the second level, where $\hat{\gamma}_{ij}$ is interpreted as an erasure if the errors in the subword $\hat{\alpha}_j$ have been only located.

The codeword

$$\gamma^*_i = (\gamma^*_{i1} \ \gamma^*_{i2} \cdots \gamma^*_{in_2})$$

or decoding discard can be obtained as the result of decoding. In the latter case go to 7.

5. The syndromes $c_{i-1,j}$, $j = \overline{1, n_2}$, are computed by means of expression (1). The numbers $t^*_{i-1,j}$, are estimated by means of the syndromes $c_{i-1,j}$ and the number t^*_{i-1} is evaluated by expression (2).
6. If $2t^*_{i-1} < d_{1i} d_{2i}$ then the codeword γ^*_i is the result; else go to 7.

7. If $T < T_i$ then $T := T+1$ and go to 2; if $T = T_i$ the decoding discard is the result.

4.3 Realized Correcting Capability

Now it will be shown that the codeword γ_i^* obtained as a result from the AK(i) algorithm equals the actual codeword γ_i if the actual number t satisfies (4).

This statement is a consequence of the following lemmas:

Lemma 1. There is at least one value of T in the AK(i) algorithm such that all errors and erasures are corrected if the words $\gamma_{sj}, s = \overline{i+1,m+1},\ j = \overline{1,n_2}$ are known and if the number of errors satisfies (4).

Proof Let e_T and τ_T be the numbers of errors and of erasures, respectively, corresponding to a value T in the word $\hat{\gamma}_i$. To correct these errors and erasures we need only to show that e_T and τ_T satisfy

$$2e_T + \tau_T < d_{2i} \tag{5}$$

for at least one $T (0 \leqslant T \leqslant T_i)$.

Let $N_s (N_s \leqslant n_2)$ be the number of subwords $\hat{\alpha}_j$ in the received word such that for each of them there are s errors. As a consequence:

$$t = \sum_{j=1}^{n_2} t_{oj} = \sum_{s=1}^{n_1} sN_s \,. \tag{6}$$

From the algorithm AK(i) it follows that the errors of a subword $\hat{\alpha}_j$ will be corrected by decoding the i-th code of the first level if the number of errors in $\hat{\alpha}_j$ is T or smaller.

Therefore

$$e_T + \tau_T = \sum_{s=T+1}^{n_1} N_s \,.$$

It is also obvious that

$$e_T \leqslant \sum_{s=d_{1i}-T}^{n_1} N_s,\quad \tau_T \geqslant \sum_{s=T+1}^{d_{1i}-T-1} N_s \,.$$

Lemma 1 is proved if the inequality

$$d_{2i} \leqslant (e_T + \tau_T) + e_T \leqslant \sum_{s=T+1}^{n_1} N_s + \sum_{s=d_1-T}^{n_1} N_s \tag{7}$$

cannot be satisfied for all $T (T = \overline{0,T_i})$ simultaneously.

Assume the contrary is true.

We distinguish between the two cases $d_{1i} = 2T_i + 1$ and $d_{2i} = 2T_i + 2$. In the first case we consider the two following expressions

$$T_i d_{2i} \leqslant \sum_{T=0}^{T_i-1} \{ \sum_{s=T+1}^{n_1} N_s + \sum_{s=d_{1i}-T}^{n_1} N_s \};$$

$$(T_i + 1)d_{2i} \leqslant \sum_{T=0}^{T_i} \{ \sum_{s=T+1}^{n_1} N_s + \sum_{s=d_{1i}-T}^{n_1} N_s \}.$$

In the second case we consider the two following expressions

$$(T_i + 1)d_{2i} \leqslant \sum_{T=0}^{T_i} \{ \sum_{s=T+1}^{n_1} N_s + \sum_{s=d_{1i}-T}^{n_1} N_s \};$$

$$(T_i + 1)d_{2i} \leqslant \sum_{T=0}^{T_i} \{ \sum_{s=T+1}^{n_1} N_s + \sum_{s=d_{1i}-T}^{n_1} N_s \}.$$

From these expressions and (7) we obtain respectively

$$(2T_i + 1)d_{2i} \leqslant 2 \sum_{T=0}^{T_i} \{ \sum_{s=T+1}^{n_1} N_s + \sum_{s=d_{1i}-T}^{n_1} N_s \} - \sum_{s=T_i+1}^{n_1} N_s - \sum_{s=d_{1i}-T_i}^{n_1} N_s, \quad (8)$$

$$(2T_i + 2)d_{2i} \leqslant 2 \sum_{T=0}^{T_i} \{ \sum_{s=T+1}^{n_1} N_s + \sum_{s=d_{1i}-T}^{n_1} N_s \} \quad (9)$$

In the first case the following inequality is obtained from (8)

$$d_{1i}d_{2i} \leqslant 2 \sum_{s=1}^{n_1} sN_s - 2 \sum_{s=d_{1i}+1}^{n_1} (s - 2T_i - 1)N_s \leqslant 2 \sum_{s=1}^{n_1} sN_s .$$

In the second case the following inequality is obtained from (9)

$$d_{1i}d_{2i} \leqslant 2 \sum_{s=1}^{n_1} sN_s - 2 \sum_{s=d_{1i}+1}^{n_1} (s - 2T_i - 2)N_s \leqslant 2 \sum_{s=1}^{n_1} sN_s .$$

As it follows from these inequalities and from (6), the number of errors satisfies

$$d_{1i}d_{2i} \leqslant 2 \sum_{s=1}^{n_1} sN_s = 2t \quad (10)$$

in both cases. Thus there is a contradiction between (10) and (4). As a consequence, we immediately obtain the proof of Lemma 1.

Lemma 2. If the number t^*_{i-1} obtained from the AK(i) algorithm satisfies

$$2t^*_{i-1} < d_{1i}d_{2i}$$

then γ^* is the correct codeword for the i-th code of the second level.

Proof. Let $\gamma^*_i = \gamma_i$. Then the syndrome $c_{i-1,j}$ is correct for all j ($j = \overline{1,n_2}$). As a consequence of the use of minimum distance decoding, the number of errors $t^*_{i-1,j}$ corresponding to the syndrome $c_{i-1,j}$ satisfies

$$t^*_{i-1,j} \leqslant t_{o,j} \; .$$

Hence

$$t^*_{i-1} = \sum_{j=1}^{n_2} t^*_{i-1,j} \leqslant \sum_{j=1}^{n_2} t_{o,j} = t$$

and therefore

$$2t_{i-1} \leqslant 2t < d_{1i}d_{2i} \; . \tag{11}$$

Now we consider the case $\gamma^*_i \neq \gamma_i$. Let $c'_{i-1,j}$, $j=\overline{1,n_2}$ and t'_{i-1} be computed through algorithm AK(i). Let $c^{\circ}_{i-1,j}$, $j=\overline{1,n_2}$, t°_{i-1} be the corresponding actual quantities.

Let α' and α° be the word which are obtained from $\hat{\alpha}$ through error correction by means of the syndromes $c'_{i-1,j}$ and $c^{\circ}_{i-1,j}$, respectively for all j ($j = \overline{1,n_2}$).

It is obvious that

$$t'_{i-1} = d(\alpha',\hat{\alpha}),$$
$$t^{\circ}_{i-1} = d(\alpha^{\circ},\hat{\alpha}), \tag{12}$$

where d(x,y) is the distance between the words x and y. Since furthermore

$$d(\alpha^{\circ},\alpha') \leqslant d(\alpha',\hat{\alpha}) + d(\alpha^{\circ},\hat{\alpha}),$$

we get

$$d(\alpha^{\circ},\alpha') - t^{\circ}_{i-1} \leqslant t'_{i-1} \; . \tag{13}$$

As it follows from (11), t°_{i-1} satisfies

$$t^{\circ}_{i-1} < \frac{d_{1i}d_{2i}}{2}$$

and therefore

$$d(\alpha^{\circ},\alpha') - \frac{d_{1i}d_{2i}}{2} < t'_{i-1} \ . \qquad (14)$$

We now estimate $d(\alpha^{\circ},\alpha')$,, which can be expressed as

$$d(\alpha^{\circ},\alpha') = \sum_{j=1}^{n_2} d(\alpha^{\circ}_j,\alpha'_j) \ . \qquad (15)$$

Remark that α°_j and α'_j must be codewords of the i-th code of the first level and satisfy

$$d(\alpha^{\circ}_j,\alpha'_j) = 0 \quad \text{or} \quad d(\alpha^{\circ}_j,\alpha'_j) \geqslant d_{1i} \ . \qquad (16)$$

But α°_j and α'_j wre both obtained from the received subword $\hat{\alpha}_j$ through an error correction procedure which makes use of the syndromes $c^{\circ}_{i-1,j}$ and $c'_{i-1,j}$, respectively. Therefore if $c^{\circ}_{i-1,j} \neq c'_{i-1,j}$ then $\alpha^{\circ}_j \neq \alpha'_j$. Moreover

$$c^{\circ}_{i-1,j} + c'_{i-1,j} = \hat{\alpha}_j H^T_{i-1} + (\gamma^{\circ}_{ij}\ \gamma_{i+1,j} \dots \gamma_{m+1,j}) +$$

$$+ \hat{\alpha} H^T_{i-1} + (\gamma'_{ij}\ \gamma_{i+1,j} \dots \gamma_{m+1,j}) =$$

$$= (\gamma^{\circ}_{ij} + \gamma'_{ij}, 0 \dots 0) \ .$$

There are d_{21} or more γ°_{ij} which are different from γ'_{ij} as a consequence of $\gamma_i \neq \gamma^*_i$. Hence there are d_{2i} or more α°_i which are different from α'_j and, as it follows from (15) and (16), $d(\alpha^{\circ},\alpha')$ satisfies

$$d(\alpha^{\circ},\alpha') \geqslant d_{2i}d_{1i} \ .$$

From this inequality and from inequality (14) we have

$$d_{1i}d_{2i} - \frac{d_{1i}d_{2i}}{2} \leqslant t'_{i-1}$$

and hence

$$2t'_{i-1} \geqslant d_{1i}d_{2i}\ .$$

Thus lemma 2 has been proved.

As a consequence of Lemmas 1 and 2, all words $\gamma_i(i=\overline{1,m})$ can be decoded correctly using AK(i) (i = m, m - 1, m - 2, . . ., 2,1) if γ_{m+1} is known and if the number of errors satisfies (4).

4.4 Decoding in the Case $b_{m+1}=0$

In this case the decoding of generalized concatenated codes is performed as follows.

i) In this case $\gamma_{m+1}=0$. Then all γ_i, $i=\overline{1,m}$, are decoded using AK(i) successively for i = m,m - 1,...,1.

ii) For $i=1$, $\alpha_{1i}=\gamma_{1j}$, and for $i=\overline{2,m+1}$ the subwords α_{ij} are computed by

(17) $$\alpha_{ij}=\gamma_{ij}+(\alpha_{1j}\ldots\alpha_{i-1,j}\ 0\ldots 0)\widetilde{H}^T_{i-1}\ .$$

If the number of errors satisfies (4) the codeword α evaluated using this procedure will be correct.

4.5 Decoding in the Case $b_{m+1}\neq 0$.

First of all the words

$$\hat{\gamma}_{m+1}=(\hat{\gamma}_{m+1,1}\ \hat{\gamma}_{m+1,2}\ \cdots\ \hat{\gamma}_{m+1,n_2})$$

are obtained by means of

$$\hat{\gamma}_{m+1,j}=\hat{\alpha}_j\widetilde{H}^T_m,\quad j=\overline{1,n_2}\ .$$

Then the codewords γ^*_{m+1} of the (m+1)-st code of the second level are obtained from $\hat{\gamma}_{m+1}$ by decoding this code.

It is obvious that if the number of errors satisfies

$$2t<d_{2,m+1}$$

then the codeword γ^*_{m+1} will be correct. After this the decoding procedure is the same as in the case $b_{m+1} = 0$.

Conclusions

When the Reed-Solomon codes with Berlekamp's decoding algorithm are used as codes of the second level, then the decoding complexity of the generalized concatenated codes grows approximately as $n^2 \log n$ with increasing block length n. Equivalently, a decoder with complexity approximately n performs the decoding in approximately $n \log n$ clock cycles.

References

[1] Dobrushin, R.L., Gelfand, S.I., Pinsker, M.S., Asymptotically Optimal Coding by Simple Schemes. The Second International Symposium on Information Theory. Abstracts of Papers. Tsahkadsor, 1971, pp. 44-46.

[2] Ziablov, V.V., Decoding Complexity of Iterative and Concatenated Codes. The Second International Symposium on Information Theory. Abstracts of Papers. Tsahkadsor, 1971, pp. 83-87.

[3] Ziablov, V.V., Pinsker, M.S., Correcting Capability and Decoding Complexity of Codes with a Small Number of Ones in the Parity-Check Matrices. The Second International Symposium on Information Theory. Abstracts of Papers. Tsahkadsor, 1971, pp. 88-91.

[4] Bloh, E.L., Ziablov, V.V., Encoding and Decoding of Generalized Concatenated Codes. The Third International Symposium on Information Theory. Abstracts of Papers, part II. Tallinn, 1973, pp. 36-40.

[5] Ziablov, V.V., Pinsker, M.S., Decoding Complexity for Low-Density Parity-Check Codes Used for Transmission over a Channel with Erasures. Problemy Peredachi Informatsii N. 1. 1974.

THE COMPLEXITY OF DECODERS (*)

John E. Savage
Division of Engineering
Brown University
Providence, Rhode Island

(*) This paper was prepared while the author was on sabbatical leave from Brown University in the Department of Mathematics, Technological University of Eindhoven, Netherlands. During this period the author was partially supported by a Guggenheim Fellowship and a Fulbright-Hays Award.

1. Introduction

The twenty-fifth anniversary of birth of Information Theory, as marked by the publication of Shannon's seminal 1948 paper [Shannon (1948)], has just recently been celebrated. These have been a most productive and a most stimulating twenty-five years and to a large extent, the goal set out by information theorists has been achieved. That goal is to find and describe the means to obtain reliable communication on noisy channels at tolerably low reduction in data rates. Algebraic and non-algebraic, block and convolutional codes have been discovered along with algebraic and non-algebraic decoding procedures which do permit reliable communication on many types of noisy channels, radio, wire and space.

The subject of this article is one that has frequently occupied the attention of those who have or would design competitive coding and decoding methods. Nonetheless, it is one that has only recently been discussed in the abstract in the information theory literature [Savage (1969, 1971)]. This subject is the complexity of decoders for error correcting codes.

Experience teaches that highly reliable communication at large data rates can only be achieved at the expense of complex encoding and decoding algorithms. It is the purpose of this article to examine meaningful and relevant measures of the complexity of decoding algorithms and to relate the values of these measures to the reliability and data rates required, as well as channel parameters. These relationships are of the form of **exchange inequalities**. For example, if a decoding algorithm is realized by a general purpose computer with random access (core) storage space of S bits which decodes a block code in T cycles to give a probability of error $P_e = 2^{-E}$ at code rate R, then a relationship such as

$$\frac{E}{E(R)} \lesssim ST$$

can be expected to hold, where E(R) is a "coding exponent" which depends on channel parameters.

This inequality requires that any exchanges of space for time or vice versa do not violate the lower bound. The lower bound, in turn, increases logarithmically with probability of error and with fixed P_e, it increases with increasing code rate R, becoming very large as R approaches channel capacity C, where C is the largest rate at which very high reliability is possible. We shall derive two different types of exchange inequality, both involving time T (measured in cycles of a sequential

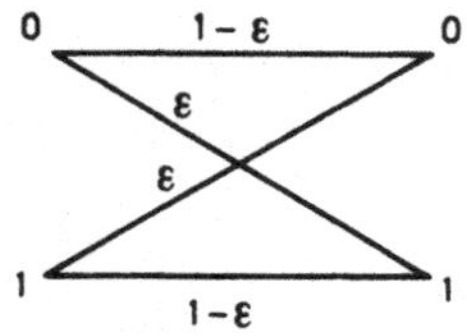

Fig. 2.1 Binary Symmetric Channel

channel. It accepts one of two inputs in $\{0,1\}$ in each unit of time and that input is transmitted correctly with probability $(1-\epsilon)$ and incorrectly with probability ϵ, $0<\epsilon<1$. Furthermore, successive transmissions are made with statistical independence.

Given a channel such as this, suppose that a probability of error is desired which is considerably less than ϵ. Then a transmission of raw data will not suffice since such produces errors with probability ϵ. A very simple and old method of reducing the error rate by adding redundancy to the message consists of repeating each message bit three times. Thus, a source output of 0 is transmitted as 000 and 1 as 111. These two triples constitute a **code**. The rate of transmission of new data is reduced by a third but the error rate is reduced also, as we shall see.

Suppose that at the receiver it is known how to group the received digits into blocks of three, corresponding to the input blocks. Then, use the following **decoding rule**: If a block of three digits contains two or more 1's decode as source digit 1, otherwise decode as 0. The following table lists all triples of received digits and the result of the decoding rule on each triple.

decoder input	output
0 0 0	0
0 0 1	0
0 1 0	0
0 1 1	1
1 0 0	0
1 0 1	1
1 1 0	1
1 1 1	1

This decoding rule does not eliminate the possibility of a decoding error, it just makes it less likely. For example, when 000 is sent, a decoding error occurs when two or more channel errors occur. The same is true when 111 is sent.

Let us evaluate the performance of this decoding rule using the **average probability of error**, P_e, as our measure of performance. If P_{e/m_i} is the probability of a decoding error when message m_i is sent, $m_i \epsilon \{0,1\}$, and $P(m_i)$ is the probability that m_i is chosen for transmission, then we define P_e by

$$P_e = \sum_{i=1}^{2} P_{e/m_i} P(m_i).$$

Given a code with more than two code words, this measure readily generalizes. As indicated, P_{e/m_i} is the probability that two or more channel errors occur. Letting 0 denote correct transmission and 1 denote an error, the following error patterns result in a decoding error when either 000 or 111 is sent: 011, 101, 110, 111. The first three each occur with probability $(1-\epsilon)\epsilon^2$ and the last with probability ϵ^3. Consequently,

$$P_{e/m_i} = 3(1-\epsilon)\epsilon^2 + \epsilon^3$$

and

$$P_e = 3(1-\epsilon)\epsilon^2 + \epsilon^3$$

also. Thus, if $\epsilon = 10^{-2}$, $P_e \cong 3 \times 10^{-4}$, which represents a reduction in the error rate by a factor of about 33. To achieve this, the data rate has been reduced by a factor of 3. Note that for this example, the probability of error is independent of $P(m_i)$, the a priori probability of message m.

With this illustration in mind, let us describe the general (block) coding problem. A source produces messages from an ensemble $\{m_1, m_2, \ldots, m_M\}$ with a priori probability $P(m_i)$, $1 \leqslant i \leqslant M$. Here,

$$1 \geqslant P(m_i) \geqslant 0$$

$$\sum_{i=1}^{M} P(m_i) = 1.$$

The messages are to be transmitted over a discrete memoryless channel (DMC) with an input alphabet $\{0,1,2,\ldots,K\text{-}1\}$ and output alphabet $\{0,1,2,\ldots,J\text{-}1\}$ where successive transmissions are statistically independent and input k is received as output j with probability $P(j/k)$ where

$$1 \geqslant P(j/k) \geqslant 0 \quad ; \quad \sum_{j=0}^{J-1} P(j/k) = 1.$$

These are known as the **channel transition probabilities.** Thus, for the BSC, $P(1/0) = P(0/1) = \epsilon$, $P(1/1) = P(0/0) = 1 - \epsilon$.

Messages are transmitted from source to receiver using **code words** $\mathbf{x}_1, \mathbf{x}_2, \ldots, \mathbf{x}_M$ (which together constitute a **code**) where each is an N-tuple over the DMC input alphabet. In our example, $M = 2$ and $\mathbf{x}_1 = (000), \mathbf{x}_2 = (111)$. The code is said to have **block length** N and to have **rate** R defined by

$$R = \frac{\log_2 M}{N} .$$

Our sample code has block length $N = 3$ and rate $R = 1/3$.

The transmission process consists of message selection by a source, a corresponding code word selection and transmission over a DMC followed by a decision in favor of some message by a decoding rule. The decoding rule partitions the set of possible received N-tuples into disjoint sets corresponding to the messages (or code words). Such a rule can be completely described by a decoding table as illustrated above) and is defined by a **decoding function** $f_D : \{0,1,\ldots,J\text{-}1\}^N \to \to \{1,2,\ldots,M\}$ which maps each possible channel output N-tuple y into a message $f_D(y) = m$, $m \in \{1,2,\ldots,M\}$. The partitioning of the space of received N-tuples into disjoint sets by the decoding function is illustrated schematically in Figure 2.2.

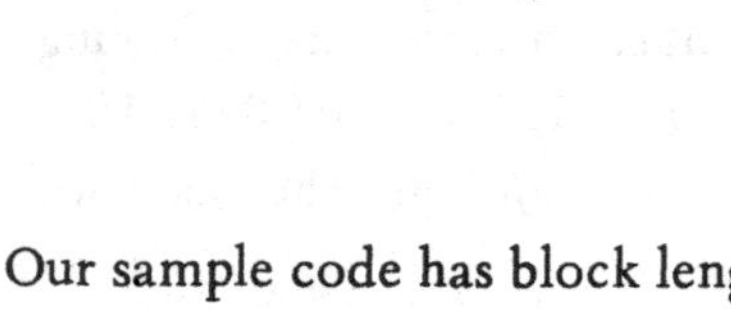

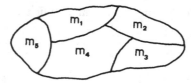

Fig. 2.2 Schematic Illustration of Decoding Regions

Given a code, the a priori probability of messages and the channel transition probabilities, it is desirable to ask for the decoding function which minimizes Pe, the average probability of error. It is fairly easy to show [Gallager (1968), p. 120] that the **maximum a posteriori** (MAP) **rule** minimizes Pe and it is defined as follows:

"Decode y as message m, that is $f_{MAP}(y) = m$, if $P(m)P(y/m) < P(m')P(y/m')$ for all m', $m' \neq m$. If there is equality for some $m'' \neq m$ assign y to either m or m''."

Here $P(y/m)$ is the probability that y is received when message m is chosen by the source and code word $\mathbf{x}_m$ transmitted. If $\mathbf{x}_m = (x_{1m}, x_{2m}, \ldots, x_{Nm})$ and $\mathbf{y} = (y_1, \ldots, y_N)$ then by the nature of the DMC

$$P(\mathbf{y}/m) = P(\mathbf{y}/\mathbf{x}_m) = \prod_{i=1}^{N} P(y_i/x_{im}).$$

When, all messages are **equiprobable**, that is, when $P(m_i) = 1/M$, the MAP rule is equivalent to the **maximum likelihood** (ML) rule defined by

"Decode $\mathbf{y}$ as message m, that is, $f_{ML}(\mathbf{y}) = m$, if $P(\mathbf{y}/m) < P(\mathbf{y}/m')$ for all m′, $m' \neq m$. If there is equality for some $m' \neq m$, assign $\mathbf{y}$ to either m or m″."

Given the a priori probability of messages and the channel transition probabilities of a DMC the (feed-back free) **coding problem** is to choose a code so that with the best decoding rule a code rate of at least R and an error probability of at most Pe are achieved. The **coding theorem** states acheivable bounds on R and Pe and the first version was stated by Shannon (1948). The following version can be found in Gallager (1968), p. 140.

Theorem 2.1. Consider a DMC with transition probabilities $\{P(j/k)\}$. There exist codes of block length N and rate (*) R for this channel such that for every message m, $1 \leqslant m \leqslant 2^{NR}$, the conditional probability of error, $P_{e/m}$, satisfies,

$$P_{e/m} \leqslant 4\ 2^{-NE_r(R)}$$

where $E_r(R)$ is the "random coding exponent" defined by

$$E_r(R) = \max_{0 \leqslant \rho \leqslant 1} \max_{Q} [E_0(\rho, Q) - \rho R]$$

and

$$E_0(\rho, Q) = -\log_2 \sum_{j=0}^{J-1} [\sum_{k=0}^{K-1} Q(k) P(j/k)^{1/(1+\rho)}]^{1+\rho}$$

and $Q = (Q(O), Q(1), \ldots, Q(K-1))$ is a probability distribution on channel inputs.

Although not explicit in this theorem, it can be shown that the exponent $E_r(R)$ satisfies $E_r(R) > 0$ for $0 \leqslant R < C$ where C is defined by

$$C = \max_{Q} \sum_{k=0}^{K-1} \sum_{j=0}^{J-1} Q(k) P(j/k) \log_2 [\frac{P(j/k)}{\sum_{k'} Q(k') P(j/k')}]$$

and is called **channel capacity**. Thus, $P_{e/m}$ as well as P_e can be made arbitrarily small for any code rate R less than channel capacity. This is the **principal theorem** of information theory, the coding theorem.

(*) It is assumed that R is such that 2^{NR} is an integer.

The channel capacity of the BSC is easily computed and it is

$$C_{BSC} = 1 + \epsilon \log_2 \epsilon + (1-\epsilon)\log_2(1-\epsilon)$$

and is convex down in ϵ, as shown in Figure 2.3.

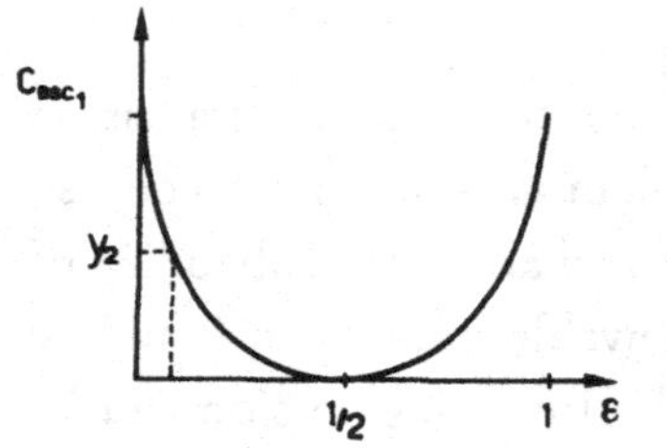

Fig. 2.3 Channel Capacity of the BSC

Given that good codes exist, the challenge of coding theory is to find some . Codes of small and moderate lengths were discovered over the lifetime of information theory but only recently [Justesen (1972)] was a "decent" procedure given for constructing good, arbitrarily long codes.

In addition to the coding theorem, there are converses which state lower bounds on the probability of error acheivable at a given rate with a given block length. Unlike the coding theorem, however, the converse depends on the a priori probability of messages and code words. We shall assume that they are equiprobable. The following "sphere-packing bound", so called because of the method of proof, is quoted in Gallager (1968) and established in Shannon, Gallager and Berlekamp (1967).

Theorem 2.2. For any code of block length N, rate R on a DMC,

$$Pe \geqslant 2^{-N\{E_{sp}[R-o_1(N)]+o_2(N)\}}$$

where

$$E_{sp}(R) = \sup_{\rho>0} [\max_Q E_0(\rho, Q) - \rho R]$$

and $o_1(N)$ and $o_2(N)$ can be taken as

$$o_1(N) = \frac{3}{N} + K\frac{\log_2 N}{N}$$

$$o_2(N) = \frac{3}{N} + \frac{\sqrt{2}}{N}\log_2\left(\frac{e^2}{P_{min}}\right).$$

They decrease with increasing N and P_{min} is the smallest nonzero channel transition probability.

The exponents defined in these two theorems, namely, the "random coding exponent", $E_r(R)$, and the "sphere-packing exponent, $E_{sp}(R)$, are shown schematically in Figure 2.4. They agree in the interval $R_{CRIT} \leqslant R \leqslant C$.

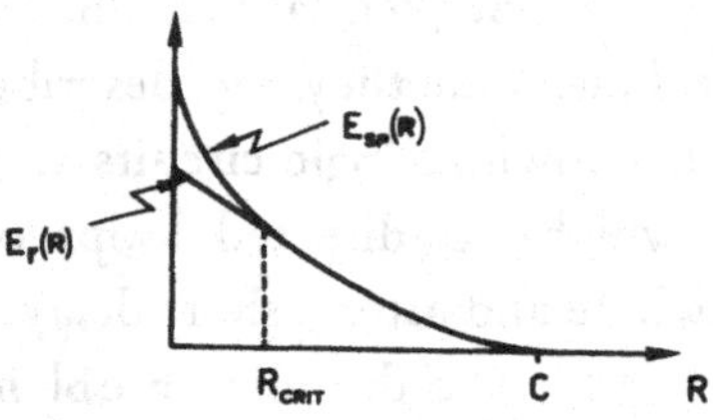

Fig. 2.4 The Random Coding and Sphere-Packing Exponents.

Improvements on the upper and lower bounds to Pe have been obtained and the gap between the exponents has been closed somewhat [Gallager (1968)].

Up to this point, we have described what is known as the feed-back free coding problem. It is also possible to use a feed-back channel, when available, to aid in the coding process and perhaps simplify the encoders and decoders. It is known [Gallager (1968)] that the sphere-packing bound stated above also applies to the BSC when feed-back is allowed. Also for all DMC's, feed-back does not change the value of channel capacity and for many it does not permit a larger coding exponent for rates in the range $R_{CRIT} \leqslant R \leqslant C$. When feed-back is used, the encoding process depends on data received on the feed-back channel, so in this case we cannot identify a fixed set of code words.

In the feed-back case, the data sent over the feed-back channel (whether it is noisy or not) depends on the data received on the forward channel. Thus, the decision reached by a decoder must be made solely on data received over the forward channel and such decisions can also be characterized by a decoding function.

3. The Decoding Problem

Machines of one type or another are required to "compute" the decoding functions or to "decode". We shall examine machine models and try to determine how much time and machine complexity is required of these machines to decode with high reliability.

As models for decoding machines, or more simply, **decoders**, we take the machines known as **sequential machines** which are discussed in Section 5. Sequential machines have internal states, receive inputs and produce outputs. They execute cycles of computation and in any one cycle an input and the current state are used to compute an output and a successor state. These are very general models and capture the essentials of today's decoding machines.

Sequential machines which execute one cycle will be called **combinational machines** and they are described in the next section. Combinational machines are well known as logic circuits and they consists of interconnections of logic elements in which no directed loops occur. Binary inputs are given to a combinational machine and after a short delay, binary outputs are generated.

The **decoding problem** is this: to determine the "size", X, and number of cycles, T, required by sequential machine decoders which decode block codes of rate at least R and achieve a probability of error of at most Pe on discrete memoryless channels. Feed-back may or may or may not be allowed. The way "size" is measured will be determined by the complexity measure used. We shall use three different measures and for each we shall find that, in general, many values of X and T are good or best, that is, it appears that good tradeoffs between X and T are possible.

As indicated in the previous section, the decoding functions are maps $f_D : \{0,1,...,J-1\}^N \rightarrow \{1,2,...,M\}$, where J is the size of the channel output alphabet, N is the block length and M is the number of message, one of which is to be transmitted with each block. Such functions can only be "computed" by a machine if the domain and range of the functions are compatible with the input and output alphabets of the machine. Because almost all machines used today are binary, without loss of much generality, we assume that all decoding functions are defined as binary functions.

Assumption: A binary decoding function (BDF) $f_D^* : \{0,1\}^{aN} \rightarrow \{0,1\}^b$ is a binary function obtained from a decoding function $f_D : \{0,1,...,J-1\}^N \rightarrow \{1,2,...,M\}$ by applying representation functions $E_J : \{0,1,...,J-1\} \rightarrow \{0,1\}^a, E_M : \{1,2,...,M\} \rightarrow \{0,1\}^b$ to the domain and range of f_D so that

$$f_D^*(E_J(y_1), E_J(y_2),..., E_J(y_N)) = E_M(f_D(y_1,...,y_N))$$

for all $y_i \in \{0,1,...,J-1\}$, $1 \leqslant i \leqslant N$. Furthermore, E_J and E_M must give distinct representations to each element of $\{0,1,...,J-1\}$ and $\{1,2,...,M\}$, resoectively. The definition of f^* on points in $\{0,1\}^{aN}$ which are not the image of points in $\{0,1,...,J-1\}^N$ is arbitrary.

The reader will note that the decoding function defined in Table 2.1 is a BDF. It is common to use as E_J and E_M, mappings of the integers into the standard binary representation of integers.

It is clear that a BDF f_D^* may not be completely defined by the representation functions E_J and E_M and the decoding function f_D. Also, the maps E_J and E_M may have a large effect on the "complexity" of f_D. The lower bounds derived in Section 7 take all of this into account and the bounds apply to all maps E_J and E_M and to every way of completing f_D^*.

4. Combinational Complexity

Combinational machines realize or compute Boolean functions. A function f is Boolean if its domain is the set of binary n-tuples, for some n, and its range is {0,1}; this is denoted by $f: \{0,1\}^n \to \{0,1\}$. Four important Boolean functions are the AND, OR, NOT and EXCLUSIVE OR functions, denoted as $f_{\cdot} : \{0,1\}^2 \to \{0,1\}$, $f_+ : \{0,1\}^2 \to \{0,1\}$, $f_- : \{0,1\} \to \{0,1\}$ and $f_\oplus : \{0,1\}^2 \to \{0,1\}$ and defined by the following tables.

x_1	x_2	$f_{\cdot}(x_1,x_2)$
0	0	0
0	1	0
1	0	0
1	1	1

x_1	x_2	$f_+(x_1,x_2)$
0	0	0
0	1	1
1	0	1
1	1	1

x_1	x_2	$f_\oplus(x_1,x_2)$
0	0	0
0	1	1
1	0	1
1	1	0

x_1	$f_-(x_1)$
0	1
1	0

These four functions and the projection functions $p_i(x_1,x_2,\ldots,x_n) = x_i$, called **variables**, can be composed, using functional composition, to "compute" other Boolean functions $f : \{0,1\}^n \to \{0,1\}$. We illustrate this by example and then give a formal description of algorithms which use only functional composition.

A very important class of Boolean functions is the set of **minterm functions**. An example of a minterm function on three variables is $m_{101}(x_1,x_2,x_3)$, shown below.

x_1	x_2	x_3	$m_{101}(x_1, x_2, x_3)$
0	0	0	0
0	0	1	0
0	1	0	0
0	1	1	0
1	0	0	0
1	0	1	1
1	1	0	0
1	1	1	0

It has value 1 on only one point of its domain, $(x_1,x_2,x_3) = (1,0,1)$, and it can be described with functional composition as

$$m_{101}(x_1,x_2,x_3) = x_1 \cdot (\bar{x}_2 \cdot x_3)$$

where we use the shorthand notation $\cdot$ for $f_\cdot$ and $-$ for f_-. In general, a **miterm** function on n variables is a Boolean function which has value 1 on exactly one point of its domain. There are 2^n minterm functions in n variables and each can be realized with functional composition of the variables and AND and NOT. A general function $f: \{0,1\}^n \rightarrow \{0,1\}$ can then be computed by OR'ing a set of minterm functions corresponding to the points in the domain of f on which it has value 1. For example, if $f(x_1,x_2,x_3)$ has value 1 only on 011, 010 and 101, then we can write it as

$$f(x_1,x_2,x_3) = \bar{x}_1 \cdot (x_2 \cdot x_3) + (\bar{x}_1 \cdot (x_2 \cdot \bar{x}_3) + x_1 \cdot (\bar{x}_2 \cdot x_3))$$

where $+$ is shorthand for f_+. We conclude that every Boolean function can be realized with operations from the **basis** $\Omega = \{+,\cdot,-\}$, and we say, that this basis is **complete.** Any basis in which AND, OR and NOT can be realized must also be complete, as is true, for example of $\Omega = \{+,-\}$.

The three operations $+$, $-$ and $\oplus$, where $\oplus$ is shorthand for $f_\oplus$, are commutative $[x_1 \circ x_2 = x_2 \circ x_1]$ and associative $[x_1 \circ (x_2 \circ x_3) = (x_1 \circ x_2) \circ x_3]$. Furthermore, certain identities hold between them which are often useful in simplifying formulas, such as the formulas given above. Some useful identities follow:

distributivity $x_1 \cdot (x_1 + x_3) = x_1 \cdot x_2 + x_2 \cdot x_3$

$x_1 \cdot (x_2 \oplus x_3) = x_1 \cdot x_2 \oplus x_1 \cdot x_3$

$x_1 + (x_2 \cdot x_3) = (x_1 + x_2) \cdot (x_1 + x_3)$.

De Morgan's rules $\overline{x_1 + x_2} = \bar{x}_1 \cdot \bar{x}_2$

$\overline{(x_1 \cdot x_2)} = \bar{x}_1 + \bar{x}_2$.

Absorption rules $x_1 + \bar{x}_1 = 1$; $x_1 + x_1 = x_1$

$x_1 \cdot \bar{x}_1 = 0$; $x_1 \cdot x_1 = x_1$

$x_1 + x_1 \cdot x_2 = x_1 \cdot (x_1 + x_2) = x_1$

substitution of constants $x_1 + 0 = x_1$, $x_1 + 1 = 1$

$x_1 \cdot 0 = 0$, $x_1 \cdot 1 = x_1$

$x_1 \oplus 0 = x_1$, $x_1 \oplus 1 = \bar{x}_1$.

An absorption rule applied to the function f of the preceding paragraph permits the simplification of its formula to $f(x_1,x_2,x_3) = \bar{x}_1 x_3 + x_1 \bar{x}_2 x_3$.

Now we define "computation chains", a class of algorithms which use only functional composition.

Definition 4.1. Given a set $\Omega = \{h_i \mid h_i : \{0,1\}^{n_i} \rightarrow \{0,1\}\}$ of Boolean functions, called the **basis**, and a **data set** $\Gamma = \{x_1, x_2, \ldots, x_n, 0, 1\}$ of the n variables and two constant functions, a k-step **chain** $\beta = (\beta_1, \beta_2, \ldots, \beta_k)$ is an ordered set of k steps $\beta_1, \beta_2, \ldots, \beta_k$ in which either $\beta_j \in \Gamma$ or

$$\beta_j = (h_i ; \beta_{j_1}, \ldots, \beta_{j_{n_i}})$$

in which case, $1 \leqslant j_r < j$ for $1 \leqslant r \leqslant n_i$ and $h_i \in \Omega$. Steps of the first type are called **data steps** and the others are called **computation steps.**

With each step is associated a function $\tilde{\beta}_j$ where $\tilde{\beta}_j = \beta_j$ if $\beta_j \in \Gamma$ and

$$\tilde{\beta}_j = h_i(\tilde{\beta}_{j_1}, \ldots, \tilde{\beta}_{j_{n_i}})$$

otherwise. Composition is the only rule used here and h_i is applied only to preceding steps. A chain is said to **compute** $f_1, f_2, \ldots, f_m$ where $f_\ell : \{0,1\}^n \rightarrow \{0,1\}$, if there exist m steps $\beta_{i_1}, \beta_{i_2}, \ldots, \beta_{i_m}$ such that $\tilde{\beta}_{i_r} = f_r$ for $1 \leqslant r \leqslant m$.

A basis Ω is **complete** if for every set $\{f_1, \ldots, f_m\}$ of Boolean functions, there exists a chain over Ω which computes the set.

Examples of complete bases are $\Omega_1 = \{+, \cdot, ^-\}$, $\Omega_2 = \{+, ^-\}$, $\Omega_3 = \{\cdot, ^-\}$ and $\Omega_4 = \{\oplus, \cdot\}$. The reader can demonstrate that Ω_4 is not complete when the constant functions are not in the data set.

Example 4.1. The following is a chain for the function $f(x_1,x_2,x_3) = \bar{x}_1 \cdot x_2 + x_1 \cdot \bar{x}_2 \cdot x_3$

j	1	2	3	4	5	6	7	8	9
β_j	x_1	x_2	x_3	$(-;\beta_1)$	$(-;\beta_2)$	$(\cdot;\beta_2,\beta_4)$	$(\cdot;\beta_1,\beta_5)$	$(\cdot;\beta_3,\beta_7)$	$(+;\beta_6,\beta_8)$

and $\tilde{\beta}_6 = \bar{x}_1 \cdot x_2$, $\tilde{\beta}_9 = f$.

A directed graph G can be associated with each chain by associating labeled nodes with each step and with an edge drawn from a node associated with β_ℓ to a node associated with β_j if $\tilde{\beta}_\ell$ is used to compute $\tilde{\beta}_j$. The node labels are from Ω

and the order of edges into a node must be preserved unless the Boolean function at such a node is commutative. The graph of the above chain is shown in Figure 4.1.

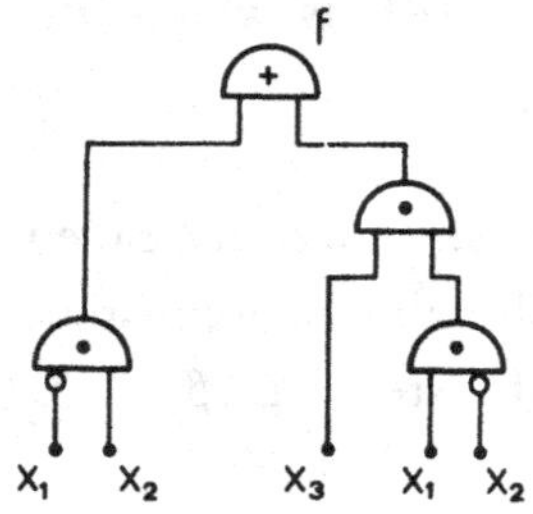

Fig. 4.1 A graph of a Chain

In such graphs, small circles denote NOT and labels represent the previously defined operations. Nodes labeled with variables are **source nodes** and the others are computation nodes. These graphs are called **combinational machines** (or switching circuits) because an actual logic circuit can be constructed by replacing edges with wires, source nodes with input terminals and computation nodes with electrical logic elements.

We identify two **complexity measures** on combinational machines, combinational complexity and delay complexity. If the set $\{f_1,\ldots,f_m\}$ of m Boolean functions all have domain $\{0,1\}^n$ call the set $f: \{0,1\}^n \rightarrow \{0,1\}^m$, $f = (f_1,\ldots,f_m)$.

Definition 4.2. The **combinatorial complexity** of the function $f: \{0,1\}^n \rightarrow \{0,1\}^m$ over Ω, $C_\Omega(f)$, is the minimum number of computation steps required to compute the associated set of m Boolean functions.

In some sense, $C_\Omega(f)$ is a measure of the "cost" of a combinatorial machine.

The graph of a chain has a **depth** which is the length of the longest directed path in the graph. The depth of the graph in Figure 4.1 is three.

Definition 4.3. The **delay complexity** of $f: \{0,1\}^n \rightarrow \{0,1\}^m$ over Ω, $D_\Omega(f)$, is the depth of the smallest depth graph of a chain for f over Ω.

Delay complexity is a measure of the minimum propagation time of a combinational machine for f. The procedure that is generally followed for computing the value of $C_\Omega(f)$ or of $D_\Omega(f)$ is to derive a lower bound and an upper bound to each. Upper bounds are usually constructive while the lower bounds are usually not constructive.

Lower bounds are often derived by stating conditions on functions and from these deriving lower bounds. One such type of condition is the following: Let $f: \{0,1\}^n \rightarrow \{0,1\}$ and let

$$f \Big|_{\substack{x_j = c_j \\ j \in J}}, J \subset \{1,2,\ldots,n\}$$

denote the subfunction of f obtained by assigning variable x_j, $j \in J$, the value $c_j \in \{0,1\}$. The $P^{(n)}_{p,q}$ condition is

"$f \in P^{(n)}_{p,q}$ if for all subsets $J \subset \{1,2,...,n\}$ of p elements, the number of distinct subfunctions

$$f \Big| \begin{array}{l} x_j = c_j \\ j \in J \end{array}$$

obtained by ranging over all values of $c_j, j \in J$, is at least q"

The functions in $P^{(n)}_{1,2}$ depend on each variable. It has been shown that

Theorem 4.2. Let $\Omega = \{+,\cdot,\oplus,^-\}$ and let $C^*_\Omega(f)$ be $C_\Omega(f)$ minus the number of NOT's. Then*

$$\begin{aligned} f \in P^{(n)}_{1,2} &\Rightarrow C^*_\Omega(f) \geq n-1 \quad &, \quad n \geq 1 \\ f \in P^{(n)}_{2,3} &\Rightarrow C^*_\Omega(f) \geq n \quad &, \quad n \geq 3 \\ f \in P^{(n)}_{3,5} &\Rightarrow C^*_\Omega(f) \geq \lceil (7n-4)/6 \rceil, \quad & n \geq 5. \end{aligned}$$

Proofs of these results can be found in Harper and Savage (1973) and Harper, Hsieh and Savage (1975). The reader may also wish to consult Harper and Savage (1972) and Savage (1976) for further results on lower bounds.

It is thought that the classes $P^{(n)}_{p,q}$ may be sufficiently structured that nonlinear lower bounds to combinational complexity may be derived. This is important because most functions $f: \{0,1\}^n \to \{0,1\}$ have $C_\Omega(f)$ on the order of $2^n/n$ [Savage (1971)].

Theorem 4.3. Let $\Omega = \{+,\cdot,\oplus,^-\}$. If $f: \{0,1\}^n \to \{0,1\}$ then

$$\begin{aligned} D_\Omega(f) &\geq \log_2[C_\Omega(f) + 1] \\ &\leq n + \lceil \log_2 n \rceil . \end{aligned}$$

A proof of the first fact is quite simple. Any binary graph of depth d can have at

(*) $\lceil x \rceil$ is the "ceiling" function which is the smallest integer $\geq x$.

most $1+2+2^2+\ldots+2^{d-1} = 2^d-1$ computation nodes. If $C_\Omega(f)$ such nodes are required, $2^d \geqslant C_\Omega(f)+1$. The second inequality follows from the minterm realization of functions. There are at most 2^n minterms of a function, which can be OR'ed with depth n, and each minterm can be computed in depth $\lceil \log_2 n \rceil$.

Chains or equivalently, combinational machines, have been studied in this section. Chains are also called "straight-line algorithms" because execution of them proceeds in a data independent fashion without looping and branching. Furthermore, the arguments of any operator in the chain are the results of preceding computations. Thus, chains form a restricted class of programs with operators of restricted type, namely, Boolean operators.

This completes our limited discussion of complexity measures on combinational machines.

5. Sequential Machines

Sequential machines have internal states, receive inputs and produce outputs. On the basis of a current state and a current input, a successor state and an output are generated. If S, I and 0 denote the sets of states, inputs and outputs, respectively, then the action of changing state is captured by a function $\delta: S \times I \rightarrow 0$, called the **transition function**, and that of producing an output is captured by a function $\lambda: S \times I \rightarrow 0$, called the **output function.** Then, formally, a **sequential machine** S is a five-tuple, $S = \{S, I, \delta, \lambda, 0\}$ where the entries have the properties and meanings indicated. The interval required for a change of state and production of an output is called a cycle.

The output function λ is shown as depending on both the current state and current input. For historial reasons, these are called **Mealey machines** while machines for which λ does not depend on the current input, just the current state, are called **Moore machines.** It can be shown that every Moore machine is equivalent to a Mealey machine and vice versa, where equivalence is defined on the basis of the input-output behavior. The Moore machine suggests that the output remains constant except during a state transition, so it is the model which we adopt here.

It is often useful to show the transition and output functions in a state diagram. In Figure 5.1 is shown the state diagram of a machine which does addition modulo 3. State q_i corresponds to a sum of i modulo 3 and a transition from q_i to $q_{(i+1) \bmod 3}$ occurs where the input is 1 and a transition from q_i back q_i occurs where the input is 0. The branch labels denote inputs and associated with each state is an

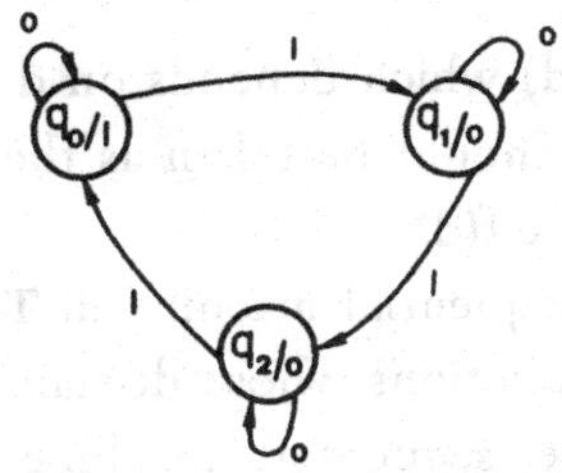

Fig. 5.1 State Diagram of a Sequential Machine

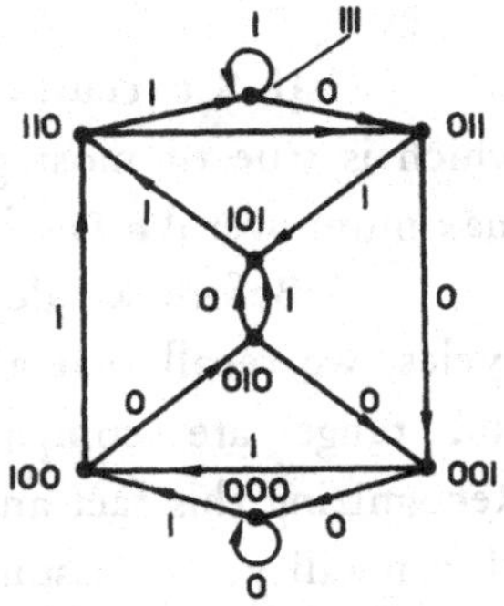

Fig. 5.2 State Diagram of a 3-Stage Shift-Register.

output, as indicated. Here q_0 is the only state with an output of 1 which denotes that in state q_0, the machine recognizes a sum of zero modulo 3.

The transition and output function of this sequential machine are given in tabular form below.

σ	u	$\delta(\sigma,u)$
q_0	0	q_0
q_0	1	q_1
q_1	0	q_1
q_1	1	q_2
q_2	0	q_2
q_2	1	q_0

σ	$\lambda(\sigma)$
q_0	1
q_1	0
q_2	0

In general, both state diagrams and tabular definitions of sequential machines are very large and must be avoided. For example, a simple feed-back shift-register which holds the last three binary inputs has eight states and the state diagram shown in Figure 5.2 (output lablels are not given).

An m-stage shift register has 2^m internal states. Thus, more compact descriptions of sequential machines are needed. These are often in the form of a circuit diagram for a machine or a circuit or formula description of the transition and output functions.

Now let us investigate what it means for a sequential to "compute" a function. This is important since decoders, which are sequential machines, compute decoding functions.

We make one basic assumption about the number of cycles executed by a sequential machine S to compute a function $f: D \to R$ of **domain** D and range R.

Assumption. If S "computes" $f: D \to R$, then the number of cycles T of S used to "compute" f(d) is the same for all $d \in D$.

If S executes a number of cycles to compute f(d) which depends on d, which is true of most general purpose computers, then T should be taken as the maximum over $d \in D$ of the number of cycles used to compute f(d).

Before we define the functions computed by a sequential machine in T cycles, we recall that a machine can only compute those functions whose domain and range are compatible with the sets describing the sequential machine. Recognizing this fact and the binary nature of today's machines, without much loss of generality, we assume that S, I and 0 are represented as sets of binary tuples through one-to-one mappings $E_S : S \to \{0,1\}^s$, $E_I : I \to \{0,1\}^c$, $E_O : O \to \{0,1\}^d$. The functions δ, λ are mapped onto binary functions δ^*, λ^* defined by

$$\delta^*(E_S(\sigma),E_I(u)) = E_S(\delta(\sigma,u)),\ \sigma \in S,\ u \in I$$

$$\lambda^*(E_S(\sigma)) = E_O(\lambda(\sigma)),\ \sigma \in S.$$

If δ^* and λ^* are not defined on all points of $\{0,1\}^s \times \{0,1\}^c$ and $\{0,1\}^s$, respectively, they are completed in some way. The resulting sequential machine is called a **binary sequential machine** (BSM).

If the BSM $S=<\{0,1\}^s, \{0,1\}^c, \delta^*, \lambda^*, \{0,1\}^d>$ executes T cycles, then it is characterized by a function $G_{S,T} : \{0,1\}^{s+Tc} \to \{0,1\}^{Td}$ which maps the initial state and a sequence of T input words into a sequence of T output words. In terms of δ^*, λ^*, the function $G_{S,T}$ is defined as follows: Let $\delta^{*(t)} : \{0,1\}^{s+tc} \to \{0,1\}^s$ be defined by

$$\delta^{*(1)} = \delta^*$$

$$\delta^{*(t)}(\sigma,u_1,u_2,\ldots,u_t) = \delta^*(\delta^{*(t-1)}(\sigma,u_1,\ldots,u_{t-1}),u_t),\ t \geq 2$$

where $\sigma \in \{0,1\}^s$ and $u_i \in \{0,1\}^c$, so that $\delta^{*(t)}$ maps the initial state and t inputs into the state entered by the machine after receiving these inputs. Then,

$$G_{S,T}(\sigma,u_1,\ldots,u_T) = (v_1,v_2,\ldots,v_T)$$

where

$$v_t = \lambda^*(\delta^{*(t)}(\sigma,u_1,\ldots,u_t))$$

defines $G_{S,T}$.

We say that S **computes** $G_{S,T}$ in T cycles. We also say that S **computes** all those binary functions obtained from $G_{S,T}$ by fixing some or none of the variables of $\sigma, u_1, \ldots, u_T$ and by erasing some or none of the components of $v_1, v_2, \ldots, u_T$ without changing the order of the unerased components. Thus, for example S might be used to compute f by putting into its initial state σ, all of the information required, giving inputs $u_i = (0,0,\ldots,0)$, and erasing all outputs except for v_T. Another way in which S might be used to compute f is to fix σ and to supply all of the necessary information through the inputs $u_1, \ldots, u_T$.

It is now clear what it means for S to compute f in T cycles. If several sequential machines are interconnected and together they "compute" f, we ask that they each execute a fixed number of cycles on each point in the domain of f. A function similar to $G_{S,T}$ could be defined for such a collection of machines, although it could be difficult to do so. Then, the definition of "f is computed by a collection of sequential machines" could be made precise,as is done above for the single sequential machine.

We are now ready to state a fundamental computational inequality which relates the "sizes" of sequential machines and the number of cycles which they execute to the combinational complexity of the function which is computed.

Theorem 5.1. Let $f: \{0,1\}^n \rightarrow \{0,1\}^m$ be computed by a collection of p interconnected binary sequential machines

$$S_i = < \{0,1\}^{s_i}, \{0,1\}^{a_i}, \delta_i^*, \lambda_i^*, \{0,1\}^{b_i} >, \quad 1 \leq i \leq p,$$

where S_i executes T_i cycles. Then, the following inequality must be satisfied

$$C_\Omega(f) \leq \sum_{i=1}^{p} C_\Omega(\delta_i^*, \lambda_i^*) T_i = W$$

where Ω is any complete basis and C_Ω is the combinational complexity measure.

This is a simple but fundamental inequality because it explicitly limits the way in which machine size can be traded for time. We should interpret $C_\Omega(\delta_i^*, \lambda_i^*)$ as the equivalent number of logic elements needed by S_i and $C_\Omega(\delta_i^*, \lambda_i^*) T_i$ as the equivalent number of uses of logic elements. If one use of one logic element constitutes one unit of computational work, then W defined in the theorem is the **computational work** done by the collection of machines. Thus, f can be computed by machines with memory only if an amount of computational work is done which is no less than the minimum number of logic uses required in the absence of memory. This is a suggestive way of looking at computation.

The proof of Theorem 5.1 is quite easy and the idea behind it is captured in Figure 5.3. Shown in Figure 5.3(a) is a schematic diagram of a sequential machine

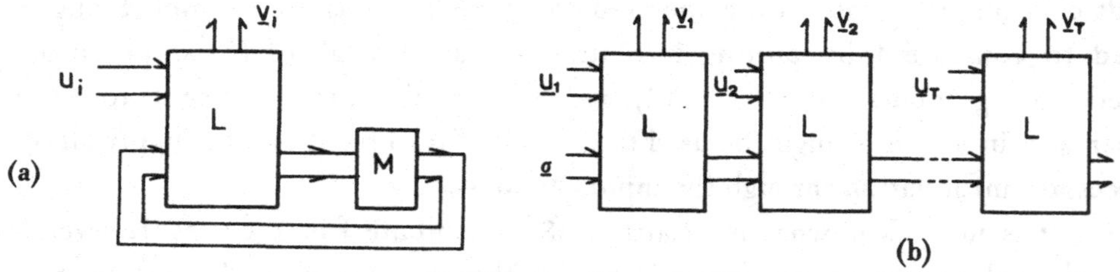

Fig. 5.3 Schematic Illustration of Proof of Theorem 5.1.

S showing a logic unit L which computes output words and successor states which are stored temporarily in a memory unit M. Then, L realizes δ^* and λ^*. In figure 5.3(b) is shown schematically a combinational machine consisting of T copies of L. This machine accepts the same inputs as the sequential machine and computes the same outputs. Since the cascade may not be the optimal realization of $G_{S,T}$ by a combinational machine, we have

$$C_\Omega(G_{S,T}) \leqslant C_\Omega(L)T.$$

But f can be computed from a combinational machine for $G_{S,T}$. Therefore,

$$C_\Omega(f) \leqslant C_\Omega(G_{S,T}) \leqslant C(L)T$$

and the inequality holds for a single sequential machine. A similar "unwinding" can be carried out when more than one machine is used to compute f.

A particularly important class of sequential machines are the general purpose computers. We take as a simple example of such machines a pair of sequential machines called a central processing unit (CPU) and a random access memory (RAM). The CPU is expected to have several registers, be capable of arithmetic and logical operations on words and testing and branching on words. The CPU has access to the outside world and can receive and transmit one c-bit binary word in each cycle. The RAM holds P words each of b-bits and has a storage capacity of S = Pb bits. It can only communicate with the CPU and in each cycle no action is taken or a store or fetch operation of one b-bit word is possible.

The transition and output function of the RAM, namely, δ^*_{RAM}, λ^*_{RAM}, are so defined that over the basis Ω consisting of all 2-input Boolean functions we have

$$S - b \leqslant C_{\Omega}(\delta^*_{RAM}, \lambda^*_{RAM}) \leqslant 6S$$

(See Savage (1972)). Therefore, the following Theorem holds.

Theorem 5.2. Let $f: \{0,1\}^n \to \{0,1\}^m$ be computed by the type of general purpose computer described above in T cycles with storage space S. Then, the following inequality must be satisfied

$$C_{\Omega}(f) \leqslant (C_{CPU} + 6S)T$$

when Ω is the set of all 2-input Boolean functions and C_{CPU} is $C_{\Omega}(\delta^*_{CPU}, \lambda^*_{CPU})$ and δ^*_{CPU} and λ^*_{CPU} describe the central processing unit.

Often, S is much larger than C_{CPU} so that this theorem then states that the space-time product ST must be large if a complex function is to be computed. More information on this type of inequality and the role it plays in computing in general can be found in Savage (1972).

6. Program complexity

In the preceding section we developed a type of computational inequality which must hold if a function is to be computed by an interconnected set of sequential machines. The method of proof of this result is interesting and suggests a way of deriving other computational inequalities. Each binary sequential machine S was characterized by a set of functions $\{G_{S,T} \mid T = 1,2,3,...\}$. which describe the input-output behavior with T cycles of computation, $T = 1,2,3,...$. Then, a straight-line program or chain of length $C_{\Omega}(\delta^*, \lambda^*)T$ was designed which computed $G_{S,T}$ and which simulated T cycles of computation on S. Thus, the computational inequality was established with a "proof by simulation". Because the program was not necessarily of shortest length, the inequality followed.

The inequality derived in this section follows from the same method of proof except that the programs allowed are of the most general type, that is, programs for universal Turing machines. The inequality derived will depend on machine parameters in a very different way than does the inequality derived above.

A **Turing machine** consists of two types of components, a finite number of potentially infinite tape units on which heads can move only to adjacent squares in each unit of time and a control unit which is a finite state sequential machine and which controls the action of the heads. No external inputs are given to the control unit during execution and all of the information required for a computation, namely, instructions and data, must be stored initially on the nonblank portion of the tapes. The tapes are ruled into cells and each cell can hold only one of finitely many characters. The control is put into a fixed initial state and the tape heads given initial positions and a computation then proceeds as follows: The symbols under the tape heads are read and used as input to the control unit. It then computes a successor state and outputs which are used to write into the cells under the heads and then to reposition the heads. If a head is moved, it is only allowed to move to an adjacent cell.

Many models for computational procedures have been introduced in which access is available to a potentially infinite store, but when the machines are limited on the distance they can jump from one word in the store to another in each unit of time, none has been found which can compute functions which cannot be computed by Turing machines. Thus, the Turing machine is indeed a very general model for computational procedures.

Every finite function, that is, a function of finite domain and range, can be computed by some Turing machine because the table of values for the function could be stored in the control unit of such a machine. However, not all functions of infinite domain can be computed on Turing machines. The interested reader should see Hopcroft and Ullman (1969) for more detailed information on Turing machines. In fact, the famous "halting function", which given a description for an arbitrary Turing machine and an arbitrary initial tape configuration returns the value "true" if that machine with that input halts and "false" otherwise, is not computable by any Turing machine. A function of infinite domain which is computable on a Turing machine but not on a finite state sequential machine is the "palindrome function" which has value 1 on a string of binary digits if the string is a palindrome, that is, it can be written as $\omega\omega^R$ where ω^R is the string ω in reverse order, and value 0 otherwise.

There are Turing machines which are **universal** in the sense that they can simulate the input-output behavior of arbitrary Turing machines given a suitable desscription of such machines. General purpose computers, given the potential for infinite storage capacity, are potentially universal.

For our second type of computational inequality, we use a **universal 1-tape Turing machine** U with a single, semi-infinite tape, that is a tape on which adjacent cells can be numbered 1,2,3,..., as shown in Figure 6.1.

The tape alphabet Σ is finite and assumed to have a single "blank" symbol B. The tape is initially blank and a computation is carried out only after a finite string of non-blank symbols is written on the tape. Furthermore, we assume without loss of generality that 0 and 1 are in Σ.

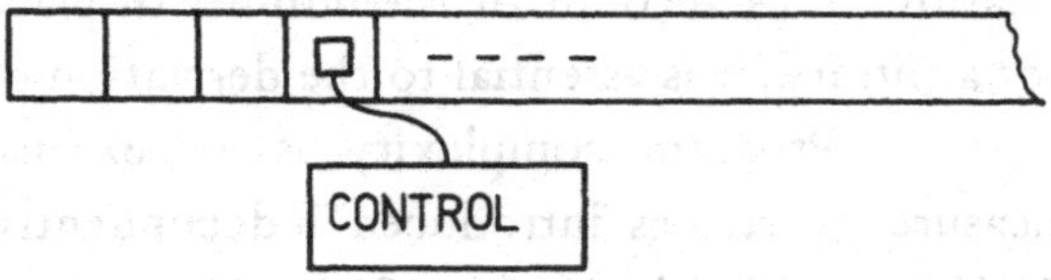

Fig. 6.1 Schematic Diagram of 1-Tape Turing Machine

We now introduce the notion of a "program" for such a machine so that we may introduce a new measure of complexity and a computational inequality. A **program** for f: $\{0,1\}^n \to \{0,1\}^m$ on U is

1) a string s(**x**) over $(\Sigma - \{B\}) \cup \{x_1, x_2, ..., x_n\}$, that is, a string whose entries are non-blank tape symbols or distinct variables of f,

such that

2) each x_i occurs once or not at all in s(**x**)

and

3) for each $\mathbf{c} \in \{0,1\}^n$, s(**c**) placed left-adjusted on the non-blank portion of the tape of U results in a halting computation with f(**c**) stored left-adjusted on the non-blank portion of the tape.

We assume that every finite function can be computed by programs of this type on the 1-tape universal Turing machine U. That such machines U exist can easily be shown [Hopcroft and Ullman(1969)].

We observe that a program is not a string over the tape alphabet alone and that it contains "instructions" as well as place markers for data. The fact that there is at most one occurrence of each variable and the absence of a requirement on the placement of variables in the program is consistent with current programming practice.

Definition 6.1. The **program complexity** of f: $\{0,1\}^n \to \{0,1\}^m$ on the 1-tape universal Turing machine U, $I_U(f)$ is the length (number of characters) of the shortest length program for f on U.

This measure of the complexity of functions is quite different from the combinational complexity measure. The full power of the "instruction set" of the most general type of computing machine is available, that is, looping, branching and recursion are available as well as functional composition, etc. In addition, the measure takes account of the number of places reserved for variables. This feature is not arbitrary; it is essential to the derivation of computational inequalities.

Program complexity is an extension to functions of the information measure on strings introduced independently by Solomonoff (1964), Kolmogorov (1968) and Chaitin (1966, 1970). Their measure is exactly the program complexity of functions $f: \{0,1\}^{o} \to \{0,1\}^{m}$, that is, the length of the shortest length string, without variables, to generate a given binary string.

Returning to the sequential machines which will be used to compute functions, we now derive a computational inequality relating the "size" of these machines and the time they use to compute a function. The inequality will follow from the construction of program for U which simulates T cycles of computation on a binary sequential machine $S = \langle \{0,1\}^{s}, \{0,1\}^{c}, \delta^*, \lambda^*, \{0,1\}^{d} \rangle$ or which computes the function $G_{S,T}$ described in the preceding section. Such a program can be constructed from a knowledge of δ^*, λ^* and of the variables for the initial state σ and the T input words $\mathbf{u}_1, \mathbf{u}_2, \ldots, \mathbf{u}_T$. For example, a program $s_G(\sigma, \mathbf{u}_1, \ldots, \mathbf{u}_T)$, could be constructed consisting of s_{SIM} which tells U how to simulate a sequential machine, $s_{\delta,\lambda}$ which tells U how to compute δ^*, λ^* and s_{DATA} which contains $\sigma, \mathbf{u}_1, \ldots, \mathbf{u}_T$. Namely, we could write

$$s_G(\sigma, \mathbf{u}_1, \ldots, \mathbf{u}_T) = s_{SIM} \cdot s_{\delta,\lambda} \cdot s_{DATA}(\sigma, \mathbf{u}_1, \ldots, \mathbf{u}_T)$$

where $\cdot$ denotes concatenation of strings. Here, only the last sub-program depends on the variables.

The reader can convince himself that such a construction is possible. Thus, if $\ell(s)$ denotes the **length** of the program $s(\mathbf{x})$, it follows that

$$\ell(s_G) = \ell(s_{SIM}) + \ell(s_{\delta,\lambda}) + \ell(s_{DATA}).$$

But $I_U(G_{S,T})$ is the length of the shortest length program for $G_{S,T}$, so

$$I_U(G_{S,T}) \leq \ell(s_{SIM}) + \ell(s_{\delta,\lambda}) + K^*(s + Tc)$$

where K^* is a constant and the last term is an upper bound to $\ell(s_{DATA})$. Since the number of data bits is $s + Tc$ and a few cells may have to be reserved as boundary indicators for each bit, the upper bound applies.

If S is used to compute f in T cycles, f is obtained from $G_{S,T}$ by fixing some or none of variables of $G_{S,T}$ and by suppressing some or none of the components in its range, without changing the order of the unerased components. We ask that the value of $G_{S,T}$, namely, $(v_1, v_2, \ldots, v_T)$ contain the necessary information to determine which bits to suppress and that the rule be a simple one. For example, the rule may be that whole words v_i are suppressed if the first bit is 1 and otherwise the first bit is suppressed and the unerased bits are strung together. Then, a simple program is appended to one for $G_{S,T}$ to create a program for f.

Theorem 6.1. Let $f: \{0,1\}^n \rightarrow \{0,1\}^m$ be a function computed in T cycles on the binary sequential machine $S = \langle \{0,1\}^s, \{0,1\}^c, \delta^*, \lambda^*, \{0,1\}^d \rangle$. Then there exist constants $K_1, K_2 > 0$ independent of s, c, T and f but dependent on U such that

$$I_U(f) \leqslant K_1(s + Tc) + K_2 + \ell(s_{\delta,\lambda})$$

where $I_U(f)$ is the program complexity of f and $\ell(s_{\delta,\lambda})$ is the length of a program describing the transition and output functions of S.

The interpretation of this inequality is that to compute f, the amount of information $I_U(f)$ that must be given to U should not exceed a quantity proportional to the amount of information that is given to S as data either initially or while running plus the amount of information required to describe S.

The inequality can also be applied to the general purpose computer described in the preceding section. If the initial state of the CPU is always fixed while that of the RAM may be changed, we have that the initial state of the sequential machine is described by s = S bits. Furthermore, the machine receives and sends c = d bit words per cycle. The transition and output functions of the combined machine are easily described and the length of programs for these will grow slowly with b, c and S. Thus, we assume that $\ell(s_{\delta,\lambda})$ can be absorbed into the constants K_1 and K_2 above and have

Theorem 6.2. If $f: \{0,1\}^n \rightarrow \{0,1\}^m$ is computed in T cycles on a general-purpose computer with storage capacity S bits and input words of size c, then there are constants $K_1, K_2 > 0$ independent of S, T, c and f but dependent on U such that

$$I_U(f) \leqslant K_1(S + Tc) + K_2$$

where $I_U(f)$ is the program complexity of f on the universal 1-tape Turing machine U.

It can be seen from this inequality and that of Theorem 5.2, that functions can be computed on general computers only if several lower bounds on space-time are simultaneously satisfied. For more on program complexity and computational inequalities see Savage (1973).

7. Lower Bounds to the Complexity of Decoding Functions

The computational inequalities derived in the preceding sections give importance to the two measures of complexity, combinational complexity, the minimum number of operation steps in a straight-line algorithm for a function, and program complexity, the length of the shortest length general program for a function. Particularly important are the relationships which must hold between space S, time T and other mchine parameters on general purpose computers (GPC's).

Fig. 7.1 Storage-Time Tradeoff Boundaries

In Figure 7.1 is shown schematically, the lower limits implied on S and T for GPC's by the computational inequalities of Theorem 5.2 and 6.2. Since both inequalities must hold, the shaded region is inaccessible, that is, the function f with combinational complexity $C_U(f)$ and program complexity $\Omega\ I_U(f)$ cannot be computed on a GPC with S and T in the shaded region. The overlap shown occurs when $C_\Omega(f) \ll I_U^2(f)$ a condition that is met by many functions. Also, functions $f: \{0,1\}^n \rightarrow \{0,1\}^m$ which have large combinational and program complexity (and most do) can be computed at very many values of S and T satisfying $S + Tc \leqslant K^* I_U(f)$ for some constant K^*, that is, at values of S and T close to the linear lower bound. Thus, the inequalities are not vacuous.

Now let us turn to lower bounding the complexity of decoding functions. Our problem is this

"Derive lower bounds to the combinational complexity and program complexity of the least complex decoding function which acheives a

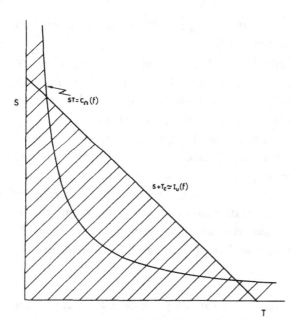

probability of error of at least $Pe = 2^{-E}$ or reliability E, at a code rate of at least R on a given discrete memoryless channel".

Implicit in this statement is the fact that feed-back may be used, if available.

Let $f_D^*: \{0,1\}^{Na} \to \{0,1\}^b$ be the binary encoding of $f_D: \{0,1,...,J-1\}^N \to \{1,2,...,M\}$ where $E_J: \{0,1,2,...,J-1\} \to \{0,1\}^a$ and $E_M: \{1,2,...,M\} \to \{0,1\}^b$ give 1-1 representations of the respective sets. By Theorem 2.2, for any given DMC there is a minimum value of block length N such that a probability of error of at most 2^{-E} and rate at least R can be achieved for codes with equally likely code words. Such bounds hold even when feed-back is available. Let $N_0 = N_0(E,R)$ be this minimum value or

$$N_0\{E_{SP}[R - o_1(N_0)] + o_2(N_0)\} \geq E$$

from Theorem 2.2 for channels **without feed-back**. For some channels, such as the BSC, the same lower bound holds with feed-back.

The program complexity of f_D^*, namely, $I_U(f_D^*)$ is lower bounded by the number of variables of f_D^* on which it depends. Since $a \geq 1$, it follows that if f_D^* provides rate R and reliability E, then

$$I_U(f_D^*) \geq N_0.$$

This is a simple lower bound but an upper bound linear in N_0 can also be derived, as well as we shall see later.

To derive a bound to $C_\Omega(f_D^*)$, more information about f_D^* is required. To say that $f_D^*(\omega_1,\omega_2,...,\omega_{Na})$ depends on a particular input variable ω_1 does not necessarily mean that that variable is used by some computation node. In fact, it may depend on this variable because a component ζ_ℓ of $f_D^*(\omega_1,\omega_2,...,\omega_{Na}) = (\zeta_1,\zeta_2,...,\zeta_b)$ equals ω_p that is, $\zeta_\ell = \omega_p$. We shall show, under a fairly weak condition, that this cannot happen when high reliability is desired unless ζ_ℓ is an output which does not change.

Assumption We assume that DMC's under consideration are **completely connected**, that is, for each input $0 \leq k \leq K-1$ and output $0 \leq j \leq J-1$, $P(j/k) > 0$. Let P_{min} be the smallest transition probability.

Consider a variable w_p of f_D^* whose value may change, that is, assume that the representation E_J is such that when applied to a channel output y_r, to give w_p, that a change in y_r causes a change in w_p. If $w_p = \zeta_\ell$, where ζ_ℓ is an output variable of f_D^*, then a change in w_p will cause a change in the value of f_D^*, that is, it causes a change in the decision by the decoder. Whatever the symbol that was sent, received as y_r and mapped into w_p, a different channel transition and a change in the decision is possible since the channel is completely connected.

Let message m be sent (with or without feed-back) and decoded correctly, let x_r be the r-th transmitted channel symbol in one such transmission and let x_r be received as y_r. With probability of at least P_{min}, x_r could be received as y_r' for which w_p is changed and a decoding error occurs. Thus, $P_{e/m} \geqslant P_{min}$ and $Pe \geqslant P_{min}$. If a smaller probability of error is required, no non-constant input variable of f_D^* can equal an output variable. Note that this argument applies whether feed-back is permitted or not.

Theorem 7.1 Consider a completely connected DMC with minimum transition probability equal to P_{min} and let Ω be the Boolean functions on two variables. If a reliability E

$$E > -\log_2 P_{min}$$

is required at rate R with an equiprobable letter source and feed-back is not used, then every decoding function f_D^* must satisfy

$$I_U(f_D^*) \geqslant N_0$$

0

$$C_\Omega(f_D^*) \geqslant \max(N_0/2, N_0 R)$$

where $N_0 = N_0(E,R)$ is the smallest integer satisfying

$$N_0\{E_{SP}(R - o_1(N_0)) + o_2(N_0)\} \geqslant E$$

where in turn $o_1(N)$ and $o_2(N)$ may be taken as the functions defined in Theorem 2.2.

Proof The lower bound to $I_U(f_D^*)$ was derived above. The lower bound to $C_\Omega(f)$ follow by observing that if

$$2^{-E} < P_{min}$$

then no non-constant input variable of f_D^* can equal an output variable. Since f_D^* depends on at least N_0 non-constant input variables, each must pass through at least one logic element. However, each element has at most two inputs so at least $N_0/2$ of them are needed. On the other hand, f_D^* has b output variables and at least $\log M_2 = NR$ distinct such variables are needed just to represent the M points in the range of f_D^* and permit rate R to be achieved. Outputs are distinct if they are the outputs of distinct logic elements. Thus, at least NR logic elements are needed. (This argument also holds when Boolean NOT's are not counted.)

Q.E.D.

For a similar lower bound to hold with feed-back, a lower bound to Pe must be derived.

In the next section, we shall derive upper bounds to these complexity measures. To see the dependence of these lower bounds on E and R, we note that the error exponent $E_{SP}(R)$ described in Theorem 2.2 behaves as $\alpha(C - R)^2$; for some constant $\alpha > 0$, for R near channel capacity C. We also note that for E large and $R < C$, N_0 is large and $o_1(N_0)$ and $o_2(N_0)$ are both small. We then have the following corollary to Theorem 7.1.

Corollary For large reliability E, under the conditions of Theorem 7.1,

$$I_U(f_D^*) \geqslant \rho(E,R) = \frac{E}{E_{SP}(R)}$$

$$C_\Omega(f_D^*) \geqslant \max(\frac{\rho}{2}, \rho R).$$

The dependence of $\rho(E,R)$ on rate R for a typical completely connected DMC is shown in Figure 7.2.

It has a non-zero value as R approaches zero and becomes very large as R approaches channel capacity.

We can conclude that the price of high reliability is large decoding complexity and that the complexity measured with either combinational complexity or program complexity grows at least linearly in E and becomes large as R approaches channel capacity.

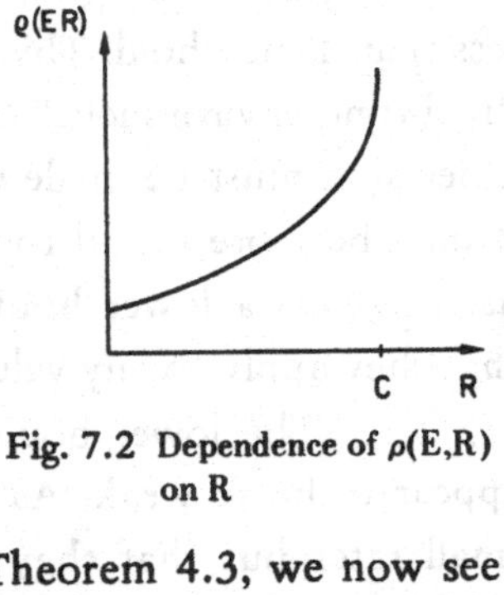

Fig. 7.2 Dependence of $\rho(E,R)$ on R

We have not said much about delay complexity, the second measure introduced in Section 4, but using Theorem 4.3, we now see that it must grow at least as fast as $\log_2(\rho + 1)$. Since delay complexity is a measure of the speed (measured in absolute terms) with which decoding can be accomplished, we also see that some finite and rather slowly growing lower bound to absolute decoding time also holds.

8. General Remarks on the Bounds and Bounding Methods

Combinational complexity and program complexity are given significance by the computational inequalities presented in Section 5 and 6. Computational work W, the measure defined in the first type of computational inequality, has not been discussed much but it does have a significance beyond that implied by the inequality.

In general, the cost of purchasing a sequential machine is roughly proportional to $C_\Omega(\delta^*, \lambda^*)$, its equivalent logical complexity, so that the cost of one unit of time on such a machine should also be proportional to this quantity, if the machine has a fixed lifetime. For accounting purpose, new equipment is usually given a fixed lifetime, so the cost to a customer for using such a sequential machine for T cycles should be proportional to $W = C_\Omega(\delta^*, \lambda^*)T$, the computational work done by the machine. It makes sense, then to introduce the efficiency measure E

$$E = \frac{C_\Omega(f)}{W} \quad .$$

It has the nice properties that (a) $E \leqslant 1$ and (b) E increases with decreasing W.

As suggested by Figure 7.1, the second type of computational inequality may prevent W from being close to $C_\Omega(f)$ at intermediate and small values of T. In other words, at any given value of T, an upper bound to efficiency which is strictly less than 1 may hold. This is an important point because decoders are often used in a "real-time environment" where data arrives continuously and at a constant rate and a decision must be made with a small delay. This suggests that for such applications, it may be more useful to lower bound W with a constraint such as this, rather than use $C_\Omega(f)$ as a lower bound. The advantage of the inequalities that have derived is that they apply to any value of T.

The lower bounds of Theorem 7.1 are a bit disappointing because they appear to be so weak. As we show in this section, the bounds are not too weak at small rates but that they do appear weak is a reflection of the state of affairs in complexity theory. There is a great deal of research under way to improve on known lower bounds but the task is not easy.

We now show that at low code rates R, we can derive upper bounds to $I_U(f^*_D)$ and $C_\Omega(f^*_D)$ which exhibit a linear dependance on reliability E. Thus, the dependence of the lower bounds on E at small rates can be improved upon by at most a constant factor.

Consider the class of binary maximal-length codes described in Gallager (1968), p. 230, which is a set of N distinct code words of length $N = 2^p - 1$, each of which is a cyclic shift of a single sequence which in turn can be generated by an p-stage feed-back shift-register. The Hamming distance (number of disagreements) between any two such code words is exactly $(N+1)/2$. This means that if such a code were used on a binary channel and no more than t errors occur, where $2t+1 \leq (N+1)/2$, then the received word is closer in Hamming distance to the transmitted code word than to any other code word and can be decoded correctly. Since $N+1 = 2^p$ is even, $t = [\,(N+1)/4\,]-1$ or fewer errors can be corrected.

From this maximal-length code choose $M \leq N$ words to form a new code. Again t or fewer errors can be corrected so define a decoding function(*) $f^*_D : \{0,1\}^N \rightarrow \{0,1\}^b$, $b = \lceil \log_2 M \rceil$, which maps a received sequence into a code word at Hamming distance t or less, if one exists, and into some one code word, otherwise. A program for a universal 1-tape Turing machine U can be constructed which consists of instructions for constructing and using the p-stage feed-back shift register to generate the M code words and instructions for comparing each of the M words with the received word in order to make a decision. The length of such a program will be linear in N for any $M \leq N$. That is, there exists a constant $K > 0$ such that

$$I_U(f^*_D) \leq KN$$

for all $N \geq 3$.

An upper bound to $C_\Omega(f^*_D)$ can also be derived. Let $\mathbf{y}$ be a received binary N-tuple and let $\mathbf{x}_1, \mathbf{x}_2, \ldots, \mathbf{x}_M$ be the M code words in the code. For each $1 \leq m \leq M$, compute an error N-tuple $\mathbf{e}_m$ where

$$e_{mi} = \begin{cases} 1 & x_{mi} \neq y_i \\ 0 & x_{mi} = y_i \,. \end{cases}$$

(*) $\lceil x \rceil$ is the "ceiling of x" which is the smallest integer $\geq$ x.

This function can be computed from the EXCLUSIVE OR operator $\oplus$ and NOT as follows:

$$e_{mi} = \overline{x_{mi} \oplus y_i}.$$

Thus, MN binary operations are used here. Now for each M, reduce $\mathbf{e}_m$ to a binary number $\mathbf{s}_m$ of $\lceil \log_2 N \rceil$ bits representing the number of 1's in $\mathbf{e}_m$ or the Hamming distance between $\mathbf{x}_m$ and $\mathbf{y}$. This last step can be realized using M copies of the counting function $f_c^{(p)}:\{0,1\}^{2p} \to \{0,1\}^p$ which maps $\mathbf{u} = (u_1,u_2,\ldots,u_{2^p})$ into $\mathbf{v} = (v_1,\ldots,v_p)$ which represents in binary notation the number of 1's in $\mathbf{u}$.

The function $f_c^{(p)}$ can be realized recursively, as shown by Muller and Preparata (1973), as follows

$$f_c^{(p)}(u_1,\ldots,u_{2p}) = \mathrm{ADD}(f^{(p-1)}(u_1,\ldots,u_{2p-1}), f_c^{(p-1)}(u_{2p-1+1},\ldots,u_{2p}))$$

where ADD $(\mathbf{w},\mathbf{w}')$ generates the sum in binary notation of the binary numbers $\mathbf{w},\mathbf{w}'$ Since this function can be realized with a number of operations linear in the length of $\mathbf{w}$ or $\mathbf{w}'$ the interested reader can show that $C_\Omega(f_c^{(p)})$ is linear in $2^p = N+1$.

The sums $\mathbf{s}_m$, $1 \leq m \leq M$, can now be compared and the smallest found which if less than or equal to t will cause $\mathbf{y}$ to be decoded as M. The function $f_{comp}: \{0,1\}^{2p} \to \{0,1\}$ defined by

$$f_{comp}(v_1,\ldots,v_p,z_1,\ldots,z_p) = \begin{cases} 1 & |\mathbf{v}| < |\mathbf{z}| \\ 0 & \text{otherwise} \end{cases}$$

can be used to compare p-tuples, where $|\mathbf{v}|$ denotes the integer represented by $\mathbf{v}$. Using the representation

$$f_{comp} = \bar{v}_p \cdot z_p + (\bar{v}_p + z_p) \cdot f'_{comp}$$

where f'_{comp} is f_{comp} on 2p-2 variables and $\cdot$,+ denote AND and OR, the reader can convince himself that

$$C_\Omega(f_{comp}) \leq 4p - 3$$

for Ω the set of all 2-input Boolean operations. With this function, a circuit can be designed which selects the m with largest s_m using on the order of Mp Boolean operations.

We conclude that there exists a constant $K^* > 0$ such that

$$C_\Omega (f_D^*) \leq K^*MN$$

for the function f_D^* described above. We next show that on the BSC with $0<\epsilon<1/4$, the probability of an error with this code decreases exponential in N. An error will not occur if t or fewer channel errors occur. Consequently,

$$Pe \leq \sum_{j=t+1}^{N} \binom{N}{j} \epsilon^j (1-\epsilon)^{N-j}$$

where $\binom{N}{j}$ is the binomial coefficient. By the standard Chernoff bounding technique [Gallager (1968), p. 127], this is readily shown to be upper bounded by

$$Pe \leq 2^{-NE(\lambda,\epsilon)}$$

where

$$E(\lambda,\epsilon) = -\lambda \log_2 \epsilon - (1-\lambda)\log_2 (1-\epsilon) + \lambda \log_2 \lambda + (1-\lambda)\log_2 (1-\lambda)$$

and

$$\lambda = (t+1)/N = (N+1)/4N \gtrsim 1/4$$

The exponent $E(\lambda,\epsilon) > 0$ as long as $\epsilon < \lambda$ or $\epsilon < 1/4$ so the reliability $E = -\log_2 Pe$ satisfies

$$E \geq NE(\lambda,\epsilon)$$

and the following theorem holds.

Theorem 8.1 There exists a code of M code words, block length N and a decoding rule f_D^* for this code such that on the BSC with crossover probability $0 < \epsilon < 1/4$

$$I_U(f_D^*) \leq K \frac{E}{E(\lambda,\epsilon)}$$

$$C_\Omega (f_D^*) \leq K^*M \frac{E}{E(\lambda,\epsilon)}$$

where K,K* > 0 are constants independent of M, N and ϵ and

$$E(\lambda,\epsilon) = -\lambda \log_2 \epsilon - (1-\lambda)\log_2(1-\epsilon) + \lambda \log\lambda + (1-\lambda)\log_2(1-\lambda).$$

Also,

$$\lambda = \frac{N+1}{4N}$$

and $E(\lambda,\epsilon) > 0$ for $0 < \epsilon < 1/4$.

We learn from this theorem that if M is fixed, that is, the rate R is small, then the combinational and program complexity of some decoding function providing probability of error $Pe = 2^{-E}$ grows no faster than linearly with the reliability E. This establishes that at low rates, the lower bounds of Theorem 7.1 are weak by at most a constant factor.

We now examine a number of well known decoding algortihms.

9. Performance of Several Known Decoding Procedures

As indicated at the end of the last section, the dependence on reliability E of the lower bounds derived in Section 7 cannot be improved substantially at low code rates. Whether such is the case at high rates remains an open question.

In this section we examine a number of known decoding algorithm and for each sketch the derivation of upper bounds to the computational work W which they require. As consequence, we not only will have upper bounds to the combinational complexity of the decoding functions which they describe, we will also have a crude measure of the cost of decoding one block with them. (See the discussion early in Section 8.) We leave it to the reader to show that the program complexity for each of the decoding functions considered is linear in block length.

We shall examine two procedures for decoding convolutional codes, namely, sequential decoding and the Viterbi algorithm, and several procedures for decoding block codes. In each case only the properties of the decoding algorithms needed to bound W are cited. The interested reader can refer to the literature for those additional details of the algorithms that he or she needs.

Both sequential decoding and the Viterbi algorithm are methods for decoding convolutional codes. Such codes are continuously encoded by sequential machines of the type shown in Figure 9.1

The machine shown is a binary shift-register which accepts one binary digit in each cycle and computes two sums, modulo -2, of subsets of the K digits in the register in each unit of time. The two sum digits are delivered to a channel in each unit of time also. More generally, such a machine might accept one of q inputs from the finite field GF(q) in each unit of time and form v sum digits over GF(q) by taking the inner product of the K digits with v K-tuples over GF(q).

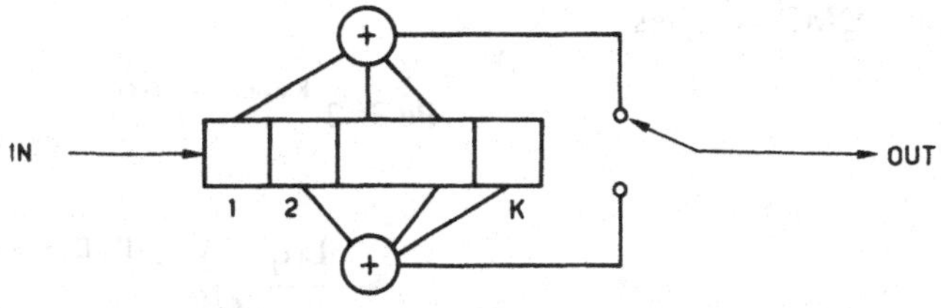

Fig. 9.1 Convolutional Encoder

Generally, convolutional codes are used to encode long sequences of information digits. If L digits are encoded, a string of L information digits followed by K-1 0's are encoded to produce a block of N = (L + K - 1)v channel digits over GF(q). Thus, the code rate is

$$R = \frac{\log_2(q^L)}{N} = \frac{L}{L+K-1}\,\frac{\log_2 q}{v} \simeq \frac{\log_2 q}{v}$$

which is approximately $(\log_2 q)/v$, as indicated, for large L.

The **Viterbi algorithm** [Viterbi (1967)] implements the maximum likelihood (ML) decoding rule for convolutional codes on a DMC (see Section 2 for the definition of maximum likelihood decoding). The algorithm executes L steps and in each step q^Ksequences of K digits over GF(q) are tested with the ML rule. By whatever method of testing used, each such step produces q^K K potentially independent symbols over GF(q). Consequently, whether each step is realized by one cycle of a complex sequential machine or many steps of a less complex machine, the work done on each step is at lest $q^K K$ (at least one logic use per symbol) and at least a total work

$$W \geqslant LKq^K$$

is needed to decode a block of L information symbols. An upper bound to W with the same dependence on L, q and K can also be derived.

Viterbi has shown that the probability of error with his algorithm on a q-ary DMC satisfies

$$Pe \geqslant 2^{-Kv[E_L(R)+0(Kv)]}$$

$$\leqslant \frac{L(q-1)}{1-q^{-\epsilon/R}} 2^{-KvE_U(R)}$$

where $\epsilon > 0$, $E_L(R)$ and $E_U(R)$ are coding exponents which are non-zero for $0<R<C$, (C is channel capacity), and $0(Kv)$ approaches zero with increasing Kv. We see that the **work W required to decode with the Viterbi algorithm grows exponentially with reliability E.** Lest the reader be discouraged from using this algorithm, let it be said that the exponential dependence is not important for E around 10, a range in which the Viterbi algorithm is as good or better than other decoding procedures for the space channel, for example.

Sequential decoding was introduced by Wozencraft [Wozencraft and Reiffen (1961)] and an improved algorithm given by Fano (1963). The several procedures known by this name involve a form of backtrack searching on the tree of possible code word sequences. (See Figure 9.2 for an example of such a tree.)

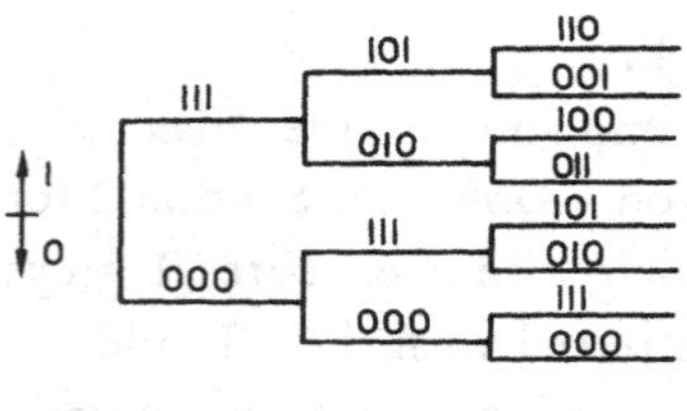

Fig. 9.2 Tree of Code Word Sequences

In effect, an encoder chooses a branch of the tree with each source digit. The decoder emulates this behavior by comparing the received sequence against branches of the tree progressing forward until a likelihood measure indicates that the chosen path is less certain to be the transmitted path. Backward steps and further attempts to move forward are then carried out.

The time at which the docoder can be fairly certain that a temporary decision on an information digit is good and not likely to be changed will fluctuate with the noise level so that, in effect, the number of computation cycles needed to decode an information symbol is a random variable. In practice, then, a large buffer capable of holding B branches of the tree and B temporary decisions on information symbols is used. The variation in the number of decoding cycles remains, however, and the buffer may occasionally fill and overflow resulting in many decoding errors or in retransmission, when possible.

The probability of overflow of the buffer has been shown from experiment and theory to behave as [Jacobs and Berlekamp (1967), Bluestein and Jordan (1963), Savage (1966A, 1966B, 1968)]

$$P_{BF} = \frac{dL}{[B(SF)]^{\alpha(R)}}.$$

where L is the number of information digits in a block, d is a small constant, SF is the speed factor (a measure of the relative speed of the decoder) and $\alpha(R)$ is the Pareto exponent. It has been shown that $\alpha(R) > 0$ for $0 < R < C$ and $\alpha(R) = 1$, at $R = R_{comp}$, a rate which is often a substantial fraction of channel capacity.

The buffer is a shift-register whose logical complexity is at best proportional to B. Thus, considering only the computational work done by it, the **work done by sequential decoders grows exponentially with reliability.** Again it must be said that for relatively small reliabilities, that is, $E \leqslant 20$, say, sequential decoding has been found to be economical, especially on the space channel.

Several decoding algorithms for block codes come under the heading of **concatenated coding and decoding** [Forney (1967), Ziv (1966, 1967)]. The codes involved are constructed from inner and outer codes. Typically, a large number of digits are broken into small groups and transmitted with coding and decoded at the receiver. This process reduces the error rate somewhat and permits fairly reliable decoding with an outer coder and decoder which operate over many of the inner blocks.

The concatenated scheme discussed by Forney (1967) uses an inner code with q code words, q a power of a prime, and an outer code which is a Reed-Solomon (1960) code of block length n, $n \leqslant q - 1$. The inner code and decoder process individual symols of the outer code and do a computational work at worst proportional to q on each symbol (the symbols would be represented as binary $[\log_2 q]$ - tuples), or at worst proportional to nq for the n symbols. Berlekamp (1968) has given an algorithm for decoding Reed-Solomon codes which uses a number of cycles T proportional to n log n and a machine whose logical complexity is also proportional to $n \log_2 n$. Thus, the Berlekamp algorithm does a work W_B proportional to

$$W_B \simeq (n \log_2 n)^2$$

With $q \cong n$, this dominates the work done by the inner decoder.

The block length of the inner code is $(\log_2 q)/R_0$ if R_0 is its rate so that the block length of the combined code is

$$N = n(\log_2 q)/R_0 .$$

The probability of error that is acheivable using a good inner code with $q \cong n$ is bounded by

$$Pe \leqslant 2^{-\frac{N}{2} E_C(R)}$$

where $E_C(R)$ is an exponent which is non-zero for $0 < R < C$. We conclude, that there exists a constant $K >,0$ such that the work done by concatenated decoder can satisfy

$$W_{CD} \leqslant K\left(\frac{ER_0}{E_C(R)}\right)^2 .$$

Therefore, **the computational work done with decoders for concatenated codes need not grow faster than E^2**. Even through this rate of growth with E is small, such codes are not in general use because at present the amount of work required at moderate values of E is too high to make them economical.

There is a moral in this last observation: Beware of arguments based exclusively on asymptotics!

The last decoding procedure considered here is for the BSC (binary symmetric channel) with feed-back [Schalkwijk (1974)] and it is used to decode what is called **multiple repetition coding** (MRC).As described here, the procedure assumes a noiseless feed-back channel, although it has been proposed that it be used in conjunction with the Viterbi algorithm when the feed-back channel is itself a BSC. A slightly different such procedure was reported in Schalkwijk (1971).

The encoding procedure has two steps. At the first step, a binary sequence of L digits is mapped into a sequence of $L_1 = L_0\ (k\text{-}1)/(k\text{-}2)$ bits where k-2 divides L_0 in such a way that neither 10^k nor 01^k occurs. (Here i^k is the k-fold repetition of i.) This can be done by inserting after each subsequence of k-2 bits one additional bit which is the complement of the last subsequence bit. Additional redundancy is added by following these L_1 bits with a tail consisting of an alternating sequence of $N\text{-}L_1$ binary digits. If

$$N - L_1 = tk$$

then it can be shown that all patterns of t or fewer errors can be corrected. Thus the code rate is

$$R = \frac{L_0}{N} = \left(\frac{k-2}{k-1}\right)\left(1 - \frac{t}{N}k\right),\ k \geq 3.$$

The second encoding step uses the feed-back channel and consists of repeating k times any bit which is seen to be received incorrectly. If during a repeat, an error is observed, the repetition process begins for the last error before resuming for preceding errors. Tail digits are dropped, as necessary, so that only N digits are sent.

The decoding process is quite simple and consists of at most t passes over the received digits, passing from last received digit to first received digit. On each pass the decoder seeks a subsequence of 10^k or 01^k which are replaced by 1 and 0, respectively. When such a subsequence is found the replacement is made and another pass begun. If no such subsequence is found, the decoding process is complete.

A decoder can then be constructed consisting of two shift-registers of length of at most N bits so that on one pass the contents of one register is examined for 10^k or 01^k and transferred to the other. If k is small relative to N, the logical complexity of the decoder is proportional to N and the number of cycles executed to no more than tN. Thus, the work done is proportional to tN^2.

The bound of Section 8 to Pe for a t-error correcting code on the BSC can be applied to MRC to show that with t proportional to N, a probability of error of

$$Pe \leq 2^{-NE(\lambda,\epsilon)}$$

can be achieved on a BSC with crossover probability ϵ when $\lambda = (t/N)$. Thus, for

$$R = \left(\frac{k-2}{k-1}\right)(1 - \lambda k)$$

the work done to decode MRC using feed-back is bounded by

$$W_{MRC} \leq K\lambda\left(\frac{E}{E(\lambda,\epsilon)}\right)^3$$

for some constant $K > 0$. With this coding and decoding method, rates close to channel capacity can be achieved. While the work done grows faster with E than it does for decoding of concatenated coding, the coefficient K may be such that at relatively large values of E, this scheme may be preferred if feed-back is available.

10. Conclusions

Coding will play an increasingly important role in the future development of communication systems. Natural resources of all kinds are scarce, including the radio spectrum, and it is quite obvious that coding will continue to be applied to reduce the cost of reliable communication.

But coding itself introduces cost, namely, the cost of decoding. We have outlined a theory of decoder complexity which in its early and abstracted form offers some promise for understanding and coping with the cost of decoding, Necessarily such a first attempt at such a theory does not permit detailed analysis but it does identify the important concepts and parameters. Computational work W is new measure of computing effort which was discovered in the first studies of decoder complexity [Savage (1971)] and it is now seen to have more general applications. We have also learned that the work required to decode grows at least linearly with reliability E and that this rate of growth can be achieved at low rates.

Computational work provides a useful measure for comparing the **asymptotic** performance of decoding strategies. We learn, for example, that in the limit of large reliability E, concatenated coding and decoding is the best known general coding method in its dependence on E.

The future can only tell what need there will be for a theory of decoder complexity. It cannot be said, however, that the problem has been overlooked or that a start has not been made in understanding it.

References

[1] Berlekamp, E.R., (1968) *Algebraic Coding*, McGraw-Hill, New York.

[2] Bluestein, G. and K.L. Jordan, (1963) Investigation of the Fano Sequential Decoding Algorithm by Computer Simulation, MIT Lincoln Laboratory, Group Report 62G-5.

[3] Chaitin, G.J., (1966) On the Length of Programs for Computing Finite Binary Sequences, *Journal of the ACM*, vol. 13, pp. 547-569.

[4] Chaitin, G.L., (1970) On the Difficulty of Computations, *IEEE Transactions on Information Theory*, vol. IT-16, No. 1, pp. 5-9.

[5] Fano, R.M., (1963) A Heuristic Discussion of Probabilistic Decoding, *IEEE Transactions on Information Theory*, vol. IT-9, pp. 64-74.

[6] Forney, G.D., (1967) *Concatenated Codes*, MIT Press, Cambridge.

[7] Gallager, R.G., (1968) *Information Theory and Reliable Communication*, Wiley and Sons, New York.

[8] Harper, L.H. and J.E. Savage, (1972) On the Complexity of the Marriage Problem, *Advances in Mathematics*, Vol. 9, No. 3, pp. 249-312.

[9] — (1973) Complexity Made Simple, *Proceedings of the International Symposium on Combinational Theory*, Rome, September 2-15.

[10] Harper, L.H., W.N. Hsieh and J.E. Savage, (1975) to appear in Journal of Theoretical Computer Science.

[11] Hopcroft, J.E. and J.D. Ullman, (1969) *Formal Languages and Their Relation to Automata*, Addison-Wesley, Reading, Mass.

[12] Jacobs, I.M. and E.R. Berlekamp, (1967) A lower Bound to the Distribution of Computation for Sequential Decoding, *IEEE Transactions on Information Theory*, Vol. IT-13, pp. 167-174.

[13] Justesen, J. (1972) A class of Constructive Asymptotically Good Algebraic Codes, *IEEE Transactions on Information Theory*, Vol. IT-18, pp. 652-656.

[14] Kolmogorov, A.N. (1968) Logical Basis for Information Theory and Probability Theory, *IEEE Transactions on Information Theory*, Vol. IT-14, pp. 662-664.

[15] Muller, D.E. and F.P. Preparata (1973) Minimum Delay Networks for Sorting and Switching, *Proceedings 7-th Princeton Conference on Information Science and Systems*, pp. 138-139.

[16] Reed, I.S. and G. Solomon, (1960) Polynomial Codes over Certain Finite Fields, *Journal of the Society for Industrial and Applied Mathematics*, Vol. 8, pp. 300-304.

[17] Savage, J.E., (1966A) Sequential Decoding — The Computation Problem, *Bell System Technical Journal*, Vol. XLV, pp. 149-175.

[18] — (1966B) The Distribution of the Sequential Decoding Computation Time, *IEEE Transactions on Information Theory*, Vol. IT-12, pp. 143-147.

[19] — (1968) Progress in Sequential Decoding, *Advances in Communication Systems*, Academic Press, pp. 149-204, edited by A.V. Balakrishman.

[20] — (1969) The Complexity of Decoders — Part I: Classes of Decoding Rules, *IEEE Transactions on Information Theory*, Vol. IT-15, pp. 689-695.

[21] — (1970) A note on the Performance of Concatenated Codes, *IEEE Transactions on Information Theory*, Vol. IT-16, pp. 512-513.

[22] — (1971) The Complexity of Decoders — Part II: Computational Work and Decoding Time, *IEEE Transactions on Information Theory*, Vol. IT-17, pp. 77-85.

[23] — (1972) Computational Work and Time on Finite Machines, *Journal of the ACM*, Vol. 9, pp. 660-674.

[24] — (1973) Bounds on the Performance of Computing Systems, Proc. 7-th Annual Symposium on Computer Science and Statistics, October 18,19, Ames, Iowa.

[25] — (1976) *The Complexity of Computing*, to be published by Wiley-Intersience, New York.

[26] Schalkwijk, J.P.M, (1971) A Class of Simple and Optimal Strategies for Block Coding on the Binary Symmetric Channel with Noiseless Feed-back, *IEEE Transactions on Information Theory*, Vol. IT-17, pp. 283-287.

[27] — (1974) A Coding Scheme for Duplex Channels, to appear in *IEEE Transactions on Communication Technology*.

[28] Shannon, C.E. (1948) A Mathematical Theory of Communication, *Bell System Tech. Journal*, Vol. 27, pp. 379-423 (Part I), pp 623-656 (Part II).

[29] Shannon, C.E., R.G. Gallager and E.R. Berlekamp, (1967) Lower Bounds to Error Probability for Coding on Discrete Memoryless Channels, *Information and Control*, Vol. 10, pp. 65-103 (Part I), pp. 522-552 (Part II).

[30] Solomonoff, R.J., (1964) A Formal Theory of Inductive Inference, *Information and Control*, Vol. 7, pp. 1-22 (Part I), pp. 224-254 (Part II).

[31] Viterbi, A,J. (1967) Error Bounds for Convolutional Codes and an Asymptotically Optimum Decoding Algorithm, *IEEE Transactions on Information Theory*, Vol. IT-13, pp. 260-269.

[32] Wozencraft, J.M. and B. Reiffen, (1961) *Sequential Decoding*, MIT Press, Cambridge.

[33] Ziv, J., (1966) Further Results on the Asymptotic Complexity of an Iterative Coding Scheme, *IEEE Transactions on Information Theory*, Vol. IT-12, pp. 168-171.

[34] — (1967) Asymptotic Performance and Complexity of a Coding Scheme for Memoryless Channels, *IEEE Transactions on Information Theory*, Vol. IT-13, pp. 356-359.

COMPLEXITY AND INFORMATION THEORY

Aldo De Luca

Laboratorio di Cibernetica del C.N.R. Arco Felice, Napoli
and
Istituto di Scienze dell'Informazione dell'Università di Salerno

General Introduction

The concept of "information" appeared in Physics in connection with the concept of "entropy". It was observed (Boltzmann, 1896) in the framework of statistical thermodynamics, that the entropy is proportional to the logarithm of the number of alternatives (or microscopic states) which are possible for a physical system knowing all the **macroscopic** information about it. The entropy gives, in other words, a measure of the total **amount of missing information** about the system.

This concept was used by many authors such as Szilard (1929) for information in physics, and Nyquist (1924) and Hartley (1928) in problems of transmission of information. Fundamental contributions in this latter direction have been given by C.E. Shannon (1948) and N. Wiener (1949).

The importance of Shannon's theory is due, without doubt, to the possibility of proving, by making use of the theory of ergodic processes, some very general and certainly non trivial theorems on problems of transmission of information. However, the validity of Shannon's theory, or better, the meaningful application of it, is confined to statistical communication theory. Many questions that have a great intuitive appeal from the information point of view, do not make sense in the framework of Shannon's theory. In spite of the several criticisms that can be made against Shannon's point of view, we stress that Shannon's approach to communication theory and his measure of information are such that the main problems that arise in the context of communication theory can be easily faced and handled, and fundamental theorems can be proved. However the success of Shannon's theory discouraged most researchers for whose problems Shannon's concept of information was neither useful nor interesting, and hampered other attempts to introduce new concepts of information.

It is quite clear that "information" is one of those intuitive concepts (or **explicanda**) of a wide "semantic halo", so that it can admit several corresponding formal concepts (or **explicata**). Moreover, the formalization of the notion of information is very much related to the "context", that is, to the utilization of this quantity whether in the theoretical or the practical sense.

For these reasons, other approaches have been proposed in the attempt to formalize and quantify the concept of information. These approaches, named semantic, pragmatic, descriptive, logic etc., are conceptually very different in spite of some analogies, even though often only formal, between the considered quantities (*). Often some concepts of information, though very deep and interesting, lack a solid mathematical frame in which one can evaluate the actual implications of these concepts or find very meaningful theorems.

Only recently, mainly through Kolmogorov's approach (Kolmogorov, 1965), has the concept of information been divorced from the context of ergodic processes and has been inserted into a completely different one, very well formalized and developed from a mathematical point of view, that is, the theory of recursive functions, or theory of algorithms.

In the Kolmogorov approach, "information" is defined in terms of algorithmic "complexity". Moreover, the concept of "randomness" appears to be related to the concept of complexity of algorithms. The intention of Kolmogorov then, is to give a "logical" basis to information theory as well as to probability theory.

The trade-off between "information" "and "algorithmic complexity" requires that one specifies exactly what "algorithm" and "complexity" of algorithms mean. For these reasons the first part of these lectures is dedicated to analyzing these concepts both on an intuitive and on a formal basis.

Historically in the same year, (1936), different researchers (Church 1936; Turing, 1936; Post, 1936) gave, independently, different formal notions of algorithm that have been proved equivalent in the sense that they yield the same class of "effectively" computable functions. Only in the last decade has a great deal of research been done first in formalizing and quantifying the notion of algorithmic-complexity and, afterwards, in developing a solid mathematical theory of complexity in the framework of algorithmic theory (Blum, 1966). Different measures of complexity of algorithms have been introduced. They can be divided into two groups. Those of the first group, which are called "static", measure, essentially, the complexity of the "description" of an algorithm, whereas those of the second group, which are called "dynamic", measure the "resources" (time, space, memory, etc.) used during the computation. Kolmogorov considers a static

(*) In the scientific literature on information theory, there is a certain lack of critical accounts of the different approaches to the notion of information, covering also some which are less known, such as the "algorithmic" or the "semantic". The reader can consult the book of Brillouin (1956) and the articles of Carnap and Bar-Hillel (1952), Kolmogorov (1965), Shannon and Weaver (1949), McKay (1950) and De Luca and Termini (1972b).

measure. However, the dynamic aspects of computation-complexity, how it is intuitive, which will appear clear in the following, can play an important role in the attempt to establish the above trade-off between the concepts of information, complexity and randomness.

ALGORITHMS AND COMPUTABILITY

1.1. Introduction

In this lecture we will give an outline of the "theory of computability" known also as the theory of "recursive functions". The main problem is to give a satisfactory formalization of the intuitive notion of "algorithm" (or "effective procedure") and a precise specification of the class of computable functions. As we said in the general introduction, different mathematical definitions have been proposed for the intuitive notion of algorithm. All of them give rise to the same class of computable functions. One of the most interesting results of the theory is the proof of the existence of **noncomputable** functions, or, which is the same, of **recursively unsolvable** decision-problems. These results have great theoretical and practical importance.

In the next section we shall first analyze the intuitive aspects of the informal notions of "algorithm" and "effectively computable" function, and then identify them with mathematically precise concepts as, for instance, Turing machines and partial recursive functions. This identification is the basic assumption of the theory, called Church's thesis. It is supported by the fact that the theorems about algorithms in any formalization of this concept can be proved in all the others that permit an "acceptable" Gödel numbering.

1.2. The informal concept of algorithm

A constant trend in the history of Mathematics, from the beginning, was that of giving "effective procedures" or "algorithms" for computing functions, deciding classes of problems, or enumerating sets of integers. Typical examples are the algorithms of the sum or of the product of integers, the Euclide algorithm for computing the greatest common divisor of two integers, the Eratostenes'"sieve" of enumerating prime numbers.

Intuitively "algorithm" means a **set of instructions** or **rules of symbolic manipulation** by which some strings of symbols (**input**) can be transformed into other strings (**output**). Characteristic features of the previous notion are the **finitude** of the set of instructions and the **deterministic** aspect of these in the sense that each step of the computation is uniquely determined by the previous ones. The computation is carried out by some "agent" that mechanically follows the rules of the algorithm in a discrete stepwise fashion, without employing "analogical" or "random" methods. A computation is over only when, after a finite number of steps any other transformation is impossible. We stress that the execution of the instructions of an algorithm has no "creative" aspects, so that it can be, in principle, carried out by a suitable (abstract) machine. Since computation theory concerns computations which are possible in principle no other constraints but to be **finite** are imposed on the size of the **input** of the machine and on the **length** of computation.

In the following we consider a universe of objects containing only a denumerable infinity of elements that are coded by integers. Let N be the set of nonnegative integers and f a function from N into N. As usual we denote the "domain" and the "range" of f by Dom f and Range f. When Dom f = N, f is said to be **total**, otherwise **partial**.

One can define a total function "computable if there exists an algorithm by which to compute it for all its arguments. This is the case for all the usual arithmetic functions. There are cases of well-defined functions for which a computing algorithm is not known (*). However, to state that a function is not computable one has to prove that there does not exist an algorithm to compute it. This requires the use of a formal notion of algorithm, that has to be adequate to the foregoing intuitive picture. However, even on an intuitive basis, one realizes that noncomputable functions have to exist since there are **continuously many** functions from integers into integers (Cantor's theorem), but there are only **denumerable many** finite sets of instructions, written in any alphabet.

Let us now, remaining on the intuitive level, establish some relationships between basic concepts of computability theory, that will become rigorous only when the notion of algorithm is formalized.

A set of integers is called "decidable" (or **recursive**) if there is an algorithm by which to decide whether any integer x does or does not belong

(*) This is the case, for instance, of the function f defined as f(n) = 1 if the Fermat equation $x^n + y^n = z^n$ has integral solutions, = 0 otherwise. Although one knows that f(1) = f(2) = 1 and that f(3) = f(4) = 0 and many other values of f, an algorithm to compute f for all values of n is not known. Fermat's conjecture asserts that f(n) = 0, for all $n \geq 3$.

to the set. Therefore defining for a set A the **characteristic function** χ_A as

$$\chi_A(x) = \begin{cases} 1, & x \in A \\ 0, & x \notin A, \end{cases}$$

one has

Proposition 1.1. A set of integers is decidable if and only if its characteristic function is computable.

A predicate P is called "computable" if there is an algorithm by which to determine for each fixed x whether P(x) is true or false. This is called the **decision problem** for the predicate P.

Proposition 1.2. A predicate is computable if and only if the set of integers $\{ x \mid P(x) \}$ for which P is true, is decidable.

The decision problem associated with P is called **recursively solvable** if and only if P is computable, otherwise it is **recursively unsolvable.**

A set of integers is called "recursively enumerable" (r.e.) if there is an algorithm by which to enumerate all its elements. One has, as it is easy to see:

Proposition 1.3. A (nonempty) set is recursively enumerable if and only if it coincides with the range of a computable function.

The fact that a set is recursively enumerable does not imply that it is decidable except in particular cases. The following proposition, whose rigorous version is called Post's theorem, gives the relation between these concepts.

Proposition 1.4. A set A is decidable if and only if A and its complement $\overline{A}$ are recursively enumerable.

It is clear that if A is a decidable set then A and $\overline{A}$ are recursively enumerable. Viceversa if A and $\overline{A}$ are recursively enumerable there are two algorithms say T_1 and T_2, by which the elements of A and $\overline{A}$, respectively, can be generated. Therefore considering a "composite machine" T formed from T_1 and T_2

that yields and sets in two lists the elements of A and $\overline{A}$ one can decide, after a finite number of steps, for any x whether x belongs to A or to $\overline{A}$. It is enough to see in which of the two lists x will appear.

As we previously said an algorithm can be regarded as a finite set E of instructions written in a certain alphabet A. A total function is computable if there is a corresponding set of instructions by which to compute it for all its arguments. However, the instructions able to compute total functions cannot be recursively enumerated. Otherwise, one could give an effective numbering

$$f_0 , f_1 ,\dots f_n ,\dots$$

of (total) computable functions denoting by f_i the function computed by the set of instructions E_i.

Following the classical Cantor's diagonal argument it would follow that the function

$$f_n (n) + 1$$

is also computable. Thus there would be an index i_o such that, for all n,

$$f_{i_o} (n) = f_n (n) + 1 \,.$$

But this gives rise to a contradiction for $n = i_o$. Therefore:

Proposition 1.5. The set of algorithms computing total functions is not recursively enumerable.

It is intuitive that the class of all algorithms can be recursively enumerated. However, an important consequence of propositions 1.5 and 1.4 is that the problem of deciding whether an arbitrary algorithm computes (or does not compute) a total function is recursively unsolvable. Proposition 1.5 is a serious barrier to developing a theory of only the total computable functions. For this reason one should consider the more general concept of **partially-computable** (p.c) functions. More precisely:

Definition 1.1. A partial function f is said to be partially computable if and only if there is an algorithm computing its value for all arguments belonging to Dom f. Outside of Dom f the computation is not defined, in the sense that it will never terminate.

1.3. The formal concepts of algorithm

As we said in the introduction, several mathematically precise concepts have been introduced to formalize the intuitive notion of algorithm. The first formal characterization of the algorithms is that of A.M. Turing (1936), who, starting from an analysis of human computational behaviour, gave a mathematical description of a class of algorithms called Turing-machines. They are abstract prototypes of digital computers in which the typical features of a computing agent that we discussed in the previous section, are drastically schematized.

A Turing machine can be "physically" visualized as a system composed of a **finite-state** device (a mechanical system capable of only a finite number of internal configurations) and a **tape** divided in squares or cells of equal size which are scanned one at a time by the device. The tape, on which are written symbols belonging to a certain alphabet, is supposed to be **potentially infinite** in the sense that more and more squares can be added to it if the computation so requires. The behavior of the machine **depends only on the internal state of the device and on the symbol** scanned on the tape according to one of the following rules (or **quadruples**)

$$\begin{array}{lllll} \text{i.} & q_i & i_k & i_h & q_s \\ \text{ii.} & q_i & i_k & R & q_s \\ \text{iii.} & q_i & i_k & L & q_s \,. \end{array}$$

The quadruple i. means that if the device is in the state q_i and scans the symbol i_k, this is replaced on the tape by i_h and the device enters in the state q_s. The quadruple ii. [iii.] means that the device goes to scan the symbol at the right [left] entering in the internal state q_s. These are the only "atomic" acts of a Turing machine which is defined as a **finite (nonempty) set of quadruples that does not contain two quadruples with the same first two symbols.** This last requirement is a consistency condition to avoid two conflicting instructions.

A Turing-machine with an **initial** internal configuration of the device and of the tape will carry out, using its quadruples, a symbolic computation that can terminate or continue forever. Representing the non-negative integers on the tape of the Turing-machine in a suitable conventional manner, it is possible to associate with it the computation of a partial function of one or more variables. The reader can consult the book of M. Davis (1958) for a rigorous mathematical presentation of Turing machines theory.

For any Turing machine, Z, we denote by $\psi_Z(x_1,...,x_n)$ the partial function of n variables associated with it. All definitions given in the previous section can be restated in terms of Turing machines. In particular a function f of one variable is said to be (Turing) partially-computable if and only if there exists a Turing machine Z such that

$$f(x) = \psi_Z(x), \quad \text{for all } x \in \text{Dom } f,$$

and when $x \notin$ Dom f the machine Z will never stop. We stress moreover that all propositions of the previous sections can be proved to be true for Turing machines in a rigorous way (Davis, 1958).

It is possible to prove (Kleene theorem; Davis 1958) that the class of partial Turing-computable functions coincides with the class of **partial recursive functions.** This class can be defined as the smallest class $\mathscr{P}$ of partial functions containing some **basis functions** and closed with respect to some **definition schemata** of new functions (**recursive operators**). Let us first define the important subclass R of $\mathscr{P}$ of total function called **primitive-recursive** functions (Péter, 1951). At one time the primitive recursive functions were considered to be the only computable functions. The class R is defined as the **smallest** class of functions such that:

i. R contains the **successor** function S, the **projection** functions U_i^n and the **zero function** N defined by

$$S(x) = x + 1, \quad U_i^n(x_1...x_n) = x_i \qquad (i = 1...n), \quad N(x) = 0 \, .$$

ii. R is closed under **composition**; that is if $f \in R$ is a function of k variables and $g_1,...,g_k \in R$ are functions of n variables, then the function defined as $f(g_1(\vec{x}),...,g_k(\vec{x}))$, where $\vec{x} \equiv (x_1...x_n)$, belongs to R.

iii. For any pair $g, h \in R$ of functions of n variables and n+ 2 variables respectively, the function f of n+ 1 variables defined as:

$$f(0,\vec{x}) = g(\vec{x})$$
$$f(y+1, \vec{x}) = h(y, \vec{x}, f(y,\vec{x}))$$

belongs to R.

The operation by which f is formed from g and h is called **primitive recursion.** The name of the class R is derived from the name of this operation. That the class of primitive recursive functions does not include all the computable functions has been proved by Ackermann (see, Péter, 1951).

The class $\mathscr{P}$ of all partial recursive functions is the smallest class of functions that in addition to requirements i. ii. iii. is also closed with respect to the application of the so-called μ-operator, associating with any total function $f \in \mathscr{P}$ of n + 1 variables the function of n variables g defined as

$$\text{iv} \quad g(\vec{x}) = \begin{cases} \min \{y \mid f(\vec{x},y) = 0\}, & \text{if there is such a } y \\ \text{undefined otherwise.} \end{cases}$$

$g(\vec{x})$ is also denoted as $\mu\, y[f(\vec{x},y) = 0]$.

Other formalizations of the intuitive notion of algorithm are the calculus of equations of Herbrand-Gödel-Kleene (1936), the **λ-calculus** of Church (1936), **Post's systems** (1936), and **Markov's normal algorithms** (1951). These formalizations, though widely different, have some common features since they are based on the concepts of **proof** (or **derivation**) and **theorem** in some **formal system**. Let us now give some general ideas about formal systems. Let A denote an **alphabet** and A* the set of all **finite strings** of symbols of A or **words** (mathematically A*, when it includes the **empty string**, is called **the free-monoid** generated by A, with respect to the operation of **concatenation** of strings). Intuitively a formal system L is determined by a (finite) set $\mathscr{A}$ of words called **axioms** and a **finite** set $\mathscr{I}$ of **inference's rules** by means of which in a mechanical way new words can be generated. One can then define a **proof** (or derivation) as a **finite** sequence of words

$$X_1, X_2, \ldots X_n,$$

where for any $i (i = 1 \ldots n)$ X_i either is an axiom or can be derived from the previous words $X_j (j < i)$ by using one of the inference rules. A word Z is called a **theorem** if there is a proof whose last word is Z. In this case one writes $\vdash_L Z$. If one considers an alphabet that contains the equal sign = and denotes by $\bar{x}$ a **code-word** representing the integer x, one can say that the word $F \in A^*$ represents the partial function f in L and only if

$$\vdash_L F\bar{x} = \bar{y} \leftrightarrow f(x) = y.$$

The pair (L,F) is said to (partially)-compute the function f. We observe that the concept of "proof" is different from the one of "computation" of algorithm, mainly by the fact that more than one word can be direct consequences of same words. Therefore a proof seems to be like a **non-deterministic computation.** This difficulty can be avoided since it is possible to show that in all considered formal systems the set of proofs is recursive and the set of theorems recursively enumerable so that there is always a mechanical procedure to yield all theorems. As stressed by Smullyan (1961): "The notions of **formal system** and **mechanical operation** are intimately connected; either can be defined in terms of the other. If we should first define a mechanical operation directly (e.g. in terms of Turing machines), we would then define a "formal" system as one whose set of theorems could be generated by such a machine... Alternatively (following the lines of Post), we can first define a formal system directly and define an operation to be "mechanical" or "recursive" if it is computable in some formal systems".

All the formal characterizations of the notion of algorithm and of effectively computable functions give rise to the same class of functions, that is the one of **partial recursive functions.** This fact is the main support for **Church's thesis** according to which the informal notion of effectively computable function can be identified with the one of partial recursive function and the notion of algorithm with any one of the previous mathematical characterizations (for instance, Turing machines).

Of course this thesis cannot be proved but only disproved by showing the existence of a function that can be effectively calculated and for which there is no algorithm to compute it in one of the above characterizations. Church's thesis is the basic assumption of computability theory, as Carnot's principle is for thermodynamics. Once Church's thesis is admitted to be true it can be used in a technical sense, for if there is evidence of the existence of a procedure to compute a function, then there must be an algorithm of any given class to do so. These kinds of proofs are called **via Church's thesis,** since they are conditioned on it. However, one should be able to prove the result directly, since if the result were false this would disprove the Church thesis. In any one of the previous mathematical characterizations it is always possible to establish a **recursively bijection** between the class of algorithms (or set of instructions) and the set of non-negative integers N (for Turing machines see Davis, 1958). In this way one can effectively associate with an algorithm an integer and, conversely, for each integer one can find the corresponding algorithm. Such a numbering of algorithms provides a numbering for partial recursive functions. In the following we denote by ϕ_i the partial recursive function of one variable associated with the i-th algorithm , and by $\mathcal{P}_n$ $[\mathcal{R}_n]$ the class of partial [total] recursive functions of n variables.

Basic theorems that can be proved in a rigorous way for each standard characterization, or in an informal manner using Church's thesis, are the following:

Theorem 1.1. (Universal algorithm). There exists an algorithm of the class capable of computing, by a suitable input, any partial recursive function.

$$\vee i \wedge x, y\, [\phi_i (\langle x, y \rangle) = \phi_x (y)]$$

where $\langle x,y \rangle$ is a recursive bijection $N \times N \rightarrow N$.

Theorem 1.2. (Iteration Theorem). There exists a recursive function of two variables by means of which one can find (in an effective manner) given any fixed pair (i,x), an algorithm of the class for computing the function of one variable $\phi_i(\langle x, y \rangle)$.

$$\vee\ s \in \mathcal{R}_2\ \wedge\ i,x,y\ [\phi_{s(i,x)}\ (y) = \phi_i\ (\langle x,y \rangle)]\ .$$

The above theorems can also be generalized to the case in which x and y are vectors.

Let us now consider an arbitrary class of machines different from the standard ones, for instance, Turing machines with 10 tapes, more heads etc. The problem that arises is when this class of machines, supposed to be capable of computing all partial recursive functions, is "equivalent" to any of the standard ones in the sense that there is an effective procedure by which to find for any machine of the class standard-instructions and viceversa.

A deep analysis of this concept has been done by Rogers (1958) who gave a rigorous definition of "acceptable class of machines" or, which is the same, of "acceptable Gödel numbering" of partial recursive functions.

A numbering is **acceptable** if and only if it satisfies the universal machine theorem and the iteration theorem. Moreover Rogers was able to prove that any two acceptable Gödel numberings are **recursively isomorphic.** In this way it is possible to develop a theory of computation whose theorems are **independent** of the acceptable class of machines.

We conclude this section by recalling two basic results of computability theory. The first result is the recursive unsolvability of the **halting-problem** for an acceptable class of machines. In other words there is no machine in the class by means of which to decide whether an arbitrary machine i of the class started with a certain input x will stop or not. The second result is the so-called recursion Kleene's theorem

Theorem 1.3. For any f $\epsilon \mathscr{R}_1$ there is an integer i such that $\phi_{f(i)} = \phi_i$.

A noteworthy consequence of theorem 1.3 is the following theorem due to Rice (1953)

Theorem 1.4. Let $\mathscr{B}$ be a subclass of $\mathscr{P}$. Then the set $\{i \mid \phi_i \epsilon \mathscr{B}\}$ is decidable if and only if $\mathscr{B} = \mathscr{P}$ or $\mathscr{B} = \emptyset$

The meaning of the theorem is that the only decidable predicates on Gödel numbers (sets of instructions) of p.r. functions, are "trivial" since they are verified by all p.r. functions or none.

COMPLEXITY OF ALGORITHMS

2.1. Introduction

We have seen in the first lecture that, in about the same year (1936) different mathematically precise concepts have been independently proposed, and later identified (Church's thesis), as equivalent explicata for the informal notion of algorithm. Moreover the same notion of "equivalence" of classes of machines became clearer by means of the formal concept of "acceptable" Gödel numbering of partial recursive functions. These were identified with the "effectively" (partially) computable functions. However the recursive unsolvability of the "halting-problem" for machines of any acceptable class, implies that it is impossible to bound a-priori the "resource" used by them during a computation. From this one realizes that any of the above classes includes algorithms capable of carrying out computations whose "complexity" exceeds any realistic limitation. Many researchers, therefore, considered the formalized notion of algorithm too wide and general to be used as a mathematical model for an actual computer. A great deal of research was concerned with finite-automata [see, for instance, Rabin and Scott, 1959] that is, machines capable of only a fixed, finite, number of states. However the class of functions "computable by finite automata" is too restricted. For the foregoing reasons a problem of great theoretical and practical relevance was that of giving some mathematical specification to the intuitive notions of "complexity of a partially-computable function", "complexity of an algorithm" and to the related ones of their measurement. These problems have been faced from several points of view so that different approaches to them exist(*). Mainly during the last decade, however, following a large amount of theoretical research on the subject, "computational complexity" became a mathematical theory that can be considered, at present, a very important part of the theory of computation. An essential contribution in this direction has been the "axiomatic approach" of M. Blum (1967a, 1967b). This author gave for any acceptable class of machines some basic requirements (or axioms) that have to be satisfied by complexity measures. These axioms, even though quite weak since they are very general, are however, sufficient to allow the proof of very important and surprising results on computational complexity(**).

2.2. The intuitive notion of algorithmic-complexity

Any attempt at a mathematical specification of the concept of "difficulty" of computing a function requires that one first clarifies what "complexity" of an algorithm is and how to measure it. Intuitively there are, at least, two **explicanda** for the algorithmic-complexity. The first concept is related to the "complexity of description" of a given algorithm (or computing program). This complexity can be identified, at a first level, with the "size" of the description itself in a given formalism (**size-complexity**). For a Turing machine a "size-measure" can be the number of quadruples defining it. In the case of some **calculi** (such as Kleene's calculus of equations) a natural measure of size is the number of symbols of specific axioms required to compute a function. However the description-complexity of an algorithm (or computing program) can be often related more to the "structure" of the description (**structure-complexity**) than to the size. To quote Borodin (1973) "... size is not the only static measure which relates to our feelings about definitional complexity. Algorithms and other "definitional constructs" also seem to exhibit a **structural** complexity. For example, we classify grammars according to the types of productions that appear and we consider programs without loops to be simpler than those with loops. In contrast to the size measure, it is not clear which structures are to be considered simple and which more complex".

(*) General reviews of computational complexity appraches, stressing their differences and relationships are in Ausiello (1970) and in Borodin (1973).

(**) An outline of the axiomatic theory of computational complexity and its recent developments may be found in Hartmanis and Hopcroft, (1971).

The second concept of complexity of an algorithm is related to the "complexity of computation" carried out by an algorithm. A measure of the complexity of a computation must depend both on the machine and on the input to it. This quantity measures the amount of "resource" used during the computation. For Turing machines computational complexity measures can be given, for instance, by the **number of steps** (time) or the **number of scanned squares** (space) required for the computation.

For obvious reasons the first complexity measures have been called "static" and the second "dynamic". We stress that the two types of algorithmic complexity can be mutually related, since it occurs fequently that one can reduce the complexity of a computation, although at the cost of increasing the size or the structure-complexity of the machine.

In practice it is often convenient, for particular functions, to measure the complexity of an algorithm by the **number** of some basic **operations** that are needed in order to carry out actual computations. For instance, in the cases of the **polynomial evaluation** or **product of matrices** one can use **multiplications** and **additions** as basic operations. For "sorting" algorithms (which make binary comparisons to determine the total order in a set of objects) an obvious measure of complexity is the number of comparisons. Many results of great practical interest have been obtained in this way. As an example it is possible to show (Borodin, 1971) that for the evaluation of a polynomial of n-th degree there is a (uniquely) optimal method of computation requiring n multiplications and n additions. However this kind of approach to complexity called also "analysis of the algorithms" cannot be considered a comprehensive and general theory of complexity since the results obtained are too much related to the particular functions and to the particular way of measuring complexity (*).

We are, on the contrary, interested in a general approach to complexity theory in which the concepts of size and computation measures have to be mathematically specified in the framework of the theory of computation trying to capture in the formal concepts the most general and relevant aspects of the intuitive notions. This formalization will be given in the next section.

Let us now, remaining at an intuitive level, analyse the concept of "difficulty" of computing a function. We recall that in any acceptable class of

(*) The reader interested in this approach may consult the book of Knuth (1969): "The art of computer programming".

machines $\{M_i\}$, for any partially computable function f, there is a denumerable infinity of indices, or **programs**, by which to compute f (Rogers, 1967).

One could think, at first sight, of measuring the difficulty of computing f by the "minimal" complexity of a program for f. While there exist always, in principle, minimal-size programs for any f, on the contrary, there are functions (which will be shown in the following sections) that have not "optimal" computation programs, in the sense that for each program for f, there is another very much faster (speed-up theorem). This fact shows that the concept of difficulty of computing a function, intended in the "dynamic" sense, cannot be made independent of the algorithms carrying out the computations. It is, therefore, more meaningful to consider a relative concept of computation-difficulty of functions that can be introduced, within a given class of machines, by a **degree of difficulty ordering** induced by the computation-complexity measures. One can define, for instance, that f is no **more difficult** to compute than g ($f \leqslant g$) if for each index i for g there is an index j for f such that Dom f $\supseteq$ Dom g and the "complexity of computation" of the program j for f is not greater than those of the program i for g, for almost all $x \in$ Dom g.

We note that such an ordering, or a more general version of it (Arbib and Blum, 1965) to be precisely defined requires a formal concept for computation-complexity. However the difficulty ordering, having fixed a computation-complexity measure such as time, generally, depends on the class of machines. For this reason and for the presence of speed-ups it is not possible to develop a theory of complexity of partially computable functions which is **independent** of the class of machines.

A great deal of research by many authors (Hartmanis and Stearns, 1965, 1969; Hartmanis, 1968; Hartmanis, Lewis II and Stearns, 1965; Hopcroft and Ullman, 1969; Hennie, 1965; Hennie and Stearns, 1966) concerned computational complexity of particular classes of machines with respect to particular, though natural, measures. The machines are **many-tape on-line** and **off-line** Turing machines (Hartmanis, Lewis II and Stearns, 1965) and the complexity measures are the **time**, **space**, and the number of **head-reversals**. The above machines are classified according to the **maximum** time T(n) or to the **maximum** space L(n) that they can use on an input of length n, before halting. Many very interesting results have been obtained on time and space bounded computations. For instance one can prove that every **context-free** language is $T(n) = n^3$ and $L(n) = [\ln n]^2$ recognizable on off-line Turing machines (Hopcroft and Ullman, 1969). These results, though very important, are however still particular for a general theory of complexity since they are tied to the particular class of machines and measures of complexity. For these reasons the previous approach to complexity-theory has been called **machine-dependent** theory.

We have previously seen that a meaningful theory of complexity of partially computable functions cannot be developed outside some fixed class of machines. However in every class of machines the results that one obtains are too specialized. What remains possible for a **general theory** of complexity is an **axiomatic theory** whose theorems are **independent** of the class of machines, in the sense that they are true for all the classes. This was the program of the **machine-independent** theory of complexity originated by M. Blum (1967 a, 1967 b).

2.3. Size of algorithms

As we said in the previous section the "size" of a machine of a given class is related to the complexity of its description. The smaller the size of an algorithm computing a given function, the higher the "efficiency" of the representation of the function, in the sense of its definition, by means of that algorithm. Having in mind some particular size-measure as the number of instructions of a computer program, one realizes that the main intuitive aspects of the notion of **size-measure** are that, for each algorithm one can effectively compute its size and, moreover, for any value of the size one may effectively generate the algorithms, finite in number, for that size. The following definition, due to M. Blum (1967 b), translates these intuitive aspects in the form of mathematical requirements for a size measure of any class $M \equiv \{M_i\}$ of machines having an acceptable numbering $\{\phi_i\}$ of p.r. functions.

Definition 2.1. A map $s : N \to N$ is called a measure of the **size of machines**, s(i) being called the size of M_i, provided that

i. $s \in \mathscr{R}_1$,

ii. there exists a recursive function giving the number of machines of each size.

No other constraints are imposed on size-measures in Blum's theory. A first consequence of definition 2.1 is that the algorithms of the class M can be **effectively enumerated by increasing size** (for algorithms of the same size the order is not important).

We consider now the problem of comparing size-measures for two classes of machines (that can also be the same). Let $M \equiv \{M_i\}$ and $\hat{M} \equiv \{\hat{M}_i\}$ be two classes of machines having the acceptable Gödel numberings $\{\phi_i\}$ and $\{\hat{\phi}_i\}$ of p.r. functions. From a theorem of Rogers (1958), as we said in Sec. 1.3, there is a recursive bijection $\alpha: N \to N$ such that $\hat{\phi}_i = \phi_{\alpha(i)}$, $i \in N$. An important consequence of this theorem is that, without loss of generality, it is always possible to assume that the classes M and $\hat{M}$ are ordered in such a way that M_i and $\hat{M}_i$ compute, for all i, the same function. Let s and ŝ denote the sizes of machines of classes M and $\hat{M}$, respectively. One has that each of the two size-measures is recursively upper-bounded by the other.

Theorem 2.1. A recursive function g exists such that

$g(s(i)) \geqslant \hat{s}(i)$ and $g(\hat{s}(i)) \geqslant s(i)$, for all i.

The proof of the theorem is easily obtained defining g as $g(x) := \max_{i \in C_x} \{s(i), \hat{s}(i)\}$, where C_x is the finite set

$$C_x \equiv \{ i \mid \text{either } s(i) \leqslant x \text{ or } \hat{s}(i) \leqslant x \}.$$

The function g is obviously recursive.

If two machines M_i and M_j of the M class have different sizes s(i) and s(j) what will occur for sizes ŝ(i) and ŝ(j) of $\hat{M}_i$ and $\hat{M}_j$? An answer to this question is given by the following

Theorem 2.2. There exists a recursive function h such that, for all i and j,

$$s(i) \leqslant s(j) \to \hat{s}(i) \leqslant h(\hat{s}(j))$$

$$h(s(i)) \leqslant s(j) \to \hat{s}(i) \leqslant \hat{s}(j).$$

This theorem is a corollary of the previous one if one assumes as function g in the theorem 2.1 the increasing function $g^*(x) = 1 + \max\{g(x), g^*(x-1)\}$ and $h = g^* \circ g^*$ (Blum, 1967b).

Let us now consider the problem of the "efficiency" of the representation of functions by the algorithms of a given class. The theorem that follows shows that in any effectively enumerable subsequence of algorithms extracted from an acceptable class M there are always "inefficient" representations.

Theorem 2.3. Let g be any recursive functions with infinite range enumerating an infinite subsequence of machines of a given class $M \equiv \{M_i\}$. Let f be any recursive function. Then there exist $i,j \in N$ such that

$$\text{i.} \quad \phi_i = \phi_{g(j)}$$

$$\text{ii.} \quad f(s(i)) < s(g(j)).$$

Proof: For any value of the size there are only finite many machines having that value. Moreover since g has an infinite range the function h defined as $h(y) := \mu j\,[f(s(y)) < s(g(j))]$ is recursive. Therefore by using Kleene's recursion theorem one has that an integer i exists such that

$$\phi_{goh(i)} = \phi_i \,.$$

It follows therefore that $\phi_i = \phi_{g(j)}$ having set $j = h(i)$, and $f(s(i)) < s(g(j))$.

A first consequence of the theorem is that the set of minimal size algorithms for p.r. functions is not recursively enumerable. If one considers only primitive recursive functions, it is possible to prove (Péter, 1951) that they are recursively enumerable. Moreover it is easy to see that one can effectively enumerate their "smallest" defining programs (defining equations). One can assume as size measure the number of letters in the defining equations. From theorem 2.3 it follows that there is a recursive primitive function such that the smallest size in the recursive primitive derivation is much larger than same general recursive program. Therefore the complexity of representations of primitive recursive functions can be considerably reduced by using general recursive programs.

The results of theorem 2.3 show also that the formalism of "sub-primitive recursive hierarchies", with which a great deal of research is concerned (Grzegorczyk, 1953; Axt, 1963; Ritchie, 1965) can be antieconomical to represent functions. These hierarchies are examples of classifications of functions obtained by using a suitable measure of the "structural complexity" of the definition-schemata (or programs) for computing functions.

We recall here that the first sub-primitive hierarchy is the one of A. Grzegorczyk, based on Ackermann's function. For any $n \geqslant 0$ let us define the following sequence $\{f_i\}$ of primitive recursive functions:

$$f_0(x,y) = x+1, \quad f_1(x,y) = x+y, \quad f_2(x,y) = x.y$$

$$f_n(x,0) = 1$$

$$f_n(x,y+1) = f_{n-1}(x, f_n(x,y)), \quad \text{for} \quad n \geqslant 3.$$

The function of three variables $\psi(n,x,y) = f_n(x,y+1) = \psi(n-1,x, \psi(n,x,y))$ is, except for the presence of variable x, just the Ackermann function.

Definition 2.2. The class G^n of the Grzegorczyk hierarchy is the smallest class containing the basis functions N,S,U_i^n and f_n, closed with respect to composition and "limited recursion".

One says that f is obtained by limited recursion if and only if it can be obtained by a recursion schema but with the additional requirement that the value of f is always upper limited by the value of some function of the class.

It is easy to prove that $G^n \subset G^{n+1}$ and that the union of the classes is just the class of primitive recursive functions. Moreover G^3 coincides with the important class of Kalmar's **elementary functions** (Péter, 1951). Other results are that $f_{n+1}(x,x)$ increases faster than any other unary function of G^n, and that G^{n+1} $(n>2)$ contains a function enumerating the functions of G^n. It is also possible to prove that all sub-primitive recursive hierarchies coincide for a sufficiently large n.

With regard to relationships of the subrecursive formalism and complexity one can prove that if a primitive recursive function is obtained from basis functions applying at most n of the operations, then $f \in G^{n+3}$. However from Blum's theorem 2.3 it follows that functions of][these subclasses of primitive recursive functions may have inefficient representations within the class itself. Concerning the complexity of computation it has been proved that for $m \geqslant 3$, $f \in G^m$ if and only if there is a Turing machine computing f such that the time (or the space) is upper bounded by a function of G^m (Cobham, 1965). In this way structural and dynamic complexity are strongly related.

2.4. Computation – complexity measures

In this section we shall analyze the formalization of the concept of measure of computation – complexity given by M. Blum (1967 a) along the lines suggested by M.O. Rabin (1960).

As we discussed in Sec. 2.2 the complexity of a computation is related to the "resource" needed by a machine during the computation. The main general aspects of a "resource – measure", used to describe the dynamic behaviour of a machine, are that this measure has to be a partially computable function of the input to the machine assuming a defined value if and only if the computation terminates. Moreover, although the "halting problem" for the machines of a given acceptable class is recursively unsolvable, one has to be able to decide for each machine of the class and for all inputs whether the amount of resource required for computation exceeds some fixed bound. To restate these intuitive requirements in mathematically precise terms let us premise a definition. We will consider a class of machines $M \equiv \{M_i\}$ having an acceptable Gödel numbering $\phi \equiv \{\phi_i\}$ of p.r. functions.

Definition 2.3. A sequence $\psi \equiv \{\psi_i\}$ of p.r. functions is called **measured** if and only if the ψ_i can be expressed, for all $i,n \in N$ as

$$\psi_i(n) = \mu\, y\, P(i,n,y)\,, \tag{2.1}$$

where P is a computable predicate of three variables; $\mu y P(i,n,y)$, for all pairs (i,n), is the minimum value of y, if there exists, for which P(i,n,y) is true. In the following we set $\psi_i(n) = \infty$ when $\psi_i(n)$ is not defined and $\psi_i(n) > m$ when $\infty \geqslant \psi_i(n) > m$.

From the iteration theorem, applied to Eq. (2.1), one has that a recursive function β exists such that $\psi_i = \phi_{\beta(i)}$ so that the indices (or programs) for the list β can be effectively generated. Moreover from a general theorem (Rogers, 1967) one derives that the map β is not onto N since there is a p.r. function that cannot be expressed in the form (2.1) whatever the computable predicate P is. For this reason the list $\{\phi_i\}$ is not a measured sequence.

The most important consequence of definition 2.3 is that for a measured sequence ψ the predicates $\psi_i(n) = m$, $\psi_i(n) < m$, $\psi_i(n) > m$ are decidable for all $i,n,m \in N$. In fact it is sufficient to compute at the most P(i,n,0), P(i,n,1), ..., P(i,n,m) to make the decision. Therefore the condition of being a measured sequence of functions seems to be an adequate formal concept for the second intuitive requirement of a "resource-measure". The first requirement is satisfied by the condition, that the domain of definition of a resource-measure has to equal that of the function which is computed.

Definition 2.4. A sequence $\Phi \equiv \{\Phi_i\}$ of p.r. functions is called a complexity-measure (or resource-measure) for ϕ (or for M) if and only if ϕ satisfies the following two axioms (*)

Axiom 1. $\text{Dom } \Phi_i = \text{Dom } \phi_i, \quad i \in N$

Axiom 2. Φ is measured.

For all $i \in N, \Phi_i$ is called the (computation) **complexity function** associated with ϕ_i.

These axioms are **independent**. For instance ϕ satisfies the first but not the second. On the contrary a sequence of identically zero functions satisfies the second axiom but not the first. Typical examples of complexity measures in the case, for instance, of Turing-machines are:

1. $\{\Phi_i(x)$ = number of steps required by the i-th Turing machine to compute $\phi_i(x)\}$.
2. $\{\Phi_i(x)$ = number of cells of the tape required by the i-th Turing machine to compute $\phi_i(x)$ if it stops, undefined otherwise$\}$.

Let us consider a Gödel numbering of p.r. functions and a complexity-measure Φ. Recalling that any other acceptable Gödel numbering $\hat{\phi}$ is recursively isomorphic to ϕ by means of a recursive permutation α of N, it is easy to verify that the list $\hat{\Phi}$, where $\hat{\Phi}_i := \Phi_{\alpha(i)}$ is a complexity measure for $\hat{\phi}$. This means that the same complexity measure can be considered, except for a recursive permutation of the indices, a complexity measure for any acceptable class of machines (**).

The following theorem states that two complexity measures are recursively related.

(*) A more general version of Blum's axioms has been introduced by Ausiello (1970) to include the case of measures that can be finite even when the computation diverges (as in the case of cycling instructions on a finite amount of tape).

(**) Some closure properties of the class of all complexity measures with respect to a given Gödel numbering have been considered in Adrianopoli and De Luca (1974).

Theorem 2.4. A function $g \in \mathscr{R}_2$ exists such that for all i and almost all n

$$g(n, \hat{\phi}_i(n)) \geqslant \phi_i(n) \quad \text{and} \quad g(n, \phi_i(n)) \geqslant \hat{\phi}_i(n).$$

Moreover it is easy to see that the recursive function g can always be taken nondecreasing in both the variables.

2.5. Basic theorems of computation complexity

In the previous section we have specified from a mathematical point of view, in the setting of recursive function theory, which are the **acceptable measures** of computation-complexity. The axioms of Blum certainly capture the main relevant features of the intuitive notion of resource-measure. Some criticism has been made, recently, by many authors on the **weakness** of the axioms. In fact the class of all acceptable measures seems to be too **wide** by including functions of a "pathological" behaviour which do not fit the "natural" models of measures such as **time** or **space** for Turing-machines. However the weakness of the axioms, as we shall see in what follows, does not prevent us from finding important results of computation-complexity which hold true for all measures.

A first proposition that can be proved by means of a very easy diagonalization method is that for any pair of recursive functions f and h there is always an index i (or program) for f (i.e. $\phi_i = f$) such that $\Phi_i(n)$ exceeds, for all n, the recursive bound h(n). This means that in the set of all programs for f in the numbering $\phi \equiv \{\phi_i\}$ there are always programs which are very "inefficient" from the computation point of view. This fact, however, says nothing about the difficulty of computing f within the class ϕ. A basic result in this latter direction is the following theorem due to Rabin (1960) showing that, whatever the complexity measure is, there are zero-one valued functions which are arbitrarily complex to compute.

Theorem 2.5. Let Φ be a complexity measure and h a recursive function. There is a zero-one function f such that for any index i for f, there results $\Phi_i(n) > h(n)$, for almost all n.

Outline of the proof: The basic idea is to define a zero-one valued function f as

$$f(n) = \begin{cases} 0 & \text{if } \phi_{s(n)}(n) = 1 \\ 1 & \text{otherwise,} \end{cases}$$

where s is a function, defined below, that "cancels" indices for computing f, since $f \neq \phi_{s(n)}$ for all n. For each n, s(n) equals the first index $j \leqslant n$ such that $\Phi_j(n) \leqslant h(n)$ and has not already been canceled before ($s(m) \neq j$, for $m \leqslant n$). If no such integer exists s(n) is undefined. It is clear that s is a partially computable function. When it is not defined one can, however, effectively know it. For this reason f is a recursive function such that any index i for f does not belong to the Range s. This implies $\Phi_i(n) > h(n)$, for almost all n.

Since the values of the function f in the theorem are zero or one only, it follows that it is not possible to recursively upper bound the complexity measures Φ_i by the sizes of the functions ϕ_i. Theorem 2.5 has been generalized by Blum (1967 a) to the case in which h is any partial recursive ϕ_j. The function f will, then have the same domain of ϕ_j and, moreover, there is an effective procedure to obtain, for each j, an index for f.

Let us now consider the problem of the "efficiency", in the computational sense, of the representations of functions in the given numbering $\{\phi_i\}$. Certainly there are functions having optimal algorithms to compute them (for instance, the **zero-function** N). However a basic theorem of Blum (1967 a) shows that there exist recursive functions which do not have "optimal programs" in the sense specified by the following

Theorem 2.6. (Speed-up). Let Φ be a complexity measure and r a recursive function of two variables. There exists a zero-one function with the property that for any index i for f there corresponds another index j for f such that

$$\phi_i(n) > r(n, \phi_j(n)), \quad \text{for almost all n.}$$

If, for instance, $r(n,m) = 2^m$, the existence of speed-up for f implies that there is an infinite sequence $j_1, j_2, j_3, \ldots$ of indices for f, such that

$$\phi_{j_2}(n) < \ln_2 \phi_{j_1}(n), \quad \phi_{j_3}(n) < \ln_2 \ln_2 \phi_{j_2}(n), \ldots..$$

for almost all n.

Some generalizations and stronger versions of theorem 2.4 have been proved by Blum, (1967 a) (**super speed-up theorem**) and by Mayer and Fisher, (1968) (**operator speed-up theorem**). The super speed-up theorem states that for any recursive g there is a zero-one function f such that for any index i for f $\Phi_i(n) > g(n)$, for almost all n. Moreover to every index i for f there corresponds an index j for f such that $\Phi_i(n) > \Phi_j(\Phi_j(n))$, for almost all n. A problem that arises is whether or not there are effective procedures for speeding up algorithms on all but a finite number of inputs. The answer to this problem is negative as has been proven by Blum (1971). We note that relationships between the speed-up theorem and similar results found in the framework of formal systems have been pointed out by Arbib (1966). An important consequence of the speed-up theorem is that one cannot classify functions in terms of minimal-complexity, since there are functions without best programs.

The problem that we consider now is whether there exist functions for which one could find, for some programs, an upper bound to the complexity. This question admits a positive solution since it is possible to prove the existence of functions for which there are upper and lower bounds to complexity. More precisely we state the following theorem, called compression theorem, due to Blum (1967 a), that generalizes similar machine-dependent results of Hartmanis and Stearns (1965).

Theorem 2.7 **(Compression theorem).** Let $\psi \equiv \{\psi_i\}$ be a measured sequence of functions. Then a recursive function h of two variables exists such that the following is true: To each partial recursive function ψ_i there corresponds a 0-1 valued partial recursive function f with Dom f = Dom ψ_i such that:

i. for any index i for f $\Phi_i(n) > \psi_i(n)$, for almost all n.

ii. there is a recursive function τ by means of which one can find, for each i, an index $\tau(i)$ for f such that

$$\Phi_{\tau(i)}(n) < h(n, \psi_i(n)), \text{ for almost all n.}$$

Viceversa if ψ is a sequence of p.r. functions in correspondence with which there exists a recursive function h such that i. and ii. are true, then ψ is measured.

A consequence of the theorem is that there is a function $\phi_{\tau(i)}$ such that $\psi_i(n) < \Phi_{\tau(i)}(n) < h(n, \psi_i(n))$, for almost all n. Therefore, as we shall see better later, if one considers for ψ_i recursive, the class $C_{\psi i}$ of the recursive functions that can be computed with an amount of resource less than or equal to $\psi_i(n)$ almost everywhere, this will be strictly contained in $C_{h * \psi_i}$ where $(h * \psi_i)(n) := h(n, \psi_i(n))$. Another implication of the compression theorem is that any function $f = \phi_{\tau(i)}$ cannot be r-speeded-up by an increasing function r that is almost everywhere greater than or equal to h.

Let us now examine, briefly, the relationships between "size" and "computation-complexity" of the machines. It occurs frequently that a "smaller" machine for computing a function requires more "computation-resource" and, on the contrary, "faster" programs of computation are of "greater" size. This latter problem is very important, since there exist functions with large speed-ups. With regard to the first problem, by theorem 2.3 we have seen that in any recursively enumerable subsequence of machines of a given (acceptable) class there exist "inefficient" size-representations of functions; that is, integers $i,j \in N$ such that $\phi_i = \phi_{g(j)}$ and $f(s(i)) < s(g(j))$.

It is possible to prove (Blum, 1967 b) that a recursive function h exists, such that, under the same hypotheses of the theorem, $\Phi_i(n) < h(n, \Phi_{g(j)}(n))$ for almost all n. This means that the computation complexity of a smaller size program i can be, almost everywhere, recursively upper bounded by the complexity of the larger size program g(j). In other words, in these cases, reducing the size the computation-complexity does not increase too much. There also exist, moreover, cases of very inefficient representations both for size and computation complexity. In order for speed-ups as stressed by Schnorr (1972), to be interesting for "programming", two conditions have to be satisfied i. the size of the faster program has to be recursively upper-bounded by the size of the lower one (that means the former has not to increase too much) ii. the number of arguments for which the speed-up does not hold has to be recursively upper bounded by the size of the slower program.

Now there are cases for which condition i. is satisfied and others for which i. is not satisfied. Schnorr proved that there do not exist functions with sufficiently large speed-ups for which both i. and ii. are satisfied.

2.6. Complexity classes

A very important problem in computational complexity theory is concerned with "complexity classes", that is classes of functions which can be computed within some recursive bound to computation-resource. As we said in Sec. 2.2, a great deal of research has been done on this subject by Hartmanis, Stearns, et al. in a **machine-dependent** context: that is, for particular classes of machines and complexity measures.

We shall consider now complexity classes and some basic theorems on them in the general framework of Blum's axiomatic theory.

Definition 2.5. Let Φ be a complexity measure. We define for any recursive function t, the complexity-class C_t^{Φ} as

$$C_t^{\Phi} \equiv \{ f \in \mathscr{R}_1 \mid \bigvee_i (\phi_i = f) \wedge \overset{\infty}{\bigwedge_n} \Phi_i(n) \leq t(n) \}.$$

That is, the class C_t^Φ contains all recursive functions for which there exists one computing program, at least, whose complexity is upper-bounded by t, for almost all n. The symbol $\overset{\infty}{\Lambda}_n$ means, as usual, "for all but a finite number " of values of n. In the following, when there are no misunderstandings, we will denote C_t^Φ simply by C_t.

Let us now interpret some of the previous theorems of computational complexity in terms of complexity-classes. From theorem 2.4 one has $C_t^\Phi \subseteq C_{t'}^{\hat{\Phi}}$ and $C_t^{\hat{\Phi}} \subseteq C_{t'}^\Phi$, where $t'(n) := g(n,t(n))$.

Rabin's theorem shows that there is an infinite number of "levels" of complexity, since for any $t \in \mathscr{R}_1$ there exists a function $f \in \mathscr{R}_1$ such that $f \notin C_t$. As a consequence of the existence of speed-ups there are functions which cannot be characterized by a definite level of complexity. For the compression-theorem one has that if $\psi \equiv \{\psi_i\}$ is a measured sequence of functions, then for every recursive function f there exists an $h \in \mathscr{R}_2$ such that $C_f \subset C_{h*f}$, with $h*f(n) := h(n,f(n))$.

This result is very important for the general problem of computational complexity theory which consists in determining how much one has to increase the computation resource in order to obtain an increase in the "computing power". From the compression theorem one knows that there are complexity levels f for which increasing the resource from f to h*f the computing power increases since at least one new function will belong to the class C_{h*f}. On the contrary the following theorem due to Trakhtenbrot (1967) and Borodin (1969) shows that there exist arbitrarily large "gaps" in the complexity range where the computing power does not increase.

Theorem 2.8 (Gap theorem). In any complexity measure, for any nondecreasing recursive function g such that for all n $g(n) \geqslant n$, there exists an increasing recursive function t such that $C_t = C_{got}$.

It is also possible to prove that $I_t = I_{got}$, where $I_t \equiv \{i \mid \phi_i \in \mathscr{R}_1$ and $\overset{\infty}{\Lambda}_n \phi_i(n) \leqslant t(n)\}$ is the set of all programs for recursive functions whose complexity is upper-bounded by t, for almost all n. It is clear that $I_t = I_{got} \rightarrow C_t = C_{got}$. Moreover an arbitrarily large t can be found to satisfy the previous theorem (**weak gap theorem**).

These theorems imply that there are arbitrarily large gaps $[t(n),\ got(n)]$ which do not contain (infinitely often) the complexity of any program for recursive functions.

A very interesting theorem due to McCreight and Meyer (1969) shows that it is possible to avoid the inconvenience of the existence of gaps by means of a suitable choice of the functions t "naming" the complexity-classes C_t.

Theorem 2.9 (Naming theorem, or honesty theorem). For any complexity measure there exists a $\gamma \epsilon \mathscr{R}_1$ such that $\{\phi_{\gamma(i)}\}$ is a measured sequence with the property that $\phi_i \epsilon \mathscr{R}_1 \rightarrow \phi_{\gamma(i)} \epsilon \mathscr{R}_1$ and $C_{\phi_i} \equiv C_{\phi_{\gamma(i)}}$.

The importance of the theorem is due to the fact that the compression-theorem guarantees that one can operate on the recursive functions of the class $\{\phi_{\gamma(i)}\}$ to eliminate computing gaps.

Therefore it seems that names for complexity-classes which do not belong to a measured list are badly chosen names for the classes themselves. Theorem 2.9 is also called the "honesty-theorem" since measured sequences of functions satisfy an important property called "honesty". More precisely

Definition 2.6. Let r be a recursive function. The p.r. function f is called r-honest if and only if there is an index i for f, such that $\phi_i(n) \leqslant r(n,f(n))$ for almost all $n \epsilon \mathrm{Dom} f$.

In other words for an "honest" function f the complexity of computation can always be upper-bounded by a recursive function of the input n and the value f(n). Now it is possible to prove (McCreight and Meyer, 1969) that the set of functions which are r-honest form a mesured sequence. Conversely for any measured sequence $\{\psi_i\}$ there is a recursive function r such that the ψ_i are r-honest.

The complexity classes do not have meaningful closure properties valid for all measures (they are not closed with respect to union and intersection) except for the very important one expressed by the following theorem (Meyer - McCreight, 1969).

Theorem 2.10 (Union theorem). Let $\{f_i\}$ be a recursively enumerable set of recursive functions such that for each i and n $f_i(n) < f_{i+1}(n)$. Then there exists a recursive function t with the property that

$$C_t = \cup_i C_{f_i} .$$

An important corollary of this theorem (see, for instance, Hartmanis and Hopcroft, 1971) is that there exists a recursive function t such that the class of functions computable on a one-tape Turing-machine in the time bound t coincides with the class of primitive recursive functions.

We wish now to consider briefly the problem of enumerating all the functions in a complexity class C_t. A natural definition of this concept is that there is an effective procedure by which one can produce integers which are programs for the functions of the class and such that for any function at least one program for it has to be produced. The two basic results known about the recursive enumerability of the complexity classes are :i. C_t^Φ is recursively enumerable in any complexity measure for sufficiently large complexity bounds and ii. there are complexity measures for which some complexity classes cannot be recursively enumerated.

The proof of result ii. (see, for instance, Hartmanis and Hopfcroft, 1971) is obtained by considering complexity measures which have a very "pathological" behaviour such that, for instance, a finite change in the values of a function can produce an enormous change in the complexity measure. This fact is strictly related to the weakness of Blum's axioms that do not exclude such measures. Therefore one of the main open problems in computational complexity theory is to analyze whether there are axioms which, added to those of Blum, could restrict the class of acceptable measures to the "natural" ones like space, time etc.

THE ALGORITHMIC APPROACH TO INFORMATION THEORY

3.1. Introduction

As we said in the general introduction the Shannon information measure and the basic code theorems on the transmission of information are meaningfully applicable essentially only in the framework of statistical communication theory. To quote Kolmogorov (1965):

> "the probabilistic approach is natural in the theory of information transmission over communications channels carrying "bulk" information consisting of a large number of unrelated or weakly related messages obeying definite probabilistic laws".

Apart from the foregoing reasons, a proper use of probabilistic information requires that one knows what "probability" means exactly and moreover when it can be applied to the description of physical phenomena.

It is well known that the "probability theory" has been formalized in 1933 by Kolmogorov, who gave a set of axioms from which it is possible to derive the probability calculus. Such an axiomatization, that essentially reduces the probability theory to measure theory, appeared to solve most of the diatribes among frequentistic, logistic and subjectivistic schools about the interpretation to be given to the term "probability". In fact, except for some cases, any "explicatum" of the intuitive notion of probability seems to obey Kolmogorov's axioms (*). However, as stressed by Kolmogorov (1963) himself, the problem of finding the bases of real applications of the theory of probability is outside the theory itself.

Historically this latter problem was initially posed in 1919 by von Mises (see Sec. 3.4) in the setting of a frequentistic interpretation of probability (Kollektiv theory). Important contributions to this theory have been given later by Wald, (1936), Ville (1939) and Church (1940) (**). Church was the first who recognized as essential, in order to give an adequate "explicatum" of the informal notion of infinite random sequence, the use of the formal concept of "algorithm". The major difficulty which arises in a theory of infinite random sequences is that the limit frequencies of the considered events cannot be calculated either a priori, since one cannot construct a random sequence, or a posteriori, since any run of real statistical experiments is of finite length.

Kolmogorov (1963, 1965) recently reproposed the same problematics but for finite sequences (that one can always construct by means of suitable algorithms) using for his treatment the notion of "complexity" of the algorithms in a "static" sense that is not dependent on the computational behaviour. The intention of Kolmogorov was that of giving an algorithmic approach to "information theory" as well as "probabilistic theory". This end will be reached by means of a suitable definition of "finite random sequence" based on two theorems, the first due to Kolmogorov (1965) and the second to Martin-Löf (1966). This theory has been extended also by Martin-Löf and other authors in the case of infinite sequences.

3.2. The Kolmogorov minimal-program complexity

In this section we shall be concerned with the Kolmogorov minimal-program complexity and its information interpretation. Historically this quantity was first introduced by Solomonoff (1964) and independently reproposed by Kolmogorov (1965). Further Chaitin (1966) defined, relative to a particular class of machines, a complexity measure for binary strings, similar to that of Kolmogorov.

Let us consider a class $\{M_i\}$ of machines having an acceptable Gödel

---

(*) The ordinary probability theory does not include, for example, the situations arising in quantum mechanics (see, for instance, Varadarajan, 1962).

(**) An excellent paper on the literature of v. Mises Kollektives is by Martin-Löf (1969).

numbering $\{\phi_i\}$ of partial recursive functions. Denoting by ℓ a recursive function satisfying the axioms of a size-measure (see definition 2.1), one can introduce for any z and pair (x, y) of integers, the following quantity

$$K_Z^{\ell}(y/x) = \begin{cases} \min \{\ell(p) \mid \phi_Z(p,x) = y\} \\ +\infty \text{ if there is no p such that } \phi_Z(p,x) = y. \end{cases}$$

We explicitly not that K_Z^{ℓ} as a function of x and y is not always defined and, generally, is not partially computable. To give an interpretation to K_Z^{ℓ} introduced in a purely formal manner, we observe that z is the Gödel number of a machine M and that x,y,p will be represented in M by strings or words of a certain alphabet A. Therefore M computes a word-function from $A^* \times A^*$ into A^*. The value of this function for the pair P,X will be denoted by M(P,X). Furthermore if we fix the ℓ-function equal to the length $|P|$ of P, we can write $K_Z^{\ell}(y/x)$ simply as $K_M(Y/X)$ and interpret it as the minimal length of an "input-program" P that one has to add to the input X of M in order to obtain the string Y. $K_M(Y/X)$ is called the conditional complexity of Y given X with respect to the algorithm M Likewise the absolute complexity $K_M(Y)$ is defined as min $\{|P| \mid M(P,X)=Y\}$ if there is a program P such that $M(P,X) = Y$; otherwise $K_M(Y) = +\infty$.

We shall also set, which is natural, $K_M(Y) = K_M(Y/\Lambda)$ where Λ is the empty string.

Let us stress that the complexity of the "description" of the machine M carrying out, for each P, the computation of the p.r. function $\phi_Z(p,x)$ depends on the fixed value of the size of M and, for all P, on the complexity of the "input-program" P that can be measured just by its length $|P|$. From the iteration theorem one has that each input program P to a machine can be eliminated and "stored" in its instructions.

Moreover from the universal-machine theorem it is easy to see that with regard to a universal machine U of the class $\{M_i\}$ it is also possible to give to $K_U(Y/X)$ the interpretation of minimal size of a machine of the class required to produce Y starting from X. We recall here that the Chaitin (1966) measure of complexity for binary strings is just defined as the minimum number of states or the shortest program of a machine belonging to a class of special Turing machines, which is required to produce the strings themselves.

The conditional complexity $K_M(Y/X)$ is not related to the computation carried out by the machine M. For this reason it is a "static" measure of the complexity of Y given X, with respect to M. For any pair X,Y the value of $K_M(Y/X)$ depends in an essential way on the algorithm M. It can also occur that the complexity is finite with respect to an algorithm and infinite with respect to another. Moreover, as we said before, for any given string Y there exists always an algorithm of the class with respect to which the conditional complexity of Y vanishes (i.e. when the input program is stored in the machine-instructions).

In order to get an "intrinsic" measure, that is machine independent, of the complexity of Y given X one can refer to a suitable class of algorithms called **asymptotically optimal.** The reason for this is given by the following basic theorem due to Solomonoff (1964) and Kolmogorov (1965).

Theorem 3.1. An algorithm Ω exists with the property that for any other machine M one has, for all X and Y,

(3.1) $$K_\Omega(Y/X) \leqslant K_M(Y/X) + c,$$

where c is a constant depending on M and Ω, but not on X and Y.

The proof is, essentially, a consequence of the existence of universal machines by means of which one can simulate at the cost of a fixed amount of input-program, the behaviour of any machine, for all X and Y.

In a similar way one has for all M and Y

(3.2) $$K_\Omega(Y) \leqslant K_M(Y) + c.$$

We observe that $K_\Omega(Y/X)$ and $K_\Omega(Y)$ are always defined for all X and Y. Moreover Eqs. (3.1) and (3.2), which we can write in the simpler form $K_\Omega(Y/X) \lesssim K_M(Y/X)$; $K_\Omega(Y) \lesssim K_M(Y)$, for all M, X, Y (where $\lesssim$ means $\leqslant$ up to an additive constant) show that apart from a constant which can always be neglected for large values of the complexity, the algorithm Ω is the "most efficient" or "optimal" for the computation of all strings Y. Therefore the algorithms Ω have been called asymptotically optimal. Furthermore if Ω_1 and Ω_1 are two such algorithms, for the same or different classes of machines, one has $|K_{\Omega_1}(Y/X)-K_{\Omega_2}(Y/X)| \leqslant c_{\Omega_1\Omega_2}$ or $K_{\Omega_1}(Y/X) \approx K_{\Omega_2}(Y/X)$, where $\approx$ means "equal up to an additive constant". For these reasons, referring to asymptotically optimal algorithms, the conditional and absolute complexity, which we shall simply denote by K(Y/X) and K(Y), give, for large values of complexity, a measure of the "intrinsic" conditional and absolute complexity of Y, respectively.

Let us now introduce for each algorithm M, the quantity:

$$I_M(Y/X) := K_M(Y) - K_M(Y/X), \tag{3.3}$$

measuring the change that one has in the minimal-program complexity of M starting from X in order to produce Y. With regard to asymptotically optimal algorithms it is possible to show, making use of theorem 3.1, that the quantity $I(Y/X) := = K(Y) - K(Y/X)$ satisfies the following properties:

i. $I(Y/X) \gtrsim 0$, i.e. $I(Y/X)$ is no less than some negative constant.
ii. $K(Y/Y) \approx 0$, and $I(Y/Y) \approx K(Y)$
iii. $I(Y/X) \leqslant K(Y)$
iv. $I(Y/X)$ generally does not equal $I(X/Y)$.

We stress that properties i., ii. and iii. apart from the indeterminacy due to the presence of an additive constant which can be neglected for high values of the complexity, correspond to those of average mutual information of probabilistic information theory (Fano, 1961). This latter quantity is, however, symmetric in X and Y.

We wish now to discuss the "information" interpretation that it is possible to give to Kolmogorov's minimal program complexity.

We have previously seen that K(Y) measures (up to a constant) for all Y the "intrinsic absolute complexity" of Y. This quantity can, then, be naturally interpreted as a measure for any string Y of the "amount of information" required to "define" Y in an effective manner. Similarly the interpretation of K(Y/X) is that of "additional information" needed to obtain Y starting from X. Finally the interpretation of I(Y/X) is the "amount of information conveyed by X about Y".

The quantities of probabilistic information theory corresponding to K(Y), K(Y/X) and I(Y/X) are, respectively, the entropy and the conditional entropy and the average mutual information. However the interpretation of K(Y), K(Y/X) and I(Y/X) in the framework of the theory of complexity of algorithms, is completely different from the probabilistic one. For instance in Shannon's information theory the question "What is the content of information of a sequence X about a sequence Y? " is meaningless.

As stressed by Kolmogorov (1968)

"The ordinary definition of entropy uses probability concepts and thus does not pertain to individual values, but to random values, i.e. the probability distributions within a given group of values... By far, not all applications of information theory fit rationally into such an interpretation of its basic concepts".

3.3. Finite random sequences

In the previous section we have introduced the Kolmogorov minimal-program complexity and considered its information interpretation. In this and in the next section we shall see that by means of Kolmogorov's measure it is possible to give an algorithmic basis also for the intuitive concept of "randomness". Roughly speaking the random elements of a population of objects, supposed to be constructable, will be those of maximal algorithmic-complexity. In this section a precise definition of finite random sequences will be given and in the next the case of infinite sequences will be considered (*).

Let us start by proving the following two general theorems on the conditional complexity, the first due to Kolmogorov (1965) and the second to Martin-Löf (1966).

Theorem 3.2 (Kolmogorov). The conditional complexity K(Y/X) satisfies for all X and Y the inequality

$$K(Y/X) \leqslant |Y| + c \tag{3.4}$$

for some constant c.

The proof of the theorem is a simple consequence of theorem 3.1. It is sufficient to assume M in Eq. (3.1) to be an algorithm computing the projection function function $\phi(p,x) = p$.

In the following without restriction of generality, all the objects that we shall consider will be binary strings. Other objects, such as natural numbers, will be always "coded" by binary strings.

Theorem 3.3 (Martin-Löf). The number of strings Y for which $K(Y/X) < \delta$, with δ positive integer, is less than 2^{δ}.

In fact the number of input programs P for which $|P| < \delta$ is given by

(*) An outline of the Kolmogorov approach to random theory may be found in Guccione and Lo Sardo (1972).

$$\sum_{j=0}^{\delta-1} 2^j = 2^\delta - 1 .$$

Therefore the number of binary strings Y that can be produced by means of the algorithm Ω is at most $2^\delta - 1$.

Let us now consider for each n the set S(n) of all 2^n binary strings of length n, and denote by X^n an arbitrary element of it. For all n and $X^n \in S(n)$ we introduce the quantity $K(X^n/n)$, that is the complexity of X^n conditioned by the knowledge of its length. Of course as we said above the integer n is represented in the binary systems by a string of length $[\ln_2 n] + 1$, where [x] denotes the greatest integer less than or equal to x. The program whose length is measured by $K(X^n/n)$ contains therefore only information about the distribution of 0's and 1's in the string X^n. Besides $K(X^n/n)$ one can consider also the absolute complexity $K(X^n)$, that is the minimal-program complexity to yield X^n without knowing its length. For this reason this second program may need, with respect to that of $K(X^n/n)$ an additional information given at the most by $[\ln_2 n] + 1$.

The difference between $K(X^n)$ and $K(X^n/n)$ which, is not very relevant for high levels of complexity, as will be cleared in the following, may be, on the contrary, considerable at low levels where the information needed to specify the 0's and 1's distribution can be less than $[\ln_2 n] + 1$.

For the above reasons in order to study the "randomness" properties of population S(n) of binary strings of length n, we shall refer to $K(X^n/n)$ more than to $K(X^n)$.

From theorems 3.2 and 3.3 it follows

Theorem 3.4. For all n, $K(X^n/n)$ satisfies the properties:

$$K(X^n/n) \leqslant n + c, \quad \text{for all } X^n \in S(n)$$
$$K(X^n/n) \geqslant n - \delta \text{ for a number of sequences greater than } 2^n(1-2^{-\delta}). \tag{3.5}$$

The fraction of sequences of S(n) for which both $(3.5)_1$ and $(3.5)_2$ hold is then greater than $1-2^{-\delta}$ $(0<\delta\leqslant n)$. For large n, increasing the value of δ, the value of the previous fraction goes to 1 so that the great majority of the sequences of S(n) has a complexity which differs from the length n only by a constant that can be neglected for high values of the complexity itself. In other words $K(X^n/n) \approx n$.

These sequences of maximal complexity are defined by Kolmogorov to be the "random" elements of the population. We list now some straightforward consequences of the previous results (Martin-Löf, 1969b) :

Corollary 3.1. For all n there exists a sequence X^n for which $K(X^n/n) \geqslant n$.

Corollary 3.2. There is no algorithm M by which to construct for infinitely many n random sequence X^n.

Corollary 3.3. The function $K(X/Z)$ is not computable.

The random sequences, as defined by Kolmogorov, cannot then be constructed in the sense that there exists no algorithm which can produce, for any n, a random sequence of length n. This fact corresponds, in the case of infinite random sequences, as considered in the Kollektiv theory of von Mises-Church, to the **irregularity** of them, that is to the impossibility of constructing them by means of any algorithm. However, in spite of these irregularities, infinite random sequences possess **statistical regularities.** For example in tossing an ideal coin the limits of the relative frequencies of heads and tails must be equal to 1/2. Therefore for the definition of finite random sequence of Kolmogorov to be adequate, one has to show that such defined random sequences are such to pass all the usual randomness tests. It has been shown by Martin-Löf (1966) that, considering a suitable and very general definition of **statistical test**, the random sequences in the sense of Kolmogorov are such to pass all "conceivable" randomness tests. In this way, besides the relationship between the concepts of "algorithm-complexity" and "information" pointed out by Kolmogorov, the other connection between the first two concepts and the one of "randomness", will appear clear.

Let us now define in a very general way the notion of **statistical test.** To quote Martin-Löf (1966): "Generally, a test is given by a prescription which, for every level of significance ϵ tell us for what observations (in our case, binary strings) the hypothesis should be rejected". The level of **significance** of the test means that the probability that the hypothesis is rejected is less than or equal to ϵ . A "randomness" test at the level $\epsilon = 2^{-m}$ ($m = 1,2....$) is, therefore, a procedure that rejects the hypothesis of randomness for those sequences which have the tested property, whose probability is $\leqslant 2^{-m}$.

Let us consider now a particular test and denote for all m = 1, 2, ... by

V_m the so called critical region at the level $\epsilon = 2^{-m}$. V_m is a set formed by all the binary strings which are rejected at the level 2^{-m}.

The very general conditions, to be satisfied by V_m are the following:

i. $V_m \supseteq V_{m+1}$ (m = 1,2...)

ii. the number of strings of length n contained in V_m is less than or equal to 2^{n-m}, for all n and m.

iii. the set $V \equiv \{ (X,m) \mid X \epsilon V_m \}$ is recursively enumerable.

The first condition means, as is natural for all tests, that if a string is rejected at level $2^{-(m+1)}$ it has to be rejected for all greater levels. The second condition has been discussed above. The third condition is due to the fact that if there exists a "prescription" by which one may reject the sequences for every level of significance, from Church's thesis it follows that an algorithm must exist by means of which one may construct the elements of V. The set V, whose sections are the critical regions V_m determines completely the test. For each test V and for any string X we introduce the quantity:

$$m_V(X) := \max \{ m \mid X \epsilon V_m \} . \tag{3.6}$$

The string X will pass the test V for all levels 2^{-m} such that $m > m_V(X)$. Therefore $2^{-m_V(X)}$ is called the critical level for the string X and the test V.

It is possible to prove (Martin-Löf, 1966) that the class of all tests is recursively enumerable.

A consequence of this fact is the following

Theorem 3.5. A test U exists with the property that for every test V

$$V_{m+c} \subseteq U_m \quad (m = 1,2,....), \tag{3.7}$$

where c is a constant depending only on U and V.

The test U is called universal. From theorem 3.5 it follows that all strings passing the test U for a certain value of significance level 2^{-m} pass any other test V neglecting a constant change in the significance level. Theorem 3.5 can be restated in terms of critical levels in the following form. For all tests

$$m_V(X) \leqslant m_U(X) + c. \tag{3.8}$$

Assuming U_o to be the set of all binary strings, one has $0 \leqslant m_U(X) \leqslant |X|$.

Let us now fix a universal test and denote simply by m(X) the critical level for X with respect to it. A small value of m(X) means a high "degree of randomness" of X. The relationship between the function m(X) and Kolmogorov's complexity is given by the following theorem of Martin-Löf (1966).

Theorem 3.6. There exists a constant c such that

$$\big| |X| - K(X/\,|X|) - m(X) \big| \leqslant c, \tag{3.9}$$

for all binary strings X.

For the random sequences in the sense of Kolmogorov $K(X/\,|X|) \approx |X|$ so that $m(X) \leqslant$ const. Therefore they will pass all statistical tests for a fixed value of the significance level. Viceversa the binary strings which pass all statistical tests for some value of significance level are just the random sequences in the sense of Kolmogorov.

Let us in particular consider the statistical test that rejects the hypothesis of randomness of a string of length n when the number n_1 of ones in it differs too much from the number of zeros n_0. One can assume as a test at the level of significance 2^{-m} the following

$$|2n_1 - n| > 2^{m/2}\sqrt{n},$$

since from the Tchebycheff inequality of probability theory (see, for instance, Cramer, 1955) one has that the number of strings of length n that pass the above test is less than or equal to 2^{n-m}. From theorem 3.6 and Eq. (3.8) it follows that $|2n_1 - n| \leqslant \text{const}\sqrt{n}$ if $K(X^n/n) \approx n$.

At the end of this section let us briefly remember a variant of the Kolmogorov concept of minimal-program complexity due to Loveland (1969 a). Let us consider a binary string X^n of length n and denote by x(r) its initial segment of length $r\,(r = 1,\dots,n)$. With respect to any algorithm M we introduce the quantity $K_M(X^n; n)$, called the **uniform complexity** of X^n given n, and defined as:

$$K_M(X^n; n) = \begin{cases} \min\{\, |P| \mid \bigwedge_{s \leqslant n} M(P,s) = x(s) \,\} \\ +\infty \quad \text{if there is no such P.} \end{cases} \tag{3.10}$$

In other words the uniform complexity measures the minimal length of a program by which the algorithm M is able to yield for all lengths, all the initial segments of the string X.

From the definition it follows that

$$K_M(X^n/n) \leqslant K_M(X^n;n) . \tag{3.11}$$

Moreover it is possible to show (Loveland, 1969 a) that the basic theorem 3.1 is still valid for the uniform complexity. Denoting simply by $K(X^n;n)$ the uniform complexity with respect to a universal algorithm, one can prove that

$$K(X^i;i) \leqslant K(X^r;r)$$

if X^i is a **prefix** of X^r. This property is not always true for Kolmogorov's complexity. There are, in fact, examples of strings whose complexity is much less than that of some initial segment. This circumstance can be considered as rather counter intuitive.

3.4. A minimal-program complexity hierarchy for infinite sequences

Until now we have considered only finite binary sequences, that is elements of A*. In this section we shall take into account, on the contrary, binary sequences of infinite length, that is elements of A^∞.

We intend to give a brief account of different definitions of **randomness** of infinite sequences that have been proposed by some authors, and their mutual relationships. Furthermore the problem of merging infinite sequences in an "algorithm-complexity" hierarchy will naturally arise. Let us denote by

$$X \equiv x_1 x_2 \ldots\ldots x_n \ldots$$

an infinite binary string and by x(n) its **initial segment** of length n. Since there is a bijection between the class of all subsets Σ of N and the class of all the strings X by the relation:

$$n \in \Sigma \quad \longleftrightarrow \quad x_{n+1} = 1 .$$

then each property of a set of integers, being for instance recursive, recursively enumerable, cohesive, etc., gives rise to a corresponding property for infinite binary strings. A **recursive** stringX is, then, such that there is an algorithm by which to construct successively all its elements. For a **recursively enumerable** string, on the contrary, there is only a procedure for yielding all the ones in it, but generally there is not an algorithm by which to decide for all n whether x_n equals 1 or 0. Furthermore it is well known that there exists a hierarchy, called **Kleene's hierarchy**

(Davis, 1958), merging "arithmetical" sets in classes (Σ_n, Π_n) of increasing complexity. In this hierarchy $\Sigma_0 \equiv \Pi_0$ is the class of recursive sets and Σ_1 the class of recursively enumerable ones. A string corresponding to a set of Π_1 is such that there is an algorithm by which one may construct all the zeros of the sequence. Going up in the hierarchy there are sequences, that cannot be constructed, more and more complex.

The "random sequences" have to be very complex in order to pass all the possible statistical tests that one can perform on them. As we said in the introduction, the first non-contradictory definition of an **infinite random** sequence was given after the work of von Mises (1919) and Wald (1936) by Church (1940).

Definition 3.1. An infinite binary sequence $X \equiv x_1 x_2 \ldots x_n \ldots$ is called Kollektiv or random in the sense of Church (C-random) if and only if the following two axioms are satisfied:

A1 **(Regularity axiom).** Denoting by $s_n = \sum_{i=1}^{n} x_i$,

$$\lim_{n \to \infty} s_n/n = 1/2.$$

A2 **(Irregularity axiom).** However one may extract from X an infinite subsequence

$$x_{n_1}, x_{n_2}, \ldots\ldots,$$

by any recursive map $f: A^* \to \{0,1\}$ that selects x_n if and only if $f(x(n)) = 1$, the limit of the frequency of the ones of the subsequences has to be 1/2.

A consequence of A2 is that X **cannot be recursive.** From the **strong law of large numbers** of probability theory and by the fact that there is only a denumerable set of selection rules, one derives that the set of C-random sequences has probability 1.

An important objection to the previous definition of randomness was made by Ville (1939) who proved the existence of Kollektives such that, for all n, $s_n/n \geqslant 1/2$, even though the probability of them is zero. For this reason the definition of von Mises-Church has a certain degree of arbitrariness. A more detailed discussion of this equation is in Martin-Löf (1969 a, 1969 b).

Other definitions of **randomness** of infinite sequences have been proposed. As observed by Schnorr (1970) they differ essentially only by the kind of the "test" that they have to pass. The following definition uses as a test the **minimal program complexity.** A similar definition has also been proposed using the Loveland uniform complexity.

Definition 3.2 (Kolmogorov). An infinite binary sequence is random (K-random) if and only if there exists a constant c such that

$$K(x(n)/n) \geqslant n\text{-}c, \quad \text{for infinite n.}$$

This definition of randomness means that if X is random there are infinite initial segments of the sequence with high complexity. It would be desiderable to define as random a sequence for which $K(x(n)/n) \geqslant n\text{-}c$, for almost all n, but there are no sequences satisfying this condition (Martin-Löf, 1971). It is possible to prove that there are C-random sequences which are not K-random.

Another notion of randomness is due to Martin-Löf and is based on the concept of a **sequential test** (Martin-Löf, 1966) that is, a natural generalization of the one given for finite sequences. Defining for the initial segment x(n) of X the critical level m(x(n)) one has :

Definition 3.3 (Martin-Löf). An infinite binary sequence is random (M.L.-random) if and only if

$$\lim_{n \to \infty} \quad m(x(n)) < +\infty .$$

We recall the objection of Schnorr (1970) to the previous concept of randomness defined on the basis of the very general concept of sequential tests, which often do not have "physical meaning". In the opinion of Schnorr one has to restrict the class of tests to the "effective" ones (i.e. the test function is computable) in terms of which a new concept of randomness for infinite binary strings has been proposed.

It is possible to prove that a K-random sequence is also M.L.-random whereas the converse is not true (Schnorr, 1970). It has also been proved (Schnorr, 1970) that there are sequences in the class $\Delta_2 \equiv \Sigma_2 \cap \Pi_2$ of Kleene's hierarchy which are M.L.-random(*).

Let us now introduce a hierarchy of minimal program complexity defining for any total function f: N→N the class C [f] containing all infinite binary strings whose Kolmogorov's (or the Loveland complexity) is upper bounded almost everywhere by the function f. Most of the results obtained are independent of the

---

(*) A similar result has been obtained by Loveland (1966) for a modified version of C-random sequences.

measure used, even though their proofs can be different in the two cases.

Theorem 3.7. All binary strings are included in the complexity class named by n+ c, where c is a suitable constant:

$$\bigvee c\,(\text{const.}) \bigwedge X \in A^{\infty}\ X \in C\,[n + c].$$

Theorem 3.8. X is recursive $\longleftrightarrow \bigvee c\,(\text{const})\ X \in C\,[c]$.

The implication → is an easy consequence of the basic theorem 3.1. The inverse implication ← has been proved by Loveland for the uniform complexity and Meyer for the Kolmogorov complexity (Loveland, 1969 a).

Theorem 3.9. X recursively enumerable $\rightarrow \bigvee c\ X \in C\,[\ln_2 n + c]$.

This result has been proved by Loveland (1969 b) for the uniform complexity but it can be extended to Kolmogorov's complexity by means of inequality (3.11).

Theorem 3.10. (Loveland). There exists a nondecreasing unbounded function E separating recursive sequences from the others

$$X \text{ is recursive } \longleftrightarrow X \in C\,[E].$$

This result has been proved by Loveland (1969 b) for the uniform complexity.

In terms of complexity classes the definition of K-random sequences (or the one of Loveland) can be rephrased in the form:

$$X \text{ is random } \longleftrightarrow \bigvee c(\text{const.})\ X \notin C\,[n\text{-}c].$$

From the above results we see that recursive sequences and random sequences are placed at low levels and high levels of the minimal-program complexity hierarchy.

However there is no-function separating in the hierarchy the random sequences from the others, as stated in the following:

Theorem 3.11. (Schnorr, 1970). There does not exist a nondecreasing unbounded function f such that A^{∞} -C [g], where g(n) = n-f(n), is precisely the set of all random sequences.

Daley (1971 a) explored the hierarchy at low (*) and high levels of complexity. He stressed that on one hand there are sequences of low complexity which possess properties of randomness, while on the other hand there are C-random sequences (that Daley calls **pseudorandom**) with small complexity.

Theorem 3.12. There is a C-random sequence such that for all f (recursive, unbounded nondecreasing) $X \in C[f]$.

As stressed by Daley (1971 a) "One might argue that such a result shows that there is very little relation between information and randomness, or that such sequences are very poor formulations of pseudo randomness, or that our complexity does not accurately reflect the information content of sequences". Concerning this, Kolmogorov (1965) first observed that there are cases in which an object can be produced with "short" programs "only as the result of computations of thoroughly unreal duration", and suggested considering minimal-program complexity with limited resources. This problem has been considered by several authors, such as Bardzin (1968), Petri (1969), Kanovic (1969), Schnorr (1971) and Daley (1973). Making use of the concept of computation-complexity measure, as formalized by Blum (see Sec. 2.4) one can define the following **bounded** (with recursive bound t) **minimal-program complexity**:

$$K^t_\Phi(Y^n/n) = \min\{\, |P| \mid U(P,n) = Y^n \text{ and } \Phi_P(n) \leqslant t(n)\},$$

where U is a **universal algorithm** ($U(P,n) = \phi_P(n)$ being $\{\phi_P\}_{P \in A^*}$ an acceptable Gödel numbering of p.r. functions) and $\{\Phi_P\}_{P \in A^*}$ is a Blum's complexity measure. Moreover in a way similar to Loveland's variant of Kolmogorov's complexity, Daley (1973) introduced a **uniform bounded minimal-program complexity**. This quantity, that is denoted by $K^t_\Phi(Y^n;n)$represents the minimal length of a program P by means of which U can produce all initial segments of Y^n, but with the condition that $\Phi_P(i) \leqslant t(i)$ $(i = 1,..,n)$. A hierarchy of infinite sequences making use of $K^t_\Phi(Y^n;n)$ can be easily constructed.

A general result that has been obtained is that a bound on the computation resources leads often to a considerable increase in minimal-program

(*) as, for instance, for *recursively approximable* sequences in the sense of Rose and Ullian (1963).

complexity. However this does not always occur. The amount of this trade-off complexity and computation resource has been studied in some cases of interest by Daley (1973). In particular Daley has shown that C -random sequences lie in high levels of the bounded minimal-program complexity hierarchy.

A LOGICAL APPROACH TO INFORMATION THEORY

4.1.Introduction

The Kolmogorov minimal-program complexity is the basis for an algorithmic approach to information theory as well to probability theory. However, as we have previously seen, the Kolmogorov measure is a "static" complexity measure, not related to the computation behaviour. For this reason it can occur that for some object it is possible to keep "low" the amount of program (information) required to produce it only by increasing the computation resources (time, space, etc.) beyond any realistic limitation. Moreover the existence of C-random sequences with "low" minimal-program complexity shows that the "dynamic" aspects of the computation cannot be ignored. On the contrary they have to play an important role in any realistic logical information theory.

In this lecture we will outline a logical approach to information theory based on "dynamic" complexity measures in the framework of the theory of "formal systems" (De Luca and Fischetti, 1973). To this end we start by considering the notion of "formal system" or "logic" in its most general formulation. In these systems the concept of "complexity-proof" measures is axiomatized, following the work of Rabin (1960), in a manner such as to preserve the most general aspects of the intuitive notion, and at the same time obtain a definition of "minimal complexity" satisfying the requirements of effective computability. This latter requirement for certain complexity measures can be fulfilled only by suitable logics. We then arrive at a definition of "quantity of information", conveyed by one object about another, that is purely logical-syntactic in its formulation. This quantity shows properties similar to mutual probabilistic information, except for the commutative property which is not valid for our measure.

A minimal-program complexity for the derivation of a theorem is, moreover, defined and the relationships with the minimal complexity of the proofs are investigated in the case of a particular class of combinatorial logics. Finally same possible applications of this approach to a logical information theory are suggested.

4.2. Complexity measures on proofs

Let A be a countable alphabet and A* the free monoid generated by A. Any subset of A* is called a language. Let $(A^*)^k$ be the cartesian product $A^* \times \ldots \times A^*$ k-times that we call the k-th power of A*. In the following we consider recursive word predicates $\mathscr{R}$ defined in suitable powers of A* (Davis, 1958).

Definition 4.1. A finite set of recursive word predicates, none of which is singulary, is a set of rules of inference (or, for short, simply "rules").

If $\mathscr{I} \equiv \{\mathscr{R}_s\}_{s=1}^{p}$ $(p \geq 1)$ is a set of rules and the predicate $\mathscr{R}(X_1,\ldots,X_k,Y)$ belongs to $\mathscr{I}$ we say that Y is a logical consequence of $X_1,\ldots,X_k$ by the rule $\mathscr{R}$, if $\mathscr{R}(X_1,\ldots X_k, Y)$ is true.

Definition 4.2. A logic $\mathscr{L}$ is a pair $\mathscr{L} \equiv (\mathscr{A},\mathscr{I})$ where $\mathscr{A}$ is a recursive set of words called axioms and $\mathscr{I}$ a set of rules.

Definition 4.3. A proof in $\mathscr{L}$ is any finite sequence of words $X_0, X_1,\ldots,X_n$ such that for any X_i $(i = 0,1,\ldots,n)$ either

i. $X_i \in \mathscr{A}$, or
ii. $X_i = X_j$ with $j < i$, or
iii. there exists a $\mathscr{R} \in \mathscr{I}$ and a set of integers $j_1,\ldots,j_s$ with $j_k < i$ $(k = 1,\ldots, s)$ such that $(X_{j_1},\ldots, X_{j_s}, X_i)$ is true.

Definition 4.4. A word W is called a **theorem** of $\mathscr{L}$ and one writes $\vdash_{\mathscr{L}} W$, if and only if there exists a proof whose last word is W.

We shall denote by $\mathscr{P}(\mathscr{L})$ and $\mathscr{F}(\mathscr{L})$ respectively, the set of all the proofs and theorems of $\mathscr{L}$. It is well known that $\mathscr{P}(\mathscr{L})$ is **recursive** where $\mathscr{F}(\mathscr{L})$ is only **recursively enumerable** (see, for instance, Davis (1958) and Rogers (1967)). By **decision problem** for the logic $\mathscr{L}$ one means the problem of deciding whether or not an arbitrary word W is a theorem of $\mathscr{L}$. By definition the decision problem for a logic $\mathscr{L}$ is recursively **solvable** if $\mathscr{F}(\mathscr{L})$ is recursive; otherwise it is recursively **unsolvable.**

Definition 4.5. Two logics $\mathscr{L} \equiv (\mathscr{A},\mathscr{I}), \mathscr{L}' \equiv (\mathscr{A}',\mathscr{I})$ are **similar** if and only if the set of rules of inference is the same for both, whereas the set of the axioms can be different.

Let $\mathscr{L} \equiv (\mathscr{A},\mathscr{I})$ be a logic and $\mathscr{B}$ a recursive set of words. We can then consider the similar logic $\mathcal{L}_{\mathcal{B}} \equiv (\mathcal{A} \cup \mathcal{B}, \mathcal{I})$ obtained from $\mathcal{L}$ by adding the words of $\mathcal{B}$ to the set $\mathcal{A}$ of the axioms. One has $\mathcal{L}_{\emptyset} = \mathcal{L}, \emptyset$ denoting the empty set, and, from definition 4.3:

$$\vdash_{\mathcal{L}} W \rightarrow \vdash_{\mathcal{L}_{\mathcal{B}}} W \tag{4.1}$$

$$\vdash_{\mathcal{L}} W \wedge \vdash_{\mathcal{L}_W} Z \rightarrow \vdash_{\mathcal{L}} Z, \tag{4.2}$$

where $\mathcal{L}_W \equiv \mathcal{L}_{\{W\}}$ is the similar logic obtained from $\mathcal{L}$ by adding the singleton W to the set of axioms. In the following, for short, we shall denote $\vdash_{\mathcal{L}_X} W$ as $\vdash_X W$. From now on we assume the set $\mathcal{B}$ to be formed only by a single sequence even though what we say can obviously be generalized when $\mathcal{B}$ is any finite set of sequences.

Let us now introduce a measure of the complexity of the proofs in a given logic $\mathcal{L}$. We denote, for any pair X,Y of words of A*, by $\mathcal{P}_Y(\mathcal{L}_X)$ the recursive set (possibly empty) of all the proofs of Y in $\mathcal{L}_X$.

Definition 4.6. A complexity measure on proofs is a partial recursive function Φ on the set of all finite sequences of words, assuming integral values, that satisfies the following axioms:

A1. Dom $\Phi \supseteq \mathcal{P}(\mathcal{L}_X)$, for all $X \in A^*$.

A2. The function $M_{\mathcal{L}}^{\Phi}$ defined as

$$M_{\mathcal{L}}^{\Phi}(X,Y,k) = \begin{cases} 1 \text{ if } \bigvee_P P \in \mathcal{P}_Y(\mathcal{L}_X) \wedge \Phi(P) \leq k, \\ 0 \text{ otherwise,} \end{cases}$$

is (total) recursive.

The meaning of A2 is that, although one generally is not able to decide whether a word Y is provable or not in $\mathcal{L}_X$, one can however decide, when Φ is a complexity measure on proofs, whether a proof of Y in $\mathcal{L}_X$, whose complexity does not exceed a fixed amount k, does exist or not.

For a given logic $\mathcal{L}$ and complexity measure on proofs Φ we define the folllowing quantity

$$K_{\mathcal{L}}^{\Phi}(Y/X) = \begin{cases} \text{minimum value of the complexity } \Phi \\ \text{of a proof of Y in } \mathcal{L}_X \text{ if } \vdash_X Y, \\ +\infty \quad \text{otherwise.} \end{cases} \tag{4.3}$$

We shall call $K^{\Phi}_{\mathcal{L}}(Y/X)$ the **conditional complexity of Y given X with respect to the complexity measure on proofs Φ and to the logic $\mathcal{L}$**. When $X \in \mathcal{A}$ we denote $K^{\Phi}_{\mathcal{L}}(Y/X)$ simply by $K^{\Phi}_{\mathcal{L}}(Y)$, and call it the **absolute complexity** of Y with respect to Φ and $\mathcal{L}$.

We emphasize that the quantity $K^{\Phi}_{\mathcal{L}}$ even though similar in definition to Kolmogorov's notion of minimal program complexity, is conceptually very different from it. In fact this latter quantity is a "static" measure related to input data and program complexity, whereas (4.3) is a "dynamic" measure related to the used resource (number of steps, length of tape, etc.). From A2 and definition (4.3) one has:

$$\begin{aligned} &\text{i.} \quad \mathrm{Dom}\ K^{\Phi}_{\mathcal{L}} \equiv \{ (X,Y) \mid \vdash_X Y \} \\ &\text{ii.} \quad K^{\Phi}_{\mathcal{L}}(Y/X) = \mu k(M^{\Phi}_{\mathcal{L}}(X,Y,k) = 1), \end{aligned} \tag{4.4}$$

so that $K^{\Phi}_{\mathcal{L}}$ is a partially computable function, which is not generally true in Kolmogorov's approach. Furthermore we note the similarity between (4.4) and Blum's axioms of computational complexity.

We observe that axiom A2 is automatically verified if one considers a logic $\mathcal{L}$ and a funtion Φ satisfying A1, such that:

A3. The set $\{P \in \mathcal{P}(\mathcal{L}_X) \mid \Phi(P) \leqslant k\}$ is finite and the number of elements of it $\nu^{\Phi}_{\mathcal{L}}(X,k)$ is a recursive function.

In fact in such a case there exists an effective procedure by which, for any X and k, one can yield all the proofs P in $\mathcal{L}_X$ of complexity less than or equal to k. From this finite set one can then extract the subsets formed by the proofs whose terminal is Y. If this subset is empty $M = 0$, otherwise $M = 1$. The function M is therefore a recursive function.

Let us consider the following examples. We assume first that the function Φ satisfying A1 is equal to the length of a proof. In this case, $K^{\Phi}_{\mathcal{L}}(Y/X)$ is the **minimal length of a proof of Y in the logic $\mathcal{L}_X$**. However in the case of a general logic the number of proofs of a fixed length can be infinite so that generally $K^{\Phi}_{\mathcal{L}}(Y/X)$ is not partial recursive. A typical feature of **combinatorial systems** (Davis, 1958) is, on the contrary, the fact that the length of a proof satisfies the requirement A3, so that $K^{\Phi}_{\mathcal{L}}$ is partially computable.

In general logics, when the alphabet A is finite, one can, however, consider as a complexity measure of a proof $P \equiv \xi_1,\ldots\ldots,\xi_n$ the length of the word $\xi_1\ldots\ldots\xi_{n-1}$ that is $\Phi(P) := \sum_{1}^{n-1}{}_i\,|\xi_i|$. In fact the number of proofs in $\mathcal{L}_X$ with complexity equal to n is less than or equal to $(2k)^n/2$, with $k = \#A$, and one can effectively generate them for all X and n. A3 is then satisfied. We call this complexity measure the **word-length** of a proof. Another complexity measure of a proof $P \equiv \xi_1,\ldots\ldots,\xi_n$, in the case of combinatorial systems, is the quantity $\Phi(P) := \max\{|\xi_i| \, i \leq n\}$, that we call the **space-length** of a proof. It is easy to verify that the axioms A1 and A2 are satisfied.

Let us now analyze some general properties of the function $K^{\phi}_{\mathcal{L}}(Y/X)$ defined in (4.3) that we shall simply denote by K(Y/X).

Property 4.1. For any pair X,Y of words of A*

$$K(Y/X) \leq K(Y). \tag{4.5}$$

This property is due to the fact that any proof in $\mathcal{L}$ is a proof in $\mathcal{L}_X$ as well, whereas the contrary is not generally true.

In order to obtain a property of K which is very meaningful for the informational interpretation of it, that we shall give later, let us make the following hypothesis on the complexity measures on the proofs Φ:

H1. $Y \in \mathcal{A} \cup \{X\}$ if and only if there exists a proof P of Y in $\mathcal{L}_X$ such that $\Phi(P) = 0$, that is
$M(X,Y,0) = 1 \longleftrightarrow Y \in \mathcal{A} \cup \{X\}$.

Such measures Φ will be called "natural". Examples of natural measures are the number of steps of a proof (i.e. the length of a proof minus one), the **word-length** of a proof and the **space-length** of a proof that we shall denote in the following by Φ^0, Φ^* and $\tilde{\Phi}$ respectively.

With regard to natural measures K satisfies the following.

Property 4.2. $K(Y/X) = 0 \longleftrightarrow Y \in \mathcal{A} \cup \{X\}$

for any pair X,Y of words of A*.

4.3. An information measure

Let us now introduce the quantity I, defined as

$$I\ (Y/X) := K(Y) - K(Y/X)\,. \tag{4.6}$$

I is a partial recursive function that is not defined when $K(Y) = +\infty$; in this case if $K(Y/X) < +\infty$ we set, as is natural, $I(Y/X) = +\infty$

From definition (4.6) I satisfies the following properties, for any $X \in A^*$ and $Y \in \mathscr{F}(\mathcal{L}_X)$:

$$\begin{array}{ll} \text{i.} & 0 \leqslant I\ (Y/X) \leqslant K\ (Y) \\ \text{ii.} & I\,(X/X) = K(X) \\ \text{iii.} & I\,(Y/X) = K\,(Y) \leftrightarrow Y \in \mathscr{A} \cup \{X\}\,. \end{array} \tag{4.7}$$

i. is a consequence of property 4.1 and ii., iii. derive from property 4.2. If $Y \in \mathscr{A}$ then $K(Y) = 0$ so that from iii. $I(Y/X) = 0$.

A further hypothesis which one can make on the complexity measures on proof Φ is:

H2. If P is a proof of Z in $\mathcal{L}_X$ and Q is a proof of Y in $\mathcal{L}_Z$, then there exists a proof Π of Y in $\mathcal{L}_X$ such that

$$\Phi(\Pi) \leqslant \Phi(P) + \Phi(Q)\,. \tag{4.8}$$

H2 is satisfied by the measures Φ^0, Φ^* and $\tilde{\Phi}$. In general, if Φ satisfies H2 one has:

Property 4.3. For any triplet X,Y,Z of words of A^*

$$K\,(Y/X) \leqslant K\,(Y/Z) + K\,(Z/X)\,. \tag{4.9}$$

This triangular inequality derives from the fact that certainly a proof of Y starting from the set of axioms $\mathscr{A} \cup \{X\}$ can be obtained by proving first a theorem Z and afterwards proving Y starting from the set of axioms $\mathscr{A} \cup \{Z\}$. Therefore, making use of definition (4.3) and the hypothesis H2, (4.9) follows. A corollary of this property is the inequality

$$K\,(Y) \leqslant K\,(X) + K\,(Y/X)\,. \tag{4.10}$$

Making use of the triangular inequality one obtains also the following property for I (Y/X)

(4.11) iv. $$I(Y/X) \leqslant K(X).$$
for $X \in A^*$ and $Y \in \mathcal{F}(\mathcal{L}_X)$.

We shall give now an interpretation of the quantities I(Y/X) and K(Y/X), formally introduced by (4.6) and (4.3) in "information terms". As we stressed in Sec. 3.2 in a probabilistic information theory the question "if X and Y are two objects, for instance two sequences, what is the content of information in X about Y? " is meaningless. In our logical approach to information theory, this question is, on the contrary, the basic one. The intuitive idea is that the intrinsic content of information in X about Y can be obtained by measuring the reduction of the minimal complexity of a proof of Y when the word X is added to the set $\mathcal{A}$ of the axioms of some logical system $\mathcal{L}$. In our framework this quantity can be just measured by I(Y/X) defined by (4.6). The numerical valuation of the foregoing content of information depends, of course, on the complexity measure of proofs Φ and on the logical system $\mathcal{L}$, that is, the set of initial data (axioms) and the rules of inference at our disposal. It is natural that I(Y/X) assumes the value $+\infty$ when $K(Y/X) < +\infty$ and $K(Y) = +\infty$ since Y cannot be proved by using the axioms of the set $\mathcal{A}$ only, while it can be proved by adding X to $\mathcal{A}$.

The properties (4.7) are the most natural ones which an "information measure" of Y given X has to satisfy.

The formal analogy between properties i. ii. iv. of I(Y/X) and those of probabilistic average mutual information suggests giving to K(Y/X) the following interpretation as "quantity of information". Intuitively the larger the complexity of a proof of a theorem in a logic, the larger becomes the amount of "data" that one needs to construct the proof or, which is the same, that comes out in performing the proof itself; K(Y/X) measures the minimal value, expressed in Φ-units, of the previous amount of data that can be interpreted as a quantity of information needed to prove Y in $\mathcal{L}_X$. If Y is not a theorem in $\mathcal{L}_X$, then, since there is no finite complexity proof of Y, we say that the quantity of information needed to prove Y is infinite. If, for instance, Φ is the word-length of a proof then K(Y/X), in the alphabet $A \equiv \{0,1\}$, is just the minimum number of binary digits that one has to write before arriving at the theorem Y of the logic $\mathcal{L}_X$. Further interpretative remarks on K(Y/X) will be postponed to next section.

In our formalism $I(Y/X) \neq I(X/Y)$. By using the triangular inequality one can, however, show that it is always true that

$$| I(Y/X) - I(X/Y) | \leq \max \{K(Y/X),\ K(X/Y)\}. \tag{4.12}$$

Definition 4.7. The word Y is said to be **independent** from the word X, with respect to the logic $\mathcal{L}$ and to the measure Φ, if and only if there exists a proof of Y in $\mathcal{L}_X$ of minimal complexity in which the word X does not appear.

In other words, to add the word X to the set $\mathscr{A}$ of the axioms does not give any "gain", in the sense of reducing the minimal complexity of a proof of Y, so that the information furnished by X about Y has to be zero. One can state and easily prove the following:

Proposition 4.1. A necessary and sufficient condition for Y to be independent of X is that $I(Y/X) = 0$.

We stress that, unlike what occurs for random variables, if Y is independent of X this **does not imply** that X is independent of Y (cfr. Eq. (4.12)).

4.4. Complexity of proofs and program complexity

In this section we introduce for logics a quantity "similar" to minimal-program complexity for algorithms. For each logic $\mathcal{L} \equiv (\mathscr{A},\mathscr{I})$, where $\mathscr{A}$ is a finite set of axioms, we define by **program-complexity** of $\mathcal{L}$ **the sum of the lenghts of its axioms** (*). Then for any $Y \in A^*$ we introduce the following quantity:

$$H_{\mathscr{I}}(Y) := \min \{ |\mathscr{A}| \ \ | \underset{(\mathscr{A},\mathscr{I})}{\vdash} Y \}, \tag{4.13}$$

having set $|\mathscr{A}| := \Sigma_i |A_i|$. We call $H_{\mathscr{I}}(Y)$ the **(absolute) minimal-program complexity** of Y with respect to the set of rules $\mathscr{I}$. It represents the minimal amount of "initial data" required to **prove** Y by means of the rules of $\mathscr{I}$. $H_{\mathscr{I}}(Y)$ is always $\leq |Y|$. In a similar way also a **conditional** minimal-program complexity $H_{\mathscr{I}}(Y/X)$ is definable.

We want now to analyze which kind of relationships exist between minimal-program complexity and minimal complexity of proofs. To this end we shall confine ourselves to considering only **combinatorial systems**, that is logics

---

(*) For our aims we shall refer to this measure only, even if more general (or axiomatic) definitions of program-complexity measures could be given.

$\Gamma \equiv (\mathscr{A},\mathscr{I})$ where $\mathscr{A}$ contains only an axiom W and each rule $\mathscr{R} \in \mathscr{I}$ depends only on two word-variables. Denoting by Q(X,Y) the predicate "Y is a direct consequence of X", we suppose that the combinatorial system is such that there exists a constant β for which

(4.14) $$Q(X,Y) \rightarrow |Y| - |X| \leqslant \beta \text{ for all } X, Y \in A^*.$$

This condition is certainly satisfied, for instance, by semi-Thue systems or Post's normal systems (Davis, 1958), which are defined by finite sets of productions as $\xi g_s \eta \rightarrow \xi \bar{g}_s \eta$ (s=1,...,p) and $g_s \xi \rightarrow \xi \bar{g}_s$ (s = 1,...,p) respectively, where $g_s, \bar{g}_s$ denote fixed words and ξ, η variable words of A^*.

A meaningful quantity for a combinatorial system $\Gamma \equiv (\{W\},\mathscr{I})$ satisfying (4.14) is, then, the minimum value of β for which (4.14) is true; this we denote by $C(\mathscr{I})$. In the above examples $C(\mathscr{I}) = \max \{\bar{g}_s - g_s \mid 1 \leqslant s \leqslant p\}$.

When $C(\mathscr{I}) > 0$ it is easy to derive the following lower bound for the complexity K^o_Γ (number of steps of a proof)

(4.15) $$K^o_\Gamma(Y) \geqslant \lceil (|Y| - |W|)/ C(\mathscr{I}) \rceil,$$

where $\lceil x \rceil$ is the minimum integer not less than x. This lower bound depends linearly on the length of the theorem Y and on the set of rules of Γ by means of the constant $C(\mathscr{I})$ only. When $C(\mathscr{I}) \leqslant 0$ a finite number of theorems can be proved from a given axiom and a lower bound to K^o_Γ can still be obtained by replacing $|Y| - |X|$ with $||Y| - |X||$ in Eq. (4.14) and $|Y| - |W|$ with $||Y| - |W||$ in Eq. (4.15). In this case the value of $C(\mathscr{I})$ will be always nonnegative.

Let $\Gamma_o \equiv (\{W_o\},\mathscr{I})$ be a combinatorial system with $|W_o|$ equal to the minimal program complexity $H_\mathscr{I}(Y)$. Then the value of lower bound to the number of steps required to prove Y of Eq. (4.15) increases to the value $\lceil (|Y| - H_\mathscr{I}(Y))/(C(\mathscr{I})\rceil$. In particular when $H_\mathscr{I}(Y) = 0$, that is the axiom is the empty string, $K^o_{\Gamma_o}(Y) \geqslant \lceil |Y| / C(\mathscr{I})\rceil$. If $H_\mathscr{I}(Y) = |Y|$ and $W_o = Y$, then $K^o_{\Gamma_o}(Y) = 0$.

Let us now make some interpretative remarks on the minimal-program complexity $H_\mathscr{I}(Y)$. This quantity does not represent the minimal-program complexity required to "define" Y in an effective manner, as in the case of minimal-program complexity for algorithms, since an infinite number of theorems can be, generally, derived in a combinatorial system $\Gamma \equiv (\{W_o\}, \mathscr{I})$ proving Y and with $|W_o| = H_\mathscr{I}(Y)$. In other words W_o is not a "code-sequence" for Y since it specifies Y only as a member of the set of theorems $\mathscr{F}(\Gamma_o)$.

In some cases by means of suitable additional rules it is possible to "select" from the set of theorems derivable in $\Gamma \equiv (\{W\}, \mathscr{I})$, at most one theorem that can be associated in a unique manner with the axiom W. These combinatorial systems become, then, computational devices or algorithms. This is the case, for instance, of semi-Thue systems that can be transformed by suitable constraints in the Markov normal-systems (Markov, 1951) which are monogenic systems. That is, for any X there is at most a word Y for which Q(X,Y) is true. One can, then, define the predicate $Comp_\Gamma(Y)$ that means "$\vdash_\Gamma Y$ and there is no-word $Z \in A^*$ such that Q(Y,Z) is true".

The minimal length $N_{\mathscr{I}}(Y)$ of a word W such that Y is "computable" in $\Gamma \equiv (\{W\}, \mathscr{I})$ has the same meaning as in Kolmogorov's theory. It is clear that $H_{\mathscr{I}}(Y) \leqslant N_{\mathscr{I}}(Y)$ and, moreover, the number of steps required to compute Y in Γ has a lower bound given by $[\,(|Y| - N_{\mathscr{I}}(Y)) / C(\mathscr{I})\,]$.

Let us now consider a set $\mathscr{I}$ of rules (as those of a universal algorithm) by means of which any word $Y \in A^*$ can be computed. Let us suppose, moreover, $A \equiv \{0,1\}$ and put a constant bound τ to the time of computation. One derives that in this case $N_{\mathscr{I}}(Y) \geqslant |Y| - C(\mathscr{I})\tau$. Setting $|Y| = n$ it is easy to prove, in a way similar to that of theorem 2.3, that the above condition will be satisfied by a number of sequences greater than $2^n(1-2^{-\tau C(\mathscr{I})})$: that is, the great majority of sequences. When $\mathscr{I}$ is the set of rules of a "universal" algorithm the sequences Y which can be computed within the time bound τ have to be "random" in the sense of Kolmogorov.

4.5. Conclusion

The logical approach to information theory presented in our last lecture aims at being a generalization of the Kolmogorov algorithmic approach, in the framework of logical-systems theory. The concept of information, in these systems, is related to the minimal "complexity" required to generate a string of symbols.

Since the complexity can be either "program-complexity" or "proof-complexity" two kind of measures of it, the first "static" and the second "dynamic", can be introduced and two measures of information can be proposed. We recall that the minimal-program complexity represents the amount of information required to "derive" a string relative to a given set of rules. However it, genrally, determines effectively the string only up to its membership to a (possibly infinite) set of theorems. The minimal proof-complexity in a given logic represents, on the contrary, the amount of information required to be sure that the string will appear in a (finite) list of theorems.

A noteworthy case is when, with respect to a given set of rules, it is possible to associate with each axiom a single string at the most. A computation is then defined and the minimal-program complexity to do it coincides with Kolmogorov's measure and the complexity of computation with a Blum's measure.

An important problem, that naturally arises, is whether in the framework of this logical approach to information theory one might, as in the case of Shannon's theory , face problems of "transmission of information" and, possibly, prove general theorems by which one may evaluate the actual implications and the relevance of the introduced concepts of information. How can a communication system in the context of this theory be envisaged? One can think of transmitting over a "channel" a "computer program" for a message instead of the message itself. In making a "balance" of the advantage of this transmission one has to take into account both the "length of program" and the "time of computation". The "structure" of the message and of the computer play, then, a relevant role in this transmission. For many purposes, when one wants to have the freedom of transmitting any message, the "receiver" can be a universal algorithm. Even though the theory is only at a preliminary stage, we think that the foregoing approach is a good frame for an analysis of the "semantic-pragmatic" levels of comunication. In other words if the "receiver" is a "decision-maker" that follows a fixed set of rules, it can be represented just as a formal system (in particular an algorithm) and the possible decisions of it depend on the set of theorems which can be derived. In this way there is the possibility, in principle, of facing the problem of how the transmitted message affects the conduct of the receiver in the desired way.

At the technical-level "non-deterministic" computations performed by a receiver can correspond to the "noise" of probabilistic information theory. We believe that the basic limitations on the transmission of information of the theory of Shannon may have a "logical structure" and an interpretation deeper than that which appears in the probabilistic context. Of course probabilistic considerations can be taken into account also in the frame of a logical information theory, but we think they do not play an essential role.

The notion of minimal-program complexity of a string, besides its implications to information theory and probability theory (see Sec. 3.3-3.4) has also relevant consequences for the **methodology of the science**, as has been stressed by Solomonoff (1964) and Chaitin (1974).

A scientific theory is the result of a process of "induction" by means of which starting from a series of observations that can be regarded as a binary string, some possibly simple theoretical schemata are made in order to derive, in a purely mechanical way the results already observed and to predict new ones. In other words the "theoretical schema" can be considered as a computer program for the observations. Chaitin (1974) observes that: "The simplicity of a theory is inversely proportional to the length of the program that constitutes it. That is to say, the best program for understanding or predicting observations is the shortest one that reproduces what the scientist has observed up to that moment. Also, if the program has the same number of bits as the observations, then it is useless, because it is too "ad hoc". If a string of observations only has theories that are programs with the same length as the string of observations, then the observations are random, and can neither be comprehended nor predicted.

...In summary, the value of a scientific theory is that it enables one to compress many observations into a few theoretical hypotheses. There is a theory only when the string of observations is not random, that is to say, when its complexity is appreciably less than its length in bits".

Let us note, however, that the complexity of a theory is also strongly related to the time of computation required in making previsions which can exceed in some cases any realistic limitations, mainly when the program-complexity of a theory is short. To quote De Luca and Termini (1972a) "Every theory, in fact, ultimately provides, in the setting of a suitable language, computational procedures allowing one to obtain previsions about the phenomena that it aims to describe. One of the main aspects of the previsions is then the time of computation which is often realistically short (this happens for instance, in the case of most linear systems), but however in some cases (as, for some non-linear or complex systems) may considerably increase and even become intolerably large.

In the latter case, the model has to be replaced by simpler ones which may however produce a lack of information about the system".

In analogy with the results of the previous section a description of very complex systems with a computational-time compressed within fixed bounds is possible only when the described phenomena are "random". In these cases, as observed by Chaitin, the theory is useless. What remains possible to do, however, is a "probabilistic description" of the systems, that can be much simpler than the "deterministic" one even though at the cost of an unavoidable loss of information about them (*).

Note added in proof.

After the delivering of these lectures several papers on the subject of computational complexity and information theory have been published. We limit ourselves to mention some of them indicating briefly the new suggested ideas.

A variant of the Kolmogorov concept of complexity has been proposed by P.C. Schnorr (J. Comput. System Sci. 7, 376-388, 1973). This quantity is called **process complexity** and represents the minimal length of the description of a string by using a sequential coding.

The attempt of applying information-theoretic computational complexity to metamathematics made by G.J. Chaitin (1974) has been continued in another paper (J. of A C M, 21, 403-424, 1974) in which the complexity of proving a given set of theorems is measured in terms of the "size" of the assumed axioms and the "length" of the proofs needed to prove the theorems. Moreover another variant of the Kolmogorov program complexity has been proposed by Chaitin (J. of A C M, 22, 329-340, 1975). Unlike the previous definitions this quantity satisfies the formal properties of the probabilistic entropy.

An attempt of studying problems of transmission of information by making use of the Kolmogorov complexity in a "linguistic" context, different from the Shannon probabilistic one, has been made by the author (Lecture notes in Computer Science, vol. 33, 103-109, Springer-Verlag, Berlin-Heidelberg-New York, 1975).

In conclusion we mention the beautiful systematic treatment of the theory of computational complexity of the functions made by G. Ausiello ("Complessità di calcolo delle funzioni", Boringhieri, Torino, 1975).

(*) A more detailed discussion on the computational aspects in making models for the analysis of complex systems can be found in De Luca and Termini (1972a).

BIBLIOGRAPHY

Algorithms and Computabiltiy

[1] Church, A., An unsolvable problem of elementary number theory, Amer. J. of Math., 58, 345-363, 1936.

[2] — "The Calculi of Lambda-conversion", Annals of Mathematics Studies, no. 6, Princeton University Press, Princeton, N.Y. 1941.

[3] Davis, M., "Computability and Unsolvability", McGraw Hill, New York, 1958.

[4] Gödel, K., On undecidable propositions of formal mathematical systems, mimeographed lecture notes, Institute for Advanced Study, Princeton, N.J., 1934.

[5] Kleene, S.C., General recursive functions of natural numbers, Matematische Annalen, 112, 727-742, 1936.

[6] — "Introduction to Metamathematics", Van Nostrand Company, Inc. Princeton, N.J. 1952.

[7] Markov, A.A., The theory of algorithms (Russian), Trudy Mathematischeskogo Instituta imeni V.A. Steklova, 38, 176-189, 1951.

[8] — "Theory of algorithms" (English transl.) Israel Program for scientific translations. Jerusalem, 1962.

[9] Péter, R., "Rekursive Funktionen". Akademiai Koadó, Budapest, 1951.

[10] Post, E.L., Finite combinatory processes-Formulation I,J. of Symbolic Logic, 1,103-105, 1936.

[11] — Formal reductions of the general combinatorial decision problem, Amer. J. Math., 65, 197-215, 1943.

[12] Rice, H.G., Classes of recursively enumerable sets and their decision problems. Trans. of Amer. Math. Soc. 74, 358-366, 1953.

[13] Rogers, H. Jr., Gödel numberings of partial recursive functions, J. of Symbolic Logic,

23, 331-341, 1958.

[14] — "Theory of Recursive Functions and Effective Computability", McGraw Hill, New York, 1967.

[15] Smullyan, R.M., "Theory of formal Systems", Annals of Mathematics Studies, no. 47, Princeton University Press, Princeton, N.J. 1961.

[16] Turing, A.M., On computable numbers, with an application to the Entscheidungs problem, Proc. of the London Math. Soc. ser. 2, 42, 230-265, 1936-1937.

Complexity of Algorithms

[17] Adrianopoli, F. and A. De Luca. Closure operations on measures of computational complexity, Calcolo, 2,1-13, 1974; (abstract) in the Notices of AMS April, 1972.

[18] Arbib,M. and M. Blum, Machine dependence of degrees of difficulty, Proc. of the Amer. Math. Soc., 16, 442-447, 1965.

[19] Arbib, M., Speed-up theorems and incompleteness theorems, in "Automata Theory" (Caianiello, E.R. ed.), Academic Press, New York, 1966.

[20] Ausiello, G., Teorie della complessità di calcolo, Calcolo, 7, 387-408, 1970.

[21] — Abstract computational complexity and cycling computations, J. Comp. and Sys. Sci. 5,118-128, 1971.

[22] Axt, P., Enumeration and the Grzegorczyk hierarchy.Zeit.Math. Logik und Grundlagen Math. 9,53-65, 1963.

[23] Blum, M., A machine independent theory of the complexity of recursive functions, J. of ACM, 14, 322-336, 1967a.

[24] — On the size of machines, Information and Control, 11, 257-265, 1967b.

[25] — On effective procedures for speeding up algorithms, J. of ACM, 18, 290-305, 1971.

[26] Borodin, A., Complexity classes of recursive functions and the existence of

complexity gaps, ACM Symp. on Theory of Computing, Marina del Rey, Calif. 1969.

[27] — Horners rule is uniquely optimal, in "Theory of machines and Computations" (Z. Kohavi and A. Paz, eds.) Academic Press, New York and London, 1971.

[28] — Computational complexity and the existence of complexity gaps, J. of ACM, 19, 158-174, 1972.

[29] — Computational complexity: Theory and practice. In "Currents in the theory of computing" (A.V.Aho, ed.) Prentice-Hall Series in automatic computation, 1973.

[30] Cobham, A., The intrinsic computational difficulty of functions, Proc. Congress for Logic, Mathematics, and Philosophy of Science. North-Holland, Amsterdam, 1964.

[31] Grzegorczyk, A., Some classes of recursive functions, Rozprawy Mat. 4, Warsaw, 1-45, 1953.

[32] Hartmanis, J. and R.E. Stearns, On the computational complexity of algorithms, Trans. of the Amer. Math. Soc., 117, 285-306, 1965.

[33] Hartmanis, J.P., P.M. Lewis II and R.E. Stearns, Classifications of computation by time and memory requirements, IFIP Congress 1965, Vol. 1, Spartan books, Washington, D.C., 1965.

[34] Hartmanis, J., Tape-reversal bounded Turing machine computations, J. of Comp. and Syst. Sci., 2, 117-135, 1968.

[35] Hartmanis, J. and R.E. Stearns, Automata - based computational complexity, Information Sciences, 1 , 173-184, 1969.

[36] Hartmanis, J. and J.E. Hopcroft, An overview of the theory of computational complexity, J. of ACM, 18, 444-475, 1971.

[37] Hennie, F.C., One tape off-line Turing machine computations, Information and Control, 8, 553-578, 1965.

[38] Hennie, F.C. and R.E. Stearns, Two-tape simulation of multi-tape Turing machines, J. of ACM, 13,533-546, 1966.

[39] Hopcroft, J.E. and J.D. Ullman, "Formal Languages and their relation to Automata"., Addison Wesley Publ. Co. 1969.

[40] Knuth, D.E., "The art of Computer programming", Addison Wesley Publ. Co. 1969.

[41] McCreight, E.M. and A.R. Meyer, Classes of computable functions defined by bounds on computation, ACM Symp. on Theory of Computing, Marina del Rey, Calif. 1969.

[42] Meyer, A.R. and P.C. Fisher, On computational speed-up, IEEE Conf. Rec. 9th Ann.Symp. on Switching and Automata theory, October, 1968.

[43] Rabin, M.O., Degrees of difficulty of computing a function and a partial ordering of recursive sets, Tech. Report 2, Hebrew University, Jerusalem, Israel, 1960.

[44] Rabin, M.O. and D. Scott, Finite automata and their decision problems, IBM J. Res.Develop., 3, 114-125, 1959.

[45] Ritchie, R.W., Classes of recursive functions based on Ackermann's function, Pacific J. of Math., 15,3 , 1965.

[46] Schnorr, C.P., Does the computational speed-up concern programming? , in "Automata, Languages and Programming" (Nivat, M. ed.) North-Holland/American Elsevier, 1972.

[47] Trakhtenbrot, B.A., Complexity of algorithms and computations, Course Notes, Novosibirsk University, 1967.

Information Theory and Complexity

[48] Bardzin, J., Complexity of programs to determine whether natural numbers not greater than n belong to a recursively enumerable set. Soviet Math. Dokl. 9,5, 1251-1254, 1968.

[49] Boltzmann, L., "Vorlesungen über Gas Theorie", Vol. 1, J.A. Barth Leipzig. 1896 English Transl. Lectures on gas theory. Berkeley Calif. 1964.

[50] Brillouin, L., "Science and Information Theory", Academic Press Inc. New York, 1956.

[51] Carnap, R. and Y. Bar-Hillel, An outline of a theory of semantic information, Tech. Rep. 247, M.I.T., Research Laboratory of Electronics, 1952, Reprinted in Bar-Hillel, Y., "Language and Information", Addison Wesley, Reading, Mass., 1962.

[52] Chaitin, G.J., On the length of programs for computing finite binary sequences, J. of ACM, 13, 547-569, 1966.

[53] — Information-theoretic computational complexity. IEEE Trans. on Information Theory IT-20, 10-15, 1974.

[54] Church, A., On the concept of a random sequence, Bull. Amer. Math. Soc. 46, 130-135, 1940.

[55] Cramer, H. "Mathematical methods of statistics", Princeton University Press, Princeton, 1945.

[56] Daley, R., Minimal-program complexity of pseudo-recursive and pseudo-random sequences. Dept. of Math. Carnegie Mellon University. Report 71-28, 1971a.

[57] — Minimal program complexity with restricted resources, University of Chicago ICR Report no. 30, 1971b.

[58] — An example of information and computation resource trade-off, J. of ACM, 20, 687-695, 1973.

[59] De Luca, A. and E. Fischetti, Outline of a new logical approach to information theory, Proc. of NATO Summer School on "New concepts and Technologies in parallel processing", Capri, 17-30 June 1973. Published by Noordhoff, Series E, n.9, Leyden, 1975.

[60] De Luca, A. and S. Termini, Algorithmic aspects in complex systems analysis, Scientia, 106, 659-671, 1972a.

[61] — A definition of a non-probabilistic entropy in the setting of fuzzy sets theory, Information and Control, 20, 301-312, 1972b.

[62] Fano, R.M., "Transmission of Information", M.I.T. Press and J. Wiley & Sons, Inc. New York and London 1961.

[63] Guccione, S. and P. Lo Sardo, Casualità ed effettività, Reprint, Istituto di Fisica Teorica dell'Università di Napoli, 1972.

[64] Hartley, R.V.L., Transmission of information, Bell Syst. Tech. J., 7, 535-563, 1928.

[65] Kanovic, M. and N.V., Petri, Some theorems of complexity of normal algorithms and computations. Soviet Math. Dokl. 10,1,233-234, 1969.

[66] Kolmogorov, A.N., Grundbegriffe der Wahrscheinlichkeitsrechnung, in Ergebnisse der Mathematik, Berlin, 1933. Reprinted: "Foundations of the theory of the Probability" 2nd ed., Chelsea, New York, 1956.

[67] — On tables of random numbers, Sankhya, 25, 369-374, 1963.

[68] — Three approaches to the quantitative definition of information, Problemy Peredachi Informatsii, 1, 3-11, 1965.

[69] — Logical basis for information theory and probability theory, IEEE Trans. on Information Theory, IT-14, 662-664, 1968.

[70] Loveland, D.W., The Kleene hierarchy classification of recursively random sequences, Trans. Amer.Math.Soc. 125, 497-510, 1966.

[71] — A variant of the Kolmogorov concept of complexity, Information and Control, 15, 510-526, 1969a.

[72] — On minimal program complexity measures, ACM Symp. on Theory of Computing, Marina del Rey, Claif. 1969b.

[73] Mac Kay, D.M., Quantal aspects of scientific information theory, Philosophical magazine, 41, 289-311, 1950.

[74] Martin-Löf, P., The definition of random sequences, Information and Control, 9,602-619, 1966.

[75] — The literature on von Mises'Kollektives revisited, Theoria, 1, 12-37, 1969.

[76] — Algorithms and randomness, Rev. Inter.Stat.Inst., 37, 265-272, 1969.

[77] — Complexity oscillations in infinite binary sequences, Z. Wahrscheinlich. verw.Geb., 19, 225-230, 1971.

[78] Nyquist, H., Certain factors affecting telegraph speed, Bell Syst. Tech.J., 3, 324, 1924.

[79] Petri, N.V., The complexity of algorithms and their operating time, Soviet Math. Dokl., 10,3,547-549, 1969.

[80] Rose, G.F. and J.S. Ullian, Approximation of functions of the integers, Pacific J. of Math. 13, 693-701, 1963.

[81] Shannon, C.E., A mathematical theory of communication, Bell Syst. Tech. J., 27, 379-423, 623-656, 1948.

[82] Shannon, C.E. and W. Weaver, "The mathematical theory of communication", University of Illinois Press, Urbana, 1949.

[83] Schnorr, C.P., A unified approach to the definition of random sequences, Math. Syst. Theory, 5, 246-258, 1970.

[84] — "Zufälligkeit und Wahrscheinlichkeit", Lectures notes in Mathematics. Springer Verlag, vol. 218, Berlin-Heidelberg-New York, Springer, 1971.

[85] Solomonoff, R.J., A formal theory of inductive inference. Part. I. Information and Control, 7,1-22,1964.

[86] Szilard, L., Über die Entropieverminderung in einem thermodynamischen System bei Eingriffen intelligenter Wesen, Zeitschrift f. Physik, 53, 840-856, 1929; Reprinted: "On decrease of entropy in thermodynamic system by the intervention of intelligent beings", Behavioral Science, 9, 301-310, 1964.

[87] Varadarajan, V.S., Probability in physics and a theorem of simultaneous observability, Comm. Pure Appl. Math., 15, 189-217, 1962.

[88] Ville, J., Étude Critique de la Notion de Collectif, Gauthier-Villars, Paris, 1939.

[89] von Mises, R., Grundlagen der Wahrscheinlichkeitsrechnung, Mathematische Zeitschrift, 5, 52-99, 1919.

[90] Wald, A., Sur la notion de collectif dans le calcul des probabilités, C.R. Acad. Sci. Paris 202, 180-183, 1936.

[91] Wiener, N., "Cybernetics", Hermann, Paris, 1948; 2nd ed., The MIT Press and J. Wiley and Sons, Inc., New York, 1961.

[92] Wiener, N., "The extrapolation, interpolation and smooting of stationary time series", John Wiley and Sons, New York, 1949.

AN INTRODUCTION TO MULTIPLE USER COMMUNICATION SYSTEMS

P.P. Bergmans

PART ONE : BROADCAST CHANNELS

Introduction

We shall first give a brief review of the concepts used in the theory of simple communication systems. These concepts will then be extended to multiple users communication systems. Finally, the notion of "superposition codes" will be introduced and we shall show how these codes dominate other transmission schemes for a very large class of multiple users channels.

Review of Simple Communication Systems

Figure 1 represents a "simple" (i.e. single-source, single-user) communication system.

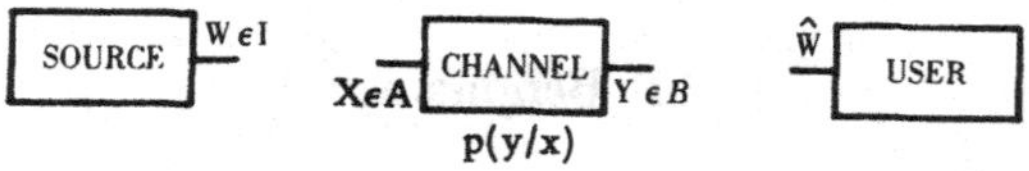

Fig. 1

The source is a "mathematical" or "equivalent" source, in the sense that its outputs are independently and identically distributed (i.i.d.), with a uniform distribution on $I = \{1,2,..M\}$. If this is not the case, source coding techniques can be used to transform an actual source to an equivalent i.i.d. memoryless source. Our goal is to transmit W to the users. To achieve this, we may use the channel n times. The input alphabet of the channel is A, and the output alphabet is B. The transition probability of the channel is given by p(y/x). Let the reproduction of W at the user's side be $\hat{W}$ (hopefully, $\hat{W} = W$).

To transmit W through the channel, we encode the source message into an n-tuple of elements of A :

$$\mathbf{x} : I \to A^n \quad (\mathbf{x}(W) \text{ encodes } W) .$$

At the decoding side, the information is recovered using a decoding function

$$g : B^n \to I \quad (g(\mathbf{y}) = \hat{W}) .$$

The information rate (in bits per channel symbol) is defined by

$$R = \frac{1}{n} \log_2 M .$$

If W is the output of the source, the probability of wrong decoding is denoted

$$\lambda(W) = \Pr [\hat{W} \neq W \text{ given that } \mathbf{x}(W) \text{ was sent}] .$$

We define the following quantity:

$$\lambda = \max_{W \in I} \lambda(W) \quad \text{(the maximum probability of error)} .$$

By definition, an (n,M,λ)-code is a code of blocklength n, size M (the number of code words), and maximum probability of error λ.

R is said to be an achievable rate if arbitrarily reliable communication can be achieved at this rate, or, mathematically, if there exists a sequence of constant rate

$$(n, 2^{nR}, \lambda^{(n)})\text{-codes}$$

such that $\lambda^{(n)} \to 0$ if n is allowed to grow without bounds.

The maximum rate at which arbitrarily reliable communication can be achieved is called the capacity C of the channel. The capacity C can also be defined in terms of information-thoretic quantities as follows.

Consider a probability distribution $p_X(x)$ on the input of the channel. This defines a probability distirbution on the output of the channel given by

$$p_Y(y) = \sum_{x \in A} p_X(x) \; p(y/x) .$$

We define the following quantities:

$$H(Y) = -E [\log_2 p_Y(Y)]$$
$$H(Y/X) = -E [\log_2 p(Y/X)]$$
$$I(Y;X) = H(Y) - H(Y/X) .$$

A well-known fact of information theory is expressed by

$$I(X;Y) = I(Y;X)$$

where $I(X;Y)$ is defined in an analogous way.
Then,

$$C' = \max_{p_X(x)} I(Y;X) .$$

For all discrete memoryless channels, we have C' = C. A coding theorem shows that C' ⩽ C, while so-called converse coding theorems show that C ⩽ C' (in fact, there are different types of converse coding theorems, but we shall not dwell on this).

A geometric interpretation of a coding-decoding scheme can easily be given, especially when $A = B$. The "code book" (the set of all code words) is represented by a set of points in A^n, one point for each source message. The decoder function partitions the space $B^n = A^n$ in disjoint decoding sets. When a code word is sent, the noise in the channel will be responsible for the fact that the received n-tuple may not coincide with the original code word. However, if the received n-tuple still falls in the right decoding set, it will be decoded properly. If the received n-tuple has been thrown out of the decoding set by the channel noise, an error will be made. This is illustrated in Fig. 2.

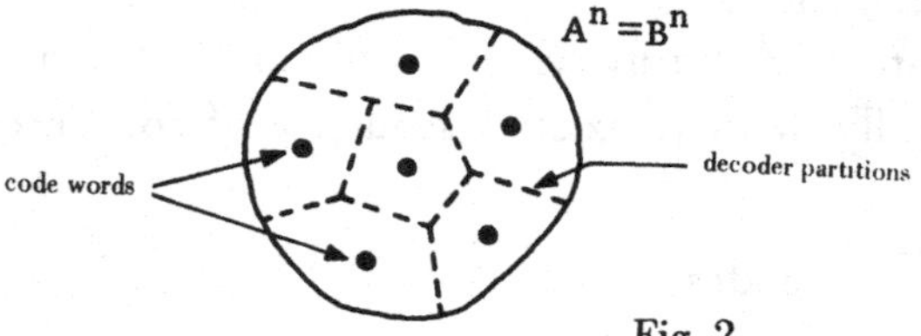

Fig. 2

Multiple Communication

The first problem in simultaneous communications was introduced by Shannon in 1961 [1]. It involves two-way communication over a channel with two input-output pairs as illustrated in Fig. 3.

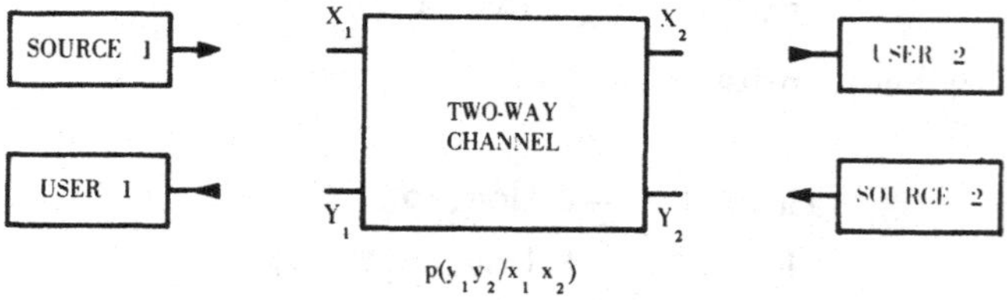

Fig. 3

The information from source 1 is intended for user 2, and the information from source 2 is intended for user 1. The encoders, at both sides of the channel, may use the information coming out of the channel at their side. This problem has not been completely solved yet, despite contributions from many authors: Jelinek [2], Libkind [3], Ahlswede [4] and Adeyemi [5].

Several other problems in simultaneous communications have been treated by various authors. In the sequel, we shall consider the broadcast channel, as introduced by T. Cover in 1972 [6]. Additional references on the broadcast channel can be found in [7 - 11].

A broadcast channel is a channel with a single input and several outputs. The transition probability of the channel is given by $p(y_1\, y_2 \,..\, y_N / x)$, where $x \in A$ is the input of the channel, and $y_i \in B_i$ is the ith output of the channel.

A broadcast channel can be used in many different communication situations. We shall restrict our attention to the so-called "one-to-one" case, represented in Fig. 4.

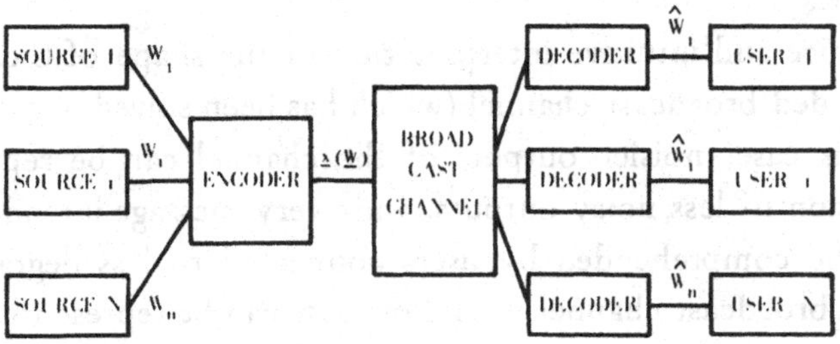

Fig. 4

Each of the sources is an i.i.d. source and the sources are independent from each other. Finally, the message from the ith source is intended for the ith user (hence the name "one-to-one"). The decoders operate independently. It is fairly obvious that in this situation, conditional dependence of the y_i's is irrelevant and that the channel can be decomposed in its marginal channel, or replaced by any other channel with same marginal transition probabilities $p_i(y_i/x)$.

The rate of the ith source is given by

$$R_i = \frac{1}{n} \log_2 M_i$$

where M_i is the number of possible messages of source i. Let $\mathbf{W} = (W_1, W_2, \ldots W_N)$ be the "composite" message. To encode this message, we use the code word $\mathbf{x}(\mathbf{W})$ from a code book of total size $M_1 M_2 .. M_N$.

The probability of error for the ith, transmission, when the composite message $\mathbf{W}$ was sent, is given by

$$\lambda_i(\mathbf{W}) = \Pr\,[\hat{W}_i \neq W_i \text{ given that } \mathbf{x}(\mathbf{W}) \text{ was sent}]\,.$$

Again, let

$$\lambda_i = \max_{\mathbf{W}} \lambda_i(\mathbf{W})$$

$\mathbf{R} = (R_1, R_2, .. R_N)$ is an achievable rate point (in N-dimensional space) if there exists a sequence of codes, with constant rate R_i for the ith transmission, such that $\lambda_i \to 0$ for all i simultaneously when $n \to \infty$. The set of all such achievable **R**'s is called the capacity region C. The genral problem of finding C has not been solved yet. Recent advances have established coding theorems and converses for a slightly different error concept (the "average" probability of error of each transmission goes to zero instead of λ). This is true for the one-to-one case as well for more complicated schemes.

A very nice and intuitive interpretation of the shape of C can be given in the case of a **degraded** broadcast channel (which has been solved in extenso, cf. [8] and [11]). In this case, noisier outputs of the channel can be represented as a statistical degradation of less noisy outputs, and every message intended for a given user can trivially be comprehended by users connected to less degraded outputs. Hence, a degraded broadcast channel with 2 outputs can be represented as in Fig. 5.

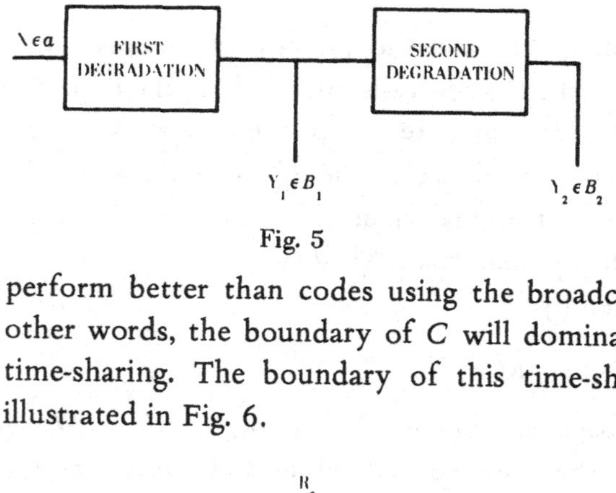

Fig. 5

For the sake of simplicity, we shall again assume that $A = B_1 = B_2$, but this is definitely not essential. Rate points on the boundary of C can be achieved for this class of broadcast channels by using so-called superposition codes. These codes will perform better than codes using the broadcast channel in a time-sharing mode. In other words, the boundary of C will dominate the set of rates achievable by simple time-sharing. The boundary of this time-sharing set is simply a straight line, as illustrated in Fig. 6.

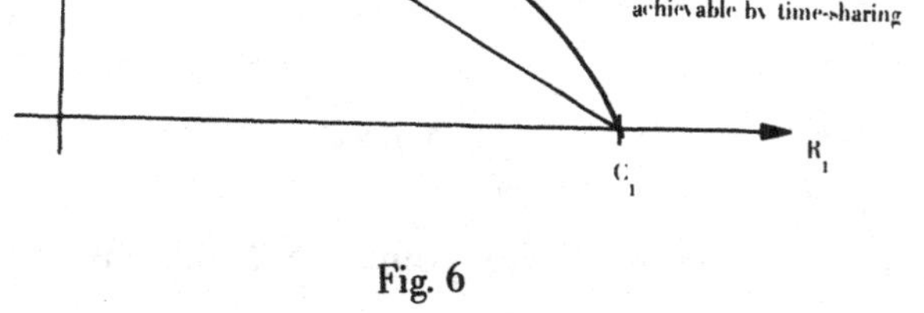

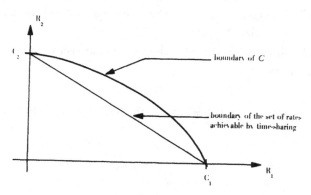

Fig. 6

In this figure, C_i is the capacity of the ith channel used as a simple channel. Obviously, $(C_1,0)$ and $(0,C_2)$ are achievable.

Superposition Codes

Let us consider the following example. We have a broadcast channel composed of a noiseless binary symmetric channel (BSC) and a BSC with cross-over probability p. We fix the following parameters:

$$n = 8$$
$$R_1 = 3/8 \text{ or } M_1 = 8$$
$$R_2 = 1/8 \text{ or } M_2 = 2 \text{ .}$$

Using time-sharing, we could reserve 3 bits for the transmission of the output of source 1 to user 1. Since this transmission is noiseless, 3 bits are just enough to transmit one of 8 possible messages ($M_1 = 8$). The remaining 5 bits can the be used to transmit one of 2 messages ($M_2 = 2$) from source 2 to user 2 over the noisy BSC. This scheme is represented in Fig. 7.

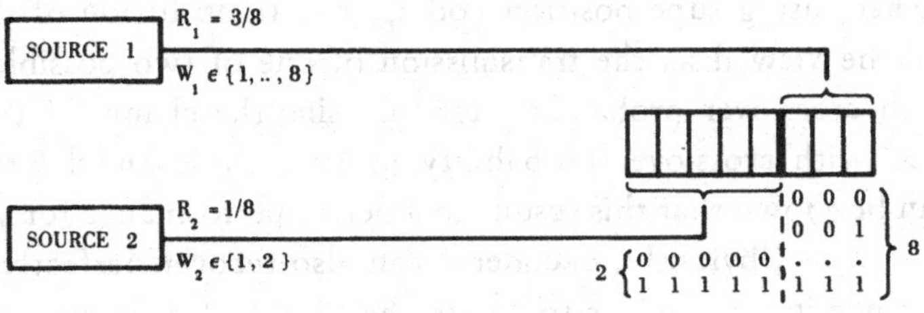

Fig. 7

We now propose another coding scheme. Suppose momentarily that we can use the 8 bits for the transmission of the output of source 2. We would need only two code words, and we would take them at a Hamming distance of 8 from each other. For example, we could use $\mathbf{u}_1 = 00000000$ and $\mathbf{u}_2 = 11111111$. Instead of shortening the $\mathbf{u}_i$'s to encode the message of source 2 like we did before, we shall keep the full length, and we shall superimpose this information, by defining the following "modified" code words:

$$\mathbf{x}(1,1) = 10000000 \qquad \mathbf{x}(1,2) = 01111111$$
$$\mathbf{x}(2,1) = 01000000 \qquad \mathbf{x}(2,2) = 10111111$$
$$\mathbf{x}(3,1) = 00100000 \qquad \mathbf{x}(3,2) = 11011111$$
$$\cdots\cdots \qquad \cdots\cdots$$
$$\mathbf{x}(8,1) = 00000001 \qquad \mathbf{x}(8,2) = 11111110$$

In words, if $W_2 = i$ and $W_1 = j$, use $\mathbf{u}_i$, and modify the jth bit.

We can represent the resulting code book in $\{0,1\}^8$ (Fig. 8).

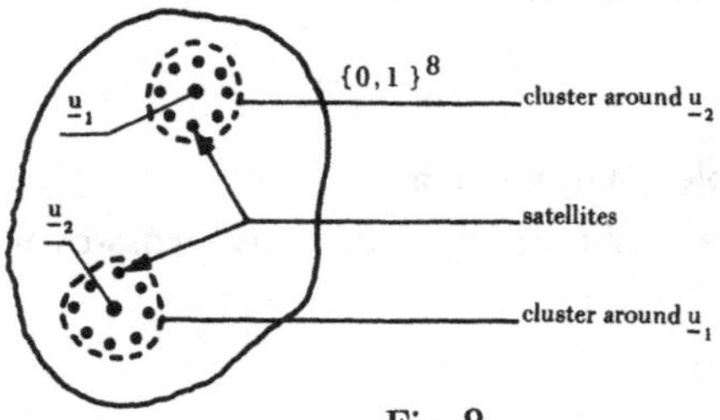

Fig. 8

The code words corresponding to the same message from source 2 are grouped in a cloud or cluster, around a cluster center (the $\mathbf{u}_i$'s in above example). The code words of a given cluster have been called the satellite code words of the cluster center.

In superimposing the message from source 1 (in other words, transmitting the satellite instead of the cluster center), we create an additional perturbation, which is more or less equivalent to passing the selected $\mathbf{u}_i$ first through a BSC with cross-over probability 1/8 (one of the 8 bits is modified), before sending it through the channel with cross-over probability p. The concatenation of 2 BSC with cross-over probabilities a and b results in a new BSC with cross-over probability given by

$$a*b = a(1-b) + b(1-a) .$$

Hence, using superposition coding, the transmission of the message from source 2 can be viewed as the transmission of one of two possible messages thorugh a BSC with cross-over probability $1/8*p$, using the channel 8 times, instead of through a BSC with cross-over probability p, using the channel 5 times as in time-sharing. It can be shown that this results in a better performance for some values of p.

Obviously, decoder 1 can also recover perfectly the message from source 1, since it is sufficient to locate the bit which is different, and since channel 1 does not introduce any perturbation.

We can now generalize the concept of superposition coding. The broadcast channel will henceforth be of the degraded type. This ensures that every information which can be received by decoder 2 can also be received by decoder 1. The decoding procedure of decoder 1 is a two-step procedure. First, the cluster center is estimated. Since this is actually the only job of decoder 2, this can also be done by decoder 1. The second step then consists in determining which of the satellites in the cluster was actually sent. This last step will actually be the one which will deliver to user 1 an estimate of the message of source 1.

As a generalization of our first example, we consider the binary symmetric broadcast channel (BSBC), with cross-over probabilties p_1 and p_2, such that (without loss of generality)

$$0 \leqslant p_1 < p_2 < \frac{1}{2} .$$

Obviously, this is a degraded broadcast channel.

Let the block length be n, and consider a number M_2 of cluster centers. Let α be the average fraction of modified bits in the superposition process (1/8 in above example). In other words, the distance between the satellites in a given cluster and the cluster center is α n. The resulting virtual BSC for transmission 2 has cross-over probability $\alpha * p_2$, and the resulting virtual capacity is given by

$$1 - h(\alpha * p_2).$$

Hence, if $R_2 = 1/n \log_2 M_2$ is smaller than above expression arbitrarily reliable communication is possible for transmission 2.

If $p_1 = 0$, we can pack

$$M_1 = \sum_{i=0}^{[n\alpha]} \binom{n}{i} \cong \binom{n}{[n\alpha]} \cong 2^{nh(\alpha)} .$$

distinguishable satellites in each cluster,

In this case, $R_1 = 1/n \log_2 M_1 = h(\alpha)$ is achievable for transmission 1, and the following rate pair is achievable:

$$\begin{cases} R_1 = h(\alpha) \\ R_2 = 1 - h(\alpha * p_2) . \end{cases}$$

If $p_1 > 0$, we cannot pack that many distinguishable (for decoder 1) satellites in each cluster, since the noise in channel 1 will perturb the transmission of the satellite. It can be shown that in this case, the following rates are achievable

$$\begin{cases} R_1 = h(\alpha * p_1) - h(p_1) \\ R_2 = 1 - h(\alpha * p_2) \end{cases}$$

α is a parameter in above expressions. By varying α it can be shown that one generates the boundary of C. These results are special cases of more general results on degraded broadcast channels. The interested reader should consult [6], [7] and [11].

Conclusions

We have introduced the notion of superposition coding as a natural coding scheme for degraded broadcast channels. It has been asserted that superposition coding yields better rates than time-sharing, in the sense that the capacity region achieved by superposition dominates the set of rates achievable by time-sharing. Rigorous proofs are available in the literature. So far, these results are valid for the asymptotic case only, since the block length must be allowed to grow without bounds. In the next paper, we show how these concepts can be used for finite n, and that superposition coding dominates time-sharing for practical communication problems.

PART TWO : SIMPLE SUPERPOSITION CODES FOR THE BROADCAST CHANNEL WITH ADDITIVE WHITE GAUSSIAN NOISE

Introduction

The reader is referred to the previous paper for the definition of the notation.

We shall now extend the notion of a broadcast channel to the continuous-amplitude discrete-time case, and we shall treat the addititve white Gaussian noise (AWGN) broadcast channel (BC) in a similar fashion. Again, we start by a brief review.

A "simple" AWGN channel is illustrated in Fig. 9. Z is the additive Gaussian noise of variance N (noise power).

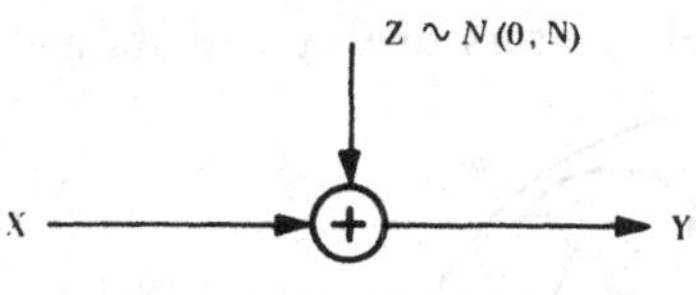

Fig. 9

Obviously, in this case, $\mathbf{x}(W) \in R^n$ (i.e. all code words are real n-tuples). Without additional constraints, it can easily be shown that the capacity of the AWGN channel is infinite. However, if we add a constraint on the maximum energy of each code word, or on the average energy of the code book, the capacity is now finite. These constraints are expressed by

$$\sum_{i=i}^{n} x_i^{\,2}(W) \leqslant nS \qquad \forall W \text{ (maximum energy)}$$

$$\frac{1}{M} \sum_{W \in I} \sum_{i=1}^{n} x_i^{\,2}(W) \leqslant nS \qquad \text{(average energy)} .$$

S is the transmission "power" (energy per dimension or per channel usage). Asymptotically, above constraints are equivalent (because of the so-called "sphere-hardening" property), and they both result in the following expression for the capacity of the AWGN channel:

$$C = \frac{1}{2} \ \ell n \left(1 + \frac{S}{N}\right) \qquad \text{(nats/transmission)} .$$

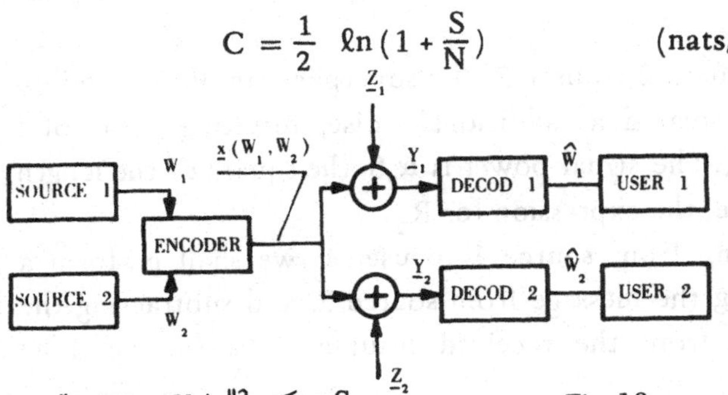

$\| \mathbf{x}(W_1 , W_2) \|^2 \leqslant nS$

Fig. 10

An AWGN BC, used in a one-to-one communication problem, as defined above, is represented in Fig. 10.

The noise vectors $\mathbf{Z}_1$ and $\mathbf{Z}_2$ have independent components, since the noise is white. However, possible component-wise dependence of $\mathbf{Z}_1$ and $\mathbf{Z}_2$ is irrelevant, again. The noise power in the two channels is N_1 and N_2 respectively.

We shall assume $N_2 > N_1$ without loss of generality.
It can be shown that the boundary of the capacity region is given by

$$\begin{cases} R_1 = \frac{1}{2} \ln\left(1 + \frac{\alpha_1 S}{N_1}\right) \\ R_2 = \frac{1}{2} \ln\left(1 + \frac{\alpha_2 S}{\alpha_1 S + N_2}\right) \end{cases} \qquad \alpha_i \geqslant 0,\ \alpha_1 + \alpha_2 = 1 .$$

The heuristic interpretation of these expressions is simple. A superposition code is used again, as depicted in Fig. 11. The cluster centers are at a distance $\sqrt{\alpha_2 nS}$ from the origin, while the satellite code words are at a distance $\sqrt{\alpha_1 nS}$ from the cluster centers. For very large n, the vector from the origin to the cluster center will be orthogonal or almost-orthogonal to most vectors from the cluster center to the satellite code word, resulting in a total energy of approximately

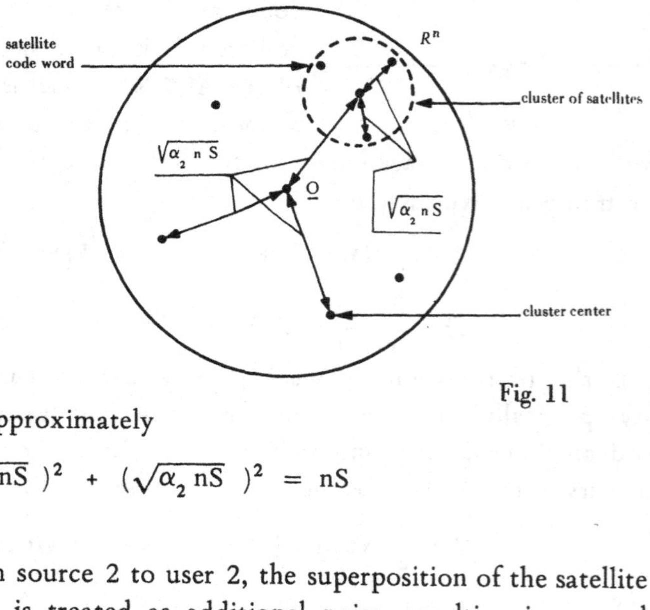

Fig. 11

$$(\sqrt{\alpha_1 nS}\ )^2 + (\sqrt{\alpha_2 nS}\ )^2 = nS$$

for each code word.

In transmitting from source 2 to user 2, the superposition of the satellite (information from source 1) is treated as additional noise, resulting in a total effective noise power of $\alpha_1 S + N_2$. The signal power is $\alpha_2 S$, the square of the length of the cluster center vectors. Hence, the expression for R_2.

However, in transmitting from source 1 to user 1, we shall perform a two-step decoding, first estimating the message from source 2, and subtracting the underlying cluster center vector from the received n-tuple. This can be done arbitrarily well (in the asymptotic sense) by decoder 2, and, a fortiori, by decoder 1, since $N_1 < N_2$. After the first step, decoder 1 recovers the information from source 1 as if this information had been transmitted with power $\alpha_1 S$ and corrupted by noise with power N_1, yielding above expression for R_1. A rigorous proof can be found in [12].

We shall now show how interesting results can be obtained in broadcast situation by superposing codes for the simple AWGN channel.

Permutation Codes for the Simple AWGN Channel

Not very many good codes are known for the AWGN channel. We shall investigate below the performance of a reasonably large class of codes, the so-called permutation codes, used in time-sharing mode and superposition mode on an AWGN BC.

Permutation codes were introduced by Slepian in 1965 [13]. There are two classes of permutation codes, but we shall be concerned here with Variant I codes only. Starting with one vector in the code book, all other vectors are generated by permuting in all possible ways the components of this vector. If all the components of the initial vector are different, we generate a code book of size $M = n!$. However, if the initial vector contains ν_1 times the number μ_1, ν_2 times the number μ_2, ..., ν_k times the number μ_k, with $\mu_1 < \mu_2 < \ldots < \mu_k$ and $\nu_1 + \nu_2 + .. + \nu_k = n$, the number of different permutations is smaller, and the size of the code book is given by

$$M = \frac{n!}{\nu_1! \, \nu_2! \ldots \nu_k!} \, .$$

Permutation codes have the following interesting properties:

1 – All vectors have the same energy.

2 – The maximum likelihood detection of a permutation code in AWGN is very simple. Consider a received vector $(Y_1, Y_2, .., Y_n)$; replace the ν_1 smallest Y_i's by μ_1, the ν_2 next smallest Y_i's by μ_2, and so on. The resulting vector is a codeword, and is the maximum likelihood estimate.

3 – From above method, it is obvious that the decoding regions do not depend on the μ's, only on the ν's. Hence, for a given energy nS, the μ's can be chosen (by computer search) to minimize the probability of error.

4 – PCM (pulse code modulation), orhtogonal codes (PPM) and biorthogonal codes are all special cases of permutation codes (the latter of Variant II codes).

Further details on permutation codes can be found in Slepian's paper.

Permutation Codes for the AWGN Broadcast Channel

In the sequel, we shall compare the performance of permutation codes used in time-sharing mode vs. superposition mode. In each of the following examples, we fix in advance the following parameters:

$$\begin{cases} n & \text{the length of one transmission (for both sources)} \\ M_1 & \text{the size of the message set of source 1} \\ M_2 & \text{the size of the message set of source 2} \\ S/N_1 & \text{the signal-to-noise ratio in channel 1} \\ S/N_2 & \text{the signal-to-noise ratio in channel 2} \end{cases}$$

To operate in time-sharing, we partition the blocklength in two intervals of length n_1 and n_2, such that $n_1 + n_2 = n$. Then, n_1 dimensions are used to transmit the output of source 1 with a permutation code of size M_1 ,which we shall denote an (n_1, M_1)-code, and n_2 dimensions are used to transmit the output of source 2, with an (n_2, M_2)-code. Let the energy of code i be E_i $(i = 1,2)$. We shall require $E_1 + E_2 = nS$. By letting E_1 vary between o and nS, with $E_2 = nS - E_1$, we generate a set of simultaneously achievable probabilities of error, using the maximum likelihood detection described above.

Except for simple cases, the probability of error of a permutation code must be calculated by numerical integration, the exact calculation involving complicate multiple integrals which cannot be solved analytically. In some cases, it is even sipler to use very long run simulations to yield the probability of error with a good degree of accuracy. Many of the results which follow were obtained by D. Arantes, of Cornell University.

Finally, the set thus obtained holds for the adopted values of n_1 and n_2 and will be different for other partitions of the total available transmission length. However, in most cases, it will not be possible to find an (n_1, M_1) permutation code and an (n_2, M_2) permutation code for all pairs n_1, n_2 such that $n_1 + n_2 = n$.

To operate in superposition mode, we now try to find an (n, M_1)

permutation code and an (n,M_2) permutation code, with energies E_1 and E_2 respectively, again such that $E_1 + E_2 = nS$. To transmit simultaneously the output of the two information sources, we select the corresponding code word in each code and simply add them together. The decoders operate as described above: decoder 2 simply attempts to recover message 2, while decoder 1 performs its operation in two steps. This decoding method is definitely suboptimal for finite blocklengths, but the results obtained in the example below and its easy implementation make it rather attractive.

By varying E_1 and E_2 within the given constraint, we can again generate the set of simultaneously achievable probabilties of error. The range of greatest interest will be for

$$0 \leqslant E_1 \leqslant \frac{nS}{5} \quad \text{(approximately)}$$

where the satellites can truly be regarded as additional noise, and where the suboptimal method of detection is very nearly optimal.

Finally we should point out that superposing two codes of energies E_1 and E_2 such that $E_1 + E_2 = nS$ will generally not result in a code in which all code words have an energy smaller than or equal to nS. The average energy of the code is still nS, however, and for the range mentioned above, the energy of each vector will be within a few percents of nS.

Example 1

Consider the following parameters :

$$\begin{cases} n & = 10 \\ M_1 & = 10 \\ M_2 & = 10 \\ S/N_1 & = 10 \text{ dB} \\ S/N_2 & = 1 \text{ dB} \end{cases}$$

In time-sharing, we use two identical permutation codes, with $n_1 = n_2 = 5$, with composition $\nu_1 = 3$ and $\nu_2 = 2$, yielding

$$M_1 = M_2 = \frac{5!}{2!\ 3!} = 10 .$$

In superposition, the two identical codes have composition $\nu_1 = 9$ and $\nu_2 = 1$, yielding the same M_1 and M_2.

The results are given in Fig. 12, where pe_i is the average probability of error for transmission i.

Example 2

This example involves better channels and codes with higher rate. We have:

$$\begin{cases} n = 10 \\ M_1 = 120 \\ M_2 = 120 \\ S/N_1 = 14 \text{ dB} \\ S/N_2 = 8 \text{ dB} \end{cases}$$

In time-sharing, we consider two identical codes, with composition $\nu_i = 1$, $i = 1, \ldots, 5$

In superposition, the codes have composition $\nu_1 = 7$ and $\nu_2 = 3$, with

$$M_1 = M_2 = \frac{10!}{7! \quad 3!} = 120$$

The curves are given in Fig. 13.

Example 3

Here we consider two different rates. Specifically, let

$$\begin{cases} n = 10 \\ M_1 = 10 \\ M_2 = 120 \\ S/N_1 = 14 \text{ dB} \\ S/N_2 = 8 \text{ dB} \end{cases}$$

The codes of examples 1 and 2, used both in time-sharing mode and in superposition mode, yield remarkable results. See Fig. 14.

Example 4

To confirm the domination of superposition codes over time-sharing

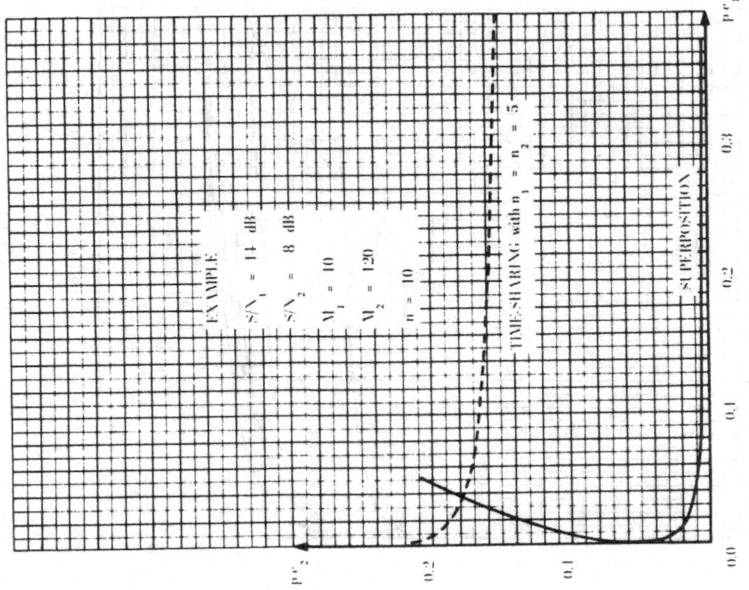
EXAMPLE
S/N1 = 14 dB
S/N2 = 8 dB
M1 = 10
M2 = 120
n = 10
TIME-SHARING with n1 = n2 = 5
SUPERPOSITION
pe1
pe2
0.0
0.1
0.2
0.3

Fig. 14

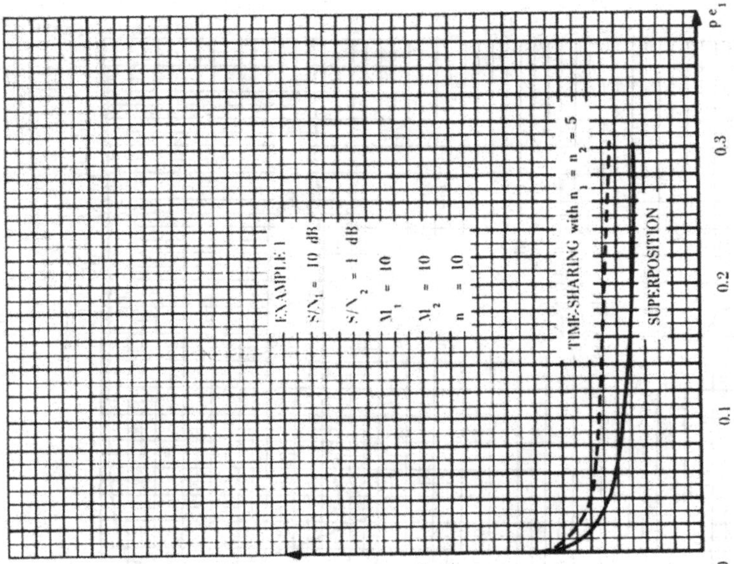
EXAMPLE 1
S/N1 = 10 dB
S/N2 = 1 dB
M1 = 10
M2 = 10
n = 10
TIME-SHARING with n1 = n2 = 5
SUPERPOSITION
pe1
pe2
0.0
0.1
0.2
0.3

Fig. 12

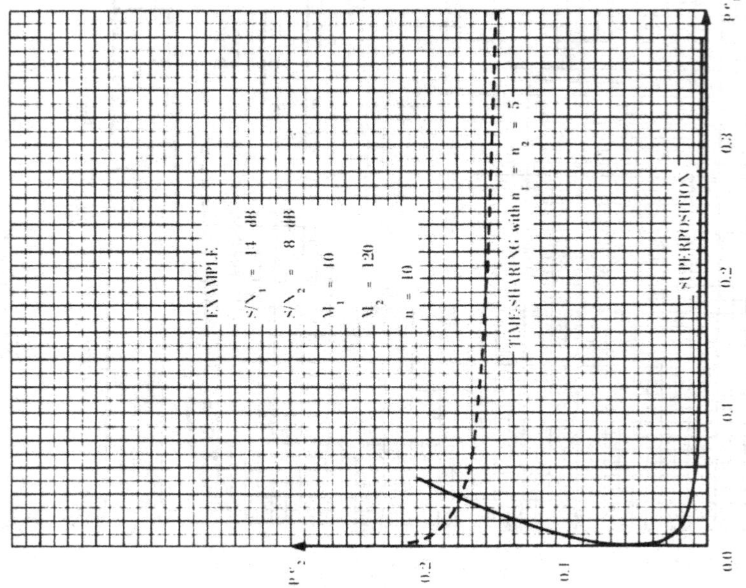
EXAMPLE
S/N1 = 14 dB
S/N2 = 8 dB
M1 = 10
M2 = 120
n = 10
TIME-SHARING with n1 = n2 = 5
SUPERPOSITION
Pe2
Pe1
0.0
0.1
0.2
0.3

Fig. 14

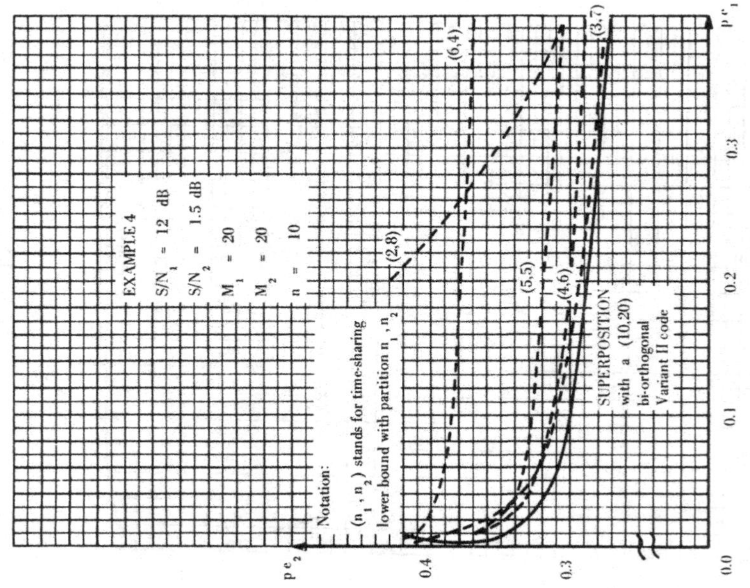
EXAMPLE 4
S/N1 = 12 dB
S/N2 = 1.5 dB
M1 = 20
M2 = 20
n = 10
Notation:
(n1 , n2) stands for time-sharing lower bound with partition n1 , n2
(6,4)
(2,8)
(5,5)
(4,6)
(3,7)
SUPERPOSITION with a (10,20) bi-orthogonal Variant II code
Pe2
Pe1
0.4
0.3
0.0
0.1
0.2
0.3

Fig. 15

codes, one should try other partitions of the blocklength (i.e. other values of n_1 and n_2). However, there are no permutations codes (n_1, M_1) and (n_2, M_2) for many possible partitions. Instead, we can use the lower bound for the probability of error for (n,M) codes (not necessarily permutation codes) on a Gaussian channel (cf. Shannon [14], Slepian [15]). With $S/N_1 = 12$ dB and $S/N_2 = -1.5$ dB, Fig. 15 shows clearly that no time-sharing code, with whatever time division, will dominate the superposition code used in this example. The low signal-to-noise ratio in the second channel is responsible for rather high values of Pe_2, but this does not invalidate the general result. A bi-orthogonal (10, 20) (Variant II) permutation code was used in superposition.

Conclusions

We have exhibited simple examples where superposition codes clearly dominate time-sharing codes, as predicted by the asymptotic results. We hope that these few cases will trigger additional research in coding applications of the results of information theory in simultaneous communications.

References

[1] Shannon, C.E., "Two-Way Communication Channels", Proc. Fourth Berkeley Symposium on Math. Stat. and Probability, 1961, Vol. I, pp. 611-644.

[2] Jelinek, F., "Coding for and Decomposition of Two-Way Channels", IEEE Trans. Inform. Theory, Vol. IT-10, No. 1, January 1964, pp. 5-17.

[3] Libkind, L.M., "Two-Way Discrete Memoryless Communication Channels", Problemy Peredachi Informatsii, Vol. 3, No. 2, 1967, pp. 37-46.

[4] Ahlswede, R., "Multi-Way Communication Channels", presented at the Second International Symposium on Information Theory, at Tsadadsor, Armenian SSR, 1971.

[5] Adeyemi, H.D., "Capacity Regions for Two-Way Channels", Ph.D. Dissertation, 1971. Dept. of Statistics, University of California, Berkeley, Calif., USA.

[6] Cover, T.M., "Broadcast Channels", IEEE Trans. Inform. Theory, Vol. IT-18, No. 1, January 1972, pp. 2-14.

[7] Bergmans, P.P., "Random Coding Theorem for Broadcast Channels with Degraded Components", IEEE Trans. Inform. Theory, Vol. IT-19, No. 3, March 1973, pp. 197-207.

[8] Gray, R.M. and Bergmans P.P., "Two Problems in Simultaneous Communications", IEEE Trans. Communications, Vol. CCM-21, No. 6, June 1973, pp. 763-768.

[9] Bergmans, P.P., "A Simple Converse for Broadcast Channels with Additive White Gaussian Noise", IEEE Trans. Inform. Theory, Vol. IT-20, No. 3, March 1974, pp. 279-280.

[10] Bergmans, P.P. and Cover T.M., "Cooperative Broadcasting", IEEE Trans. Inform. Theory, Vol. IT-20, No. 5, May 1974, pp. 317-324.

[11] Gallager, R.G., "Coding for Degraded Broadcast Channels", to appear.

[12] Bergmans, P.P., "Degraded Broadcast Channels", Ph.D. Disertation, 1972. Dept. of Electrical Engineering, Stanford University, Stanford, Cal., USA.

[13] Slepian, D., "Permutation Modulation", Proc. IEEE, VOl. 53, March 1965, pp. 228-236.

[14] Shannon, C.E., "Probability of Error for Optimal Codes in a Gaussian Channel", Bell Sys. Tech. Journal, Vol. 38, No. 3, May 1959, pp. 611-656.

[15] Slepian, D., "Bounds on Communication", Bell Sys. Tech. Journal, Vol. 42, No. 3, May 1963, pp. 681-707.

ALGORITHMS
FOR
SOURCE CODING

by

F. Jelinek, IBM Thomas J. Watson Research
Center, Yorktown Heights, N.Y., U.S.A.,

and

G. Longo, Istituto di Elettronica,
Università di Trieste, Trieste, Italy

1. The Source Coding Problem Posed

Consider an **information source** whose outputs are independent identically distributed random variables (i.i.d. r.v.s) taking values in the finite set or **alphabet** $J_c = \{0,1,\ldots,c-1\}$; a **coding alphabet** $J_d = \{0,1,\ldots,d-1\}$; and a noiseless transmission medium, or **channel**, that accepts letters from J_d as inputs and delivers them to the destination without any change (see figure 1).

The following problem is then relevant, and will be considered all along these notes:

Problem: How to encode finite-length sequences of source outputs $z_1, z_2, \ldots$ into sequences or **codewords** $s_1, s_2, \ldots$ of channel inputs so that

a) the receiver will be able to recover the sequences $z_1, z_2, \ldots$ from the corresponding sequences $s_1, s_2, \ldots$ and this without errors;

b) the smallest number of letters s is used to encode the letters z.

Fig. 1.1 Block-diagram of a noiseless communication system

A naive approach to this problem would indicate a **block-to-block** encoding as the solution, i.e. sequences $\mathbf{z}^k = z_1 \ldots z_k$ all of the same length k should be encoded into codewords $\mathbf{s}^n = s_1 \ldots s_n$ all of the same length n. Obviously this can be done provided

$$d^n \geqslant c^k > d^{n-1} ,$$

which implies that the rate R, defined as n/k log d, be approximately equal to log c*. One may ask whether this procedure is optimal. The answer is that one can do better, and precisely the following encoding procedures can be adopted:

1) **Block-to-variable length (BV) encoding**, where the codeword length n is a random variable that depends on the particular source sequence

*) Unless otherwise indicated, here and in the sequel all logarithms are to the base 2.

being encoded. The relevant parameter is now

$$\frac{R}{\log d} = \frac{E[n]}{k},$$

where E[n] stands for the average of the r.v. n.

2) **Variable length-to-block** (V B) encoding, where variable length sequences of source symbols are encoded into fixed length codewords. The quantity of interest here is

$$\frac{R}{\log d} = \frac{n}{E[k]}.$$

3) **Variable length-to-variable length** (VV) encoding, where both the length of the source sequences and of the codewords are r.v.s, and therefore one should try to minimize the ratio

$$\frac{R}{\log d} = \frac{E[n]}{E[k]}.$$

4) **Block-to-block** (BB) encoding, where, however, not all of the c^k k-length source sequences get their own codeword. Here

$$R = \frac{n}{k},$$

but an occasional decoding error may result, and specifically when the information source outputs a sequence whose codeword does not correspond to that source sequence only.

We shall be concerned with approaches 1 and 4 first, and mention approach 2 later.

The notation used in this paper is generally that of [1], and to this book the reader is referred for prerequisites as well as for a deeper treatment of certain portions of this article.

2. Block-to-Variable Length Encoding

Let us now introduce some more precise terminology and notation.

Let J_c be the set $\{0,1,..., c-1\}$, where $c \geqslant 2$ is an integer; a sequence z_1, $z_2,..., z_k$ of k symbols from J_c will be denoted by $\mathbf{z}^k$:

(2.1) $$\mathbf{z}^k \triangleq z_1 z_2 \dots z_k \qquad (z_i \in J_c).$$

Definition 2.1. A **code** (of order k) is a mapping Φ from all sequences $\mathbf{z}^k$, $z_i \in J_c$, into sequences of symbols from some alphabet J_d, i.e.

(2.2) $$\Phi(\mathbf{z}^k) = s_1(\mathbf{z}^k)\ s_2(\mathbf{z}^k) \dots s_m(\mathbf{z}^k),$$

where m depends on $\mathbf{z}^k$: $m = m(\mathbf{z}^k)$, and $s_i \in J_d$. The sequences $\Phi(\mathbf{z}^k)$ are called **codewords.**

If the sequences from J_c are of different lengths (i.e. if the code is variable-length), there must be some way for the decoder to find where a codeword ends and the next one begins. The simplest device could be the "comma insertion", where the encoder would use one of its output symbols as a codeword separation marker, or "comma". Since this symbol could not be used inside any codeword, it would be underutilized, although it has been shown [2] that the resulting increase in the average codeword length m is not necessarily too serious. We now proceed, however, to the description of the **prefix coding** method which does not use any comma and yields the shortest average codeword length $\overline{m}$.

Consider a potentially infinite **rooted tree** in which d branches leave each node, d being the size of the source coder output alphabet J_d. Let the branches stemming from each node be labelled by 0,1,..., d−1; see figure 2.1, where the first three levels of such a tree are shown in the ternary case (d = 3).

Let now each node of the tree be associated with the sequence of symbols encountered going from the root to the node itself along the tree branches. As an example, the three encircled nodes in figure 2.1 correspond to the sequences (1 0 0), (1 0), (2 2), respectively. This implies that any set of d-ary sequences is associated with a set of nodes, and therefore that any d-ary code is associated with such a set, whose nodes are referred to as **code nodes.** Note next that if the codewords are chosen in such a way that no path leading from the root to a code node ever goes through another code node, then the decoder is able to insert its own commas in the ongoing sequence of received symbols. Actually, the decoder looks at the received sequence and traces the corresponding path along the tree until a code node is reached, this is recognized as such, the decoder inserts a comma, then goes back to the root and traces the next path up to a code node, and so on.

A code whose code nodes have this property is called a prefix code, this name being a consequence of the following definition.

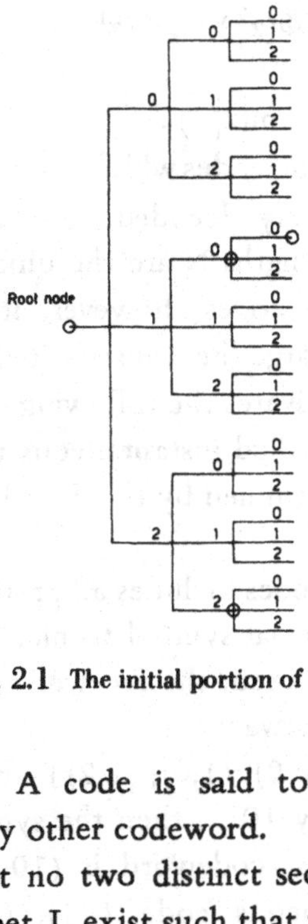

Fig. 2.1 The initial portion of a ternary tree

Definition 2.2. A code is said to satisfy the prefix condition if no codeword is a prefix of any other codeword.

This means that no two distinct sequences z_1^k, z_2^k of symbols from the source coder input alphabet J_c exist such that their codewords $\Phi(z_1^k)$, $\Phi(z_2^k)$ satisfy the following equality

$$\Phi(z_1^k) = [\Phi(z_2^k), s^h] , \tag{2.3}$$

where s^h is some sequence of $h \geq 0$ symbols from J_d. Remark that the prefix condition implies that distinct sequences have distinct codewords, which is obvious from (2.3) when $h = 0$.

The code corresponding to the encircled nodes of figure 2.1 is not prefix, since (1,0) is a prefix of (1,0,0). The code { (1,0,0), (1,0), (2,2) }, however, can always be decoded, because no codeword begins with a 0, and the receiver can always find the codeword end simply by looking one symbol ahead. This example illustrates that the prefix condition is sufficient but not necessary for the unique decipherability of the code, which is now defined:

Definition 2.3. A code Φ is **uniquely decipherable** if, given any integer n and any sequence of message blocks $\mathbf{z}_1^k, \ldots, \mathbf{z}_n^k$, the sequence

$$s^\ell = [\Phi(\mathbf{z}_1^k), \ldots, \Phi(\mathbf{z}_n^k)]$$

can be uniquely decoded (here $\ell = \sum_{i=1}^{n} m(\mathbf{z}_i^k))$.

We remark that there might be codes which for a specific N yield sequences $[\Phi(\mathbf{z}_1^k), \ldots, \Phi(\mathbf{z}_N^k)]$ that can be uniquely decoded; however, the only codes that can be decoded for all transmission lengths N are the uniquely decipherable codes.

For uniquely decipherable codes, however, it might happen that the receiver is to be told in advance how large the transmission length is going to be, and this is a shortcoming. This remark motivates the following definition.

Definition 2.4 A code is called **instantaneous** if the position of the last symbol of every codeword can be determined by the decoder without looking at any subsequent symbol.

The class of instantaneous codes includes all prefix codes and block codes, as well as those codes which use a special symbol to mark the end of a codeword. When an instantaneous code is being used, the receiver need not be told the total transmission length, and this is a great advantage.

Remark that the code {(1,0,0), (1,0), (2,2)} above is not instantaneous, and if a codeword sequence begins by 10..... then the symbol next to 0 has to be looked at to decide whether the first codeword is (100) or (10). As a further example, consider the non-instantaneous code { (1), (10), (00) }, whereby a sequence like 1000...0 can be decoded only when it is known whether the number of zeros is even or odd.

Going back to prefix codes, it is rather obvious that it si convenient to represent them by the limited portion of the rooted tree they actually use. E.g. figure 2.2 illustrates the prefix code { (00), (01), (100), (101), (1100), (1101), (1110), (1111) }, whose codewords correspond to the terminal tree nodes.

A necessary and sufficient condition for the existence of a prefix code with a given set of codeword lengths is provided by the so-called **Kraft-Szilard inequality** [3], illustrated by the following lemma.

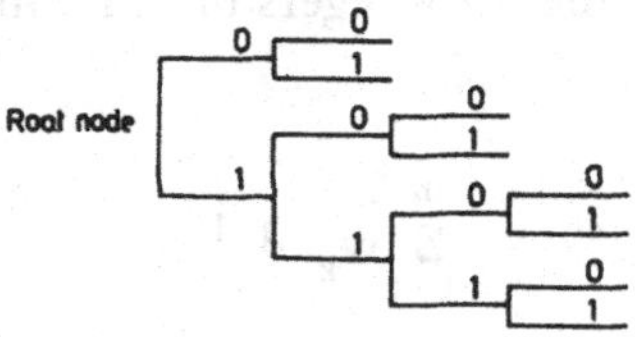

Fig. 2.2 **Tree representation of a binary prefix code**

Lemma 2.1

There exists a prefix code of size g and alphabet J_d with word lengths $m_1, m_2, \ldots m_g$ (the m_i's are positive integers) if and only if

$$\sum_{i=1}^{g} d^{-m_i} \leqslant 1. \tag{2.4}$$

Proof. (The lemma can be proved in several ways; the proof we give here exploits the structure of the tree corresponding to the code). We first prove the necessity of (2.4). Put

$$m \triangleq \max_{i=1,\ldots,g} \{m_i\}.$$

If we define the level of a node as its distance (in branches) from the root node, then a codeword of length m_i corresponds to a code node of level m_i. In a prefix code no path leading to a code node can be continued to reach any other code node, and therefore a codeword of length m prevents the existence of exactly d^{m-m_i} code nodes on level m, $m \geqslant m_i$. On the other hand, the total number of code nodes on level m whose existence is prevented by the code nodes situated at lower levels, cannot be greater than d^m, which is the total number of tree nodes on level m. Therefore

$$\sum_{i=1}^{g} d^{m-m_i} \leqslant d^m , \tag{2.5}$$

whence immediately (2.4) follows by multiplying both sides by d^{-m}.

We now prove the sufficiency of condition (2.4). We first write $\sum_{i=1}^{g} d^{-m_i}$ in a different way. If w_k is the number of integers in $\{m_1, m_2, \ldots, m_g\}$ equal to k, for $k = 1,2,\ldots,m$, then

$$\sum_{i=1}^{g} d^{-m_i} = \sum_{k=1}^{m} w_k d^{-k} ,$$

so that (2.4) implies:

$$\begin{aligned} & w_1 d^{-1} \leqslant 1 \\ & w_2 d^{-2} + w_1 d^{-1} \leqslant 1 \\ & \ldots\ldots\ldots \\ & w_m d^{-m} + \ldots + w_1 d^{-1} \leqslant 1, \end{aligned}$$

or

$$\begin{aligned} & w_1 \leqslant d \\ & w_2 \leqslant d^2 - w_1 d \\ & \ldots\ldots\ldots\ldots \\ & w_m \leqslant d^m - w_1 d^{m-1} - \ldots - w_{m-1} d^{-1} . \end{aligned} \tag{2.6}$$

The first inequality in (2.6) states that the number w_1 of codewords of length 1 is not larger than the number of tree nodes of level one, so that m_1 such nodes can be used as code nodes; the second inequality in (2.6) states that the number w_2 of codewords of length two is not larger than the number of tree nodes of level two which are available once w_1 d of them have been forbidden from being code nodes by the existence of the w_1 first-level code nodes. And so on: the other inequalities in (2.6) simply state the existence of sufficiently many available nodes at the higher levels up to the m-th. The last inequality asserts that w_m, the number of codewords of length m, is not larger than the number of tree nodes available on the m-th level once those prevented from being used by the existence of the shorter codewords have been removed. In conclusion (2.6) shows the possibility of placing the necessary code nodes in the tree at levels $m_1, m_2, \ldots, m_g$ so as to satisfy the prefix condition. Q.E.D.

We remark that the above proof is constructive, i.e. shows a way of actually constructing a code whose codewords have lengths $\{m_i\}$ $(1 \leqslant i \leqslant g)$

satisfying (2.4). Without loss of generality we may assume that $m_1 \leqslant m_2 \leqslant \dots \leqslant m_g$; then we form a d-ary rooted tree up to the m_g–th level. Next we choose any node at level m_1 and eliminate all the $d^{m_g - m_1}$ terminal nodes stemming from it; in the remaining portion of the tree we choose any surviving node at level m_2 and eliminate all the $d^{m_g - m_2}$ terminal nodes stemming from it. We continue this way until, having deleted $\sum_{i=1}^{g-1} d^{m_g - m_i}$ terminal nodes from the original tree, one is left with at least one node at level m ; one then chooses one node at this level and eliminates the (possibly) remaining terminal nodes. The resulting tree corresponds to the desired code and its codewords can be read directly by going, along the surviving branches, from the root up to each code node.

We only mention, without proof, the following more general result:

Proposition (McMillan's inequality)

The inequality

$$\sum_{i=1}^{g} d^{-m} \leqslant 1 \tag{2.4}$$

is also a necessary and sufficient condition for the existence of a uniquely decipherable code of word lengths $\{m_i\}_{i=1}^{g}$ and alphabet J_d.

So far we have been looking for the possibility of uniquely (and, possibly, instantaneously) recovering the information contained in the encoded sequence.

It is also very important to construct codes whose average codeword length $\bar{m}$ be as small as possible. The following theorem sets a precise lower bound to $\bar{m}$, which is precisely defined as follows:

$$\bar{m} = \sum_{\mathbf{z}^k} P(\mathbf{z}^k)\ m(\mathbf{z}^k)\ , \tag{2.7}$$

where $P(\mathbf{z}^k)$ is the probability of the block $\mathbf{z}^k$ and $m(\mathbf{z}^k)$ is the length of the corresponding codeword $\Phi(\mathbf{z}^k)$.

Theorem 2.1

The average length $\bar{m}$ of every uniquely decipherable code that encodes sequences of length k of symbols generated by a discrete memoryless source of entropy H(Z) satisfies the inequality

$$\bar{m} \geqslant \frac{k}{\log d}\quad H(Z)\ , \tag{2.8}$$

where d is the size of the code alphabet. On the other hand there exist prefix codes for which

$$\bar{m} < \frac{k}{\log d} H(Z) + 1. \tag{2.9}$$

Proof. Since the probabilities $P(\mathbf{z}^k)$ are given, the numbers $[\log_d P(\mathbf{z}^k)]$ are perfectly defined (here $[x]$ is the integral part of x). If we now set $m(\mathbf{z}^k) = -[\log_d P(z^k)]$, then the $m(\mathbf{z}^k)$ are the codeword lengths of a code satisfying (2.4), as it follows from observing that $-m(\mathbf{z}^k) = [\log_d \mathbf{P}(\mathbf{z}^k)] \leqslant \log_d \mathbf{P}(\mathbf{z}^k)$, and therefore

$$\sum_{\mathbf{z}^k} d^{-m(\mathbf{z}^k)} \leqslant \sum_{\mathbf{z}^k} d^{\log_d P(\mathbf{z}^k)} = \sum_{\mathbf{z}^k} \mathbf{P}(\mathbf{z}^k) = 1.$$

By Lemma 2.1, then, there exists a prefix code having these numbers as codeword lengths and such a code can be easily constructed. Its average length, according to definition (2.7) is given by

$$\bar{m} = \sum P(\mathbf{z}^k)\, m(\mathbf{z}^k) < \sum \mathbf{P}(\mathbf{z}^k)\, [-\log_d \mathbf{P}(\mathbf{z}^k) + 1] =$$

$$= \frac{k}{\log d} H(Z) + 1$$

(the last equality follows from the memoryless character of the source). This proves (2.9); to prove the first part of the theorem, we use the following inequality

$$\sum_{i=1}^{n} p_i \log \frac{p_i}{q_i} \geqslant 0, \tag{2.10}$$

where $\{p_i\}$ and $\{q_i\}$ are any two probability distributions, i.e. $p_i \geqslant 0$, $q_i \geqslant 0$, $\sum_{i=1}^{n} p_i = \sum_{i=1}^{n} q_i = 1$, which is an immediate consequence of the elementary inequality

$$\log_e x \leqslant x - 1. \tag{2.11}$$

In (2.11) equality holds iff $x = 1$, and in (2.10) equality holds iff $p_i = q_i$ $(1 \leqslant i \leqslant n)$. Now, from (2.10) it follows in particular that

$$\sum P(\mathbf{z}^k) \log \frac{\mathbf{P}(\mathbf{z}^k)}{Q(\mathbf{z}^k)} \geqslant 0, \tag{2.12}$$

where $Q(\mathbf{z}^k) \triangleq d^{-m(\mathbf{z}^k)}/\sum_{\mathbf{z}^k} d^{-m(\mathbf{z}^k)}$. Since by hypothesis $\sum_{\mathbf{z}^k} d^{-m(\mathbf{z}^k)} \triangleq K \leqslant 1$, inequality (2.12) implies

$$0 \leqslant \sum_{\mathbf{z}^k} P(\mathbf{z}^k) \log P(\mathbf{z}^k) - \sum_{\mathbf{z}^k} P(\mathbf{z}^k) \log d^{-m(\mathbf{z}^k)} + \log K ,$$

whence

$$0 \leqslant -k\,H(Z) + \sum_{\mathbf{z}^k} P(\mathbf{z}^k)\, m(\mathbf{z}^k) \log d ,$$

because $\log K \leqslant 0$. This proves (2.8) on account of (2.7). Q.E.D.

As a consequence of this theorem, we see that for every uniquely decipherable (in particular prefix) code

$$R \geqslant H(Z) , \tag{2.13}$$

where the quantity R, defined by

$$R \triangleq \frac{1}{k}\, \bar{m} \log d , \tag{2.14}$$

can be called again the rate of the code. On the other hand prefix codes exist whose rate satisfies

$$R < H(Z) + \frac{1}{k} . \tag{2.15}$$

This result provides a particularly interesting interpretation for the source entropy H(Z), in terms of the average length per source letter, $\bar{m}/k$, but leaves two very important questions open:

1) What is the encoding scheme corresponding to the lowest rate R for a given k, while preserving uniquely decipherability?
2) When k is large, how can we encode in such a way that the encoding complexity does not increase exponentially with k, as it does when a dictionary-type scheme is used?

Let us deal first with problem (1), which can be restated as follows: M d-nary codewords are to be used, having probabilities $p_1 \geqslant p_2 \geqslant \ldots \geqslant p_m$ ($\sum_{i=1}^{M} p_i = 1$); what prefix code minimizes the average codeword length $\bar{m}$?

The answer is the **Huffman code** [4]. We shall not prove its optimality here (see [5]), but shall describe the procedure which gives it for d = 2:

1. Let $q_j^0 = p_j$ $(j = 1,2,\ldots, M(0) = M)$. To each j assign the code sequence $S_j^0 = \{j\}$; set $S^0 = \{S_j^0\} = \{\{j\}\}$. Set k = 0.
2. Let $p_j' = q_j^k$; $V_j = S_j^k$ $(j = 1,2,\ldots, M(k)-2)$; $P'_{M(k)-1} = q^k_{M(k)-1} + q^k_{M(k)}$;
   $V_{M(k)-1} = \{0, S^k_{M(k)-1}\} \cup \{1, S^k_{M(k)}\}$, $M(k+1) = M(k) - 1$.
   [Here $\{0,V\}$ denotes the set of sequences V each of which has been prefixed by the digit 0.]
3. If $M(k+1) = 1$ go to 6, else continue.
4. Let $q_1^{k+1} \geqslant q_2^{k+1} \geqslant \ldots \geqslant q^{k+1}_{M(k+1)}$ be a merge of the sequence $p_1', p_2', \ldots, p'_{M(k)-2}$ with the element $p'_{M(k)-1}$ and let $S^{k+1} = \{S_j^{k+1}\}$ $(j = 1,2, \ldots, M(k+1))$ be the corresponding rearrangement of the set $\{V_j\}$
5. Set $k = k + 1$, and go to 2.
6. The set of codewords is $\{V_1\}$. The last digit of each codeword denotes the message to which the codeword corresponds.

Example 2.1.

Let M = 8, and $p_1 = p_2 = 0.25$, $p_3 = p_4 = 0.125$, $p_5 = p_6 = p_7 = p_8 = 0.0625$. Then

$$\{p_j\} = \{q_j^0\} = \{.25, .25, .125, .125, .125, .0625, .0625, .0625, .0625\}$$
$$\{q_j^1\} = \{.25, .25, .125, .125, .125, .0625, .0625\}$$
$$\{q_j^2\} = \{.25, .25, .125, .125, .125, .125\}$$
$$\{q_j^3\} = \{.25, .25, .25, .125, .125\}$$
$$\{q_j^4\} = \{.25, .25, .25, .25\}$$
$$\{q_j^5\} = \{.5, .25, .25\}$$
$$\{q_j^6\} = \{.5, .5\}$$

and

$$S^0 = \{\{1\}, \{2\}, \{3\}, \{4\}, \{5\}, \{6\}, \{7\}, \{8\}\}$$
$$S^1 = \{\{1\}, \{2\}, \{3\}, \{4\}, \{07, 18\}, \{5\}, \{6\}\}$$
$$S^2 = \{\{1\}, \{2\}, \{3\}, \{4\}, \{07, 18\}, \{05, 16\}\}$$
$$S^3 = \{\{1\}, \{2\}, \{007, 018, 105, 116\}, \{3\}, \{4\}\}$$
$$S^4 = \{\{1\}, \{2\}, \{007, 018, 105, 116\}, \{03, 14\}\}$$
$$S^5 = \{\{0007, 0018, 0105, 0116, 103, 114\}, \{1\}, \{2\}\}$$
$$S^6 = \{\{0007, 0018, 0105, 0116, 103, 114\}, \{01, 12\}\}$$
$$\{V_1\} = \{\{00007, 00018, 00105, 00116, 0103, 0114, 101, 112\}\}$$

To generalize the above algorithm to the case $d > 2$, we remark that it might be impossible to perform d merges at each step (as in the binary case) and still end up with a single set. It might therefore be necessary to have an initial merge of size $\ell < d$, and in any case $2 \leqslant \ell \leqslant d$, where l is such that if all subsequent merges are of size d, we are left with a single set at the end.

More formally, if M is the number of codewords and n indicates the total number of merges, the following must be true

$$M - [(\ell - 1) + (n-1)(d-1)] = 1 , \tag{2.16}$$

whence

$$\ell = M - (n-1)(d-1) , \tag{2.17}$$

so that if $M-(n-1)(d-1) > 1$ (n is the largest integer for which the right-faced side of (2.17) is nonnegative), then

$$\ell \equiv M \pmod{d-1} ; \tag{2.18}$$

whereas, if $M-(n-1)(d-1)$ is either 0 or 1, then

$$\begin{aligned} \ell &= (d-1) + M - (n-1)(d-1) \\ &= M - (d-1)(n-1) , \end{aligned} \tag{2.19}$$

since ℓ cannot be less than 2.
Putting togther (2.18) and (2.19), one easily gets:

$$\ell = 2 + R_{d-1}(M-2) , \tag{2.20}$$

where $R_{d-1}(M-2)$ is the remainder of the division of $M-2$ by $d-1$.

Example 2.2

As an example, let $d = 3$, $M = 8$, and let the probability distribution be the same as in Example 2.1. Since $R_2(6) = 0$, from (2.20) we get $\ell = 2$, and the procedure gives successively:

$$\{q_j^1\} = \{.25, .25, .125, .125, .125, .0625, .0625\}$$

$$\{q_j^2\} = \{.25, .25, .25, .125, .125\}$$
$$\{q_j^3\} = \{.5, .25, .25\},$$

and

$$S^0 = \{\{1\}, \{2\}, \{3\}, \{4\}, \{5\}, \{6\}, \{7\}, \{8\}\}$$
$$S^1 = \{\{1\}, \{2\}, \{3\}, \{4\}, \{07, 18\}, \{5\}, \{6\}\}$$
$$S^2 = \{\{1\}, \{2\}, \{007, 018, 15, 26\}, \{3\}, \{4\}\}$$
$$S^3 = \{\{0007, 0018, 015, 026, 13, 24\}, \{1\}, \{2\}\}$$
$$\{V_1\} = \{\{00007, 00018, 0015, 0026, 013, 024, 11, 22\}\}.$$

From what we have said so far, it is apparent that the Huffman encoding procedure relates probabilities to codewords and not source sequences to codewords. Clearly there is no way to compute the codewords from the source sequences, and consequently a codebook is necessary and its size is c^k, the number of different k-length source sequences from alphabet J_c. This is practical only if k is small. Later we shall see another encoding scheme, due to Elias, which eliminates the need for any codebook.

3. Variable Length-to-Block Encoding

Variable length-to-block encoding consists in associating variable length sequences of source outputs to codewords of constant length. It can be considered as a generalization of run length encoding, and is a technique of data compression which becomes particularly attractive for "skew sources" (a skew source is one in which the frequencies of some output letters very much exceed that of the others), or for retrieval situations which require block formatting of data. Variable length-to-block coding was first considered by Tunstall [6], who invented a constructive encoding and proved its optimality in a certain sense (see below).

Before describing the actual Variable-to-Block (VB) model, it is convenient to introduce some terminology and notation.

Let us begin by considering an arbitrary sequence w of length L(w), made up with letters from J_c:

$$w = z^1 z^2 \dots z^{L(w)} \qquad (z^i \in J_c). \tag{3.1}$$

Let W(T) denote a set of T sequences w_i:

$$W(T) = \{w_i : 1 \leq i \leq T\}.$$

If w is an arbitrary c-nary sequence and z is an element of J_c, a new c-nary sequence w may be formed appending z to w as a suffix (i.e. forming the concatenation of w and z). The following definitions describe two special properties that the set W(T) may have.

Definition 3.1. Let w_i and w_j be sequences belonging to W(T). If for every i and j $(i \neq j)$ w_i is not a prefix of w_j, then W(T) is called **proper.**

Definition 3.2. If every infinite length c-nary string has a prefix which belongs to W(T), then W(T) is called **complete.**

From now on the sequences of W(T) will be referred to as **words.**

An example of word set that is both complete and proper is W(3) = = {0, 10, 11}, for c = 2.

Now we state a series of lemmas concerning complete and proper word sets. Their proof can be found in Schneider [4].

Lemma 3.1. $W(c) = J_c$ is complete and proper and it is the only complete and proper word set of size c.

Lemma 3.2. If W(T) is a complete and proper set of size T and w* is an element of W(T), then the word set

$$W(T+c-1) = \{W(T) - w^*\} \bigcup_{z \in J_c} w^*z$$

is proper and complete and has size T+c–1. The set W(T +c–1) is said to be an **extension** of W(T) and w* is called the extending word.

Lemma 3.3. There exists a sequence

$$\{W_n\} \qquad (n = 1,2,...)$$

of complete and proper word sets, the n-th having size c + (n–1) (c–1), with $W_1 = J_c$, and being W_{n+1} an extension of W_n.

Lemma 3.4. If W(T) is a complete and proper word set of size T, then

$$T = c + n(c-1)$$

for some integer $n \geq 0$.

From now on we will restrict our attention to complete and proper word sets W(T) relative to a Constant Memoryless Source (CMS). Any infinite output sequence of a CMS can be segmented into elements of any given W(T) by the completeness property. On the other hand the properness of W(T) implies that only one partition of the sequence into elements of W(T) is possible. Thus, one may think of any source output sequence not only as an infinite string of the z's, but also of the w's. In this manner the constant memoryless c-nary information source can be thought of also as a T-nary information source, which outputs letters from the alphabet W(T), although not at a constant rate*.

The distribution $\{Q(z)\}$ on the letter alphabet J_c of the CMS induces the following product probability distribution $\{P(w)\}$ on the word alphabet W(T):

If $w = z^1 \dots z^{L(w)}$, then

$$P(w) = \prod_{i=1}^{L(w)} Q(z^i) . \tag{3.2}$$

Consider any w_i belonging to W(T), and assign a d-nary sequence $\Phi(w_i)$ to w_i with the following properties: $\Phi(w_i)$ has length $M = 2 + [\log_d T]$ (the symbol $[y]$ denotes the greatest integer less than or equal to y). The first letter of $\Phi(w_i)$ will be the digit 1 (the purpose of this tag will become clear in Section 4). The remaining $1 + [\log_d T]$ digits will be the expansion to the base d of i, the index of w_i. The collection of the ordered pairs $\{w_i, \Phi(w_i)\}$ of c-nary source words and corresponding d-nary sequences will be called a Variable Length-to-Block code defined on W(T). $\Phi(w_i)$ will be called the codeword of w_i.

A measure of the efficiency of a VB coding scheme is the compression ratio, defined as

$$\frac{M \log d}{E[L(w)]} ,$$

where $E[x]$ denotes the expected value of the random variable x. The compression ratio is seen to be proportional to the average number of code digits per source output.

Tunstall proved that for every admissible size $T = c + (n-1)(c-1)$ the following algorithm yields the complete and proper word set W(T) which minimizes

*) We remark that the properness requirement is a restriction which is not necessary for unique coding and decoding. The word set $W(6) = \{000, 001, 010, 01, 10, 1\}$ is not proper (01 is a prefix of 010), but could be used as a basis of a Variable Length-to-Block code based on the rule: "parse the source output sequence into the longest source-words found in the set". Under such a rule the sequence 1110010011... would be parsed unambiguously as (1), (1), (10), (010), (01), (1... . On the other hand, the properness restriction avoids many complications in the analysis and in particular allows a straightforward probability assignment to the source words.

the compression ratio.

Algorithm 3.1

Let $W^*(c) = J_c$ (this is a complete and proper word set of size c).

Let $W^*(c+(c-1))$ be the complete and proper word set of size $(c+(c-1))$ which is formed by extending $W^*(c)$ with the most probable word in $W^*(c)$.

Let $W^*(c+2(c-1))$ be the complete and proper word set of size $(c+2(c-1))$ which is formed by extending $W^*(c+(c-1))$ with the most probable word in $W^*(c+(c-1))$...

Continue this procedure, i.e. form $W^*(c+n(c-1))$ by extending $W^*(c+ +(n-1)(c-1))$ with its most probable word $(n = 1,2,...)$.

Proof of the Optimality of Algorithm 3.1.

Let W(T) be any complete and proper word set. Using the terminology of Lemma 3.3, let $W_n = W(T)$ and assume W_n is the extension of the word set W_{n-1} by a given word w*; then

$$E[L] \triangleq \sum_{w \in W_n} P(w)\, L(w) = \sum_{\substack{w \in W_{n-1} \\ w \neq w^*}} P(w)\, L(w) + P(w^*) \sum_z P(z)\,(L(w^*)+1) \tag{3.3}$$

$$= \sum_{w \in W_{n-1}} P(w)\, L(w) + P(w^*)\,.$$

But W_{n-1} in turn can be considered as the extension of some W_{n-2} by some appropriate word w**, and continuing this way one sees by recursion that

$$E[L] = \sum_i P(w_i)\,, \tag{3.4}$$

where the sum is over all the codewords that have been extended to form the sequence $\{W_i\}$ of word sets such that $W_n = W(T)$. Furthermore, eq. (3.3) shows that the extension of any W_{i-1} that maximizes E[L] on W_i, has to be performed by means of a word w* whose probability is maximal. If we can show that the optimal word set W(T) can be obtained by extending the optimal word set $W(T-(c-1))$, our proof will be complete.

We make use of the complete induction scheme.

First of all, the above statement is clearly true for $T = c$. We must now show that the statement is true for $T = c+n(c-1)$ when it is true for $T = c+k(c-1)$ for $k = 1,2,...,n-1$.

Let $\lambda, w_1, \ldots, w_k$ be the words that must be extended to obtain the optimal W(T), (λ is the empty word), let $\lambda, w'_1, \ldots, w'_{k-1}$ be the extension words for W(T−(c−1)) and let w'_k be the optimal extension word for W(T−(c−1)). [It is important to bear in mind that, by the induction hypothesis, the latter word set was obtained by Algorithm 3.1.] Suppose that the word indexing is such that

$$1 = P(\lambda) \geqslant P(w_1) \geqslant \ldots \geqslant P(w_{k-1}) \geqslant P(w_k)$$

(3.5)

$$1 = P(\lambda) \geqslant P(w'_1) \geqslant \ldots \geqslant P(w'_{k-1}) \geqslant P(w'_k) .$$

If W(T) is not an extension of W(T−(c−1)), then

$$\sum_{i=1}^{k-1} P(w_i) < \sum_{i=1}^{k-1} P(w'_i) , \tag{3.6}$$

but at the same time $P(w_k) > P(w'_k)$, since W(T) is optimal. But w'_k is the most probable word in W(T−(c−1)), and the only way for the last inequality to hold is that w_k already belongs to the set $\{w'_1, w'_2, \ldots, w'_{k-1}\}$. We will show that this is impossible. Let j be the least integer for which $w_j \neq w'_j$. Then, since W(T−(c−1)) is obtained by Algorithm 3.1, $P(w'_j) > P(w_j)$ and either w_j is one of the words $w'_{j+1}, \ldots, w'_k$ or $P(w_j) < P(w'_k)$, which because of (3.5) implies that $P(w_k) < P(w'_k)$, so that W(T) cannot be optimal. Then for some $\ell > j$, $w_j = w'_\ell$. Since every w_i is an extension of a lower index word, for all n $P(w_{j+1}) \leqslant P(w_{\ell+1})$, because of the optimality property of W(T−(c−1)) j now if n is such that $j + n = k$, we get $P(w_k) \leqslant P(w'_{k+m})$ for $m = \ell - j$ and a fortiori $P(w_k) \leqslant P(w'_k)$, by (3.5), thus contradicting the optimality of W(T). Q.E.D.

We remark that, unfortunately, the instrumentation complexity of any code produced by Algorithm 3.1 grows linearly with T, i.e. again exponentially with the code word length M. We shall treat this problem later.

We denote the minimal compression ratio for a given admissible T by $R_{min}(T)$, and define it as follows

$$R_{min}(T) \triangleq \min_{W(T)} \frac{M \log d}{E[L(w)]} ,$$

where the minimum is taken over all word sets of size T and M is $2 + [\log_d T]$ as before. We will show that $R_{min}(T)$ converges to the source entropy H(Z) as $T \to \infty$. Let us first prove two lemmas.

Lemma 3.5. Consider a constant memoryless source, and let W(T) be a complete and proper word set of size T defined on the source alphabet, J_c. Let $\{P(w)\}$ be the probability distribution on W(T) generated by the source distribution $\{Q(z)\}$, and define the average word length and the word entropy respectively by

$$E[L(w)] = \sum_{w \in W(T)} P(w)\, L(w)$$

$$H(W) = -\sum_{w \in W(T)} P(w) \log P(w) .$$

Then

$$H(W) = E[L(w)]\; H(Z) . \tag{3.7}$$

Proof. Let $\{W_n\}$ be the sequence corresponding to W(T) mentioned in Lemma 3.3, and define

$$E_n = \sum_{w \in W_n} P(w)\, L(w)$$

$$H_n = -\sum_{w \in W_n} P(w) \log P(w) .$$

The lemma is obviously true for $W_1 = J_c$, so we only need to prove it for W_{n+1} assuming it true for W_n. Let w* be the word that extends W_n to W_{n+1}, i.e.

$$W_{n+1} = \{ (W - w^*) \bigcup_{z \in J_c} w^* z \}.$$

Then, by the inductive hypothesis:

$$H_{n+1} = H_n + P(w^*) \log P(w^*) - \sum_{z \in J_c} P(w^*)Q(z) \log P(w^*)Q(z) =$$

$$= H_n + P(w^*)H(Z) = [E_n + P(w^*)]\, H(Z) .$$

But

$$E_{n+1} = E_n - P(w^*)L(w^*) + \sum_{z \in J_c} P(w^*)Q(z)[L(w^*) + 1] =$$

$$= E_n + P(w^*) ,$$

and therefore $H_{n+1} = E_{n+1}\, H(Z)$, which is (3.7).

Q.E.D.

Let W*(a) be the word set of size a formed by Algorithm 3.1. Define

$$\max(a) = \max_{w \in W^*(a)} P(w) ,$$

(3.8)

$$\min(a) = \min_{w \in W^*(a)} P(w) ,$$

and assume that the ordering of the source letters is such that

(3.9) $$Q(O) \leqslant Q(1) \leqslant \dots \leqslant Q(c-1) .$$

Lemma 3.6

$$\frac{\max(a)}{\min(a)} \leqslant \frac{1}{Q(O)} .$$

Proof. W*(a) was obtained from W*(a–(c–1) by extending its most probable word, let it be w*. Assume first that the least probable word of W*(a) is an extension of w*, i.e.

$$\min(a) = P(w^*)Q(O) .$$

On the other hand the most probable codeword has probability max(a) = P(w*), and therefore

$$\frac{\max(a)}{\min(a)} \leqslant \frac{P(w^*)}{P(w^*)Q(O)} = \frac{1}{Q(O)} .$$

On the other hand, if the least probable codeword of W*(a) was already a codeword of W*(a–(c–1), since max(a) ⩽ P(w*) = max(a–(c–1)) and min(a) = min (a–(c–1)), we have:

$$\frac{\max(a)}{\min(a)} \leqslant \frac{P(w^*)}{\min(a-(c-1))} = \frac{\max(a-(c-1))}{\min(a-(c-1))} ,$$

and therefore the result holds for W*(a) if it is true for W*(a–(c–1)). Since it clearly holds for W*(c), the lemma is proven by induction. Q.E.D.

We are now able to prove the following theorem.

Theorem 3.1 (Source Coding)

For the sequence of optimal variable-to-block word sets

$$\lim_{T \to \infty} R_{min}(T) = H(Z) .$$

Proof. The quantity $R_{min}(T)$ was defined to be $M \log d/E[L(w)]$, the expectation in the denominator being evaluated with respect to the probability measure P(w) induced on W*(T) by Q(z). If H(Z) = 0, the source has only one possible output, and the theorem holds, since no encoding is required. Otherwise, since $M = 2 + [\log_d T]$, by Lemma 3.5 we get

$$R_{min}(T) \;=\; \frac{(2 + [\log_d T]) \log d}{H(W)} \; H(Z) \;\geqslant\; \frac{\log d + \log T}{\log T} \; H(Z)\,. \quad (3.10)$$

On the other hand, using lemma 3.6:

$$H(W) \geqslant -\log \max(T) \geqslant \log Q(O) - \log \min(T) \geqslant \log Q(O) + \log T$$

(the last inequality follows from being $\min(T) \leqslant 1/T$), and again by lemma 3.5:

$$R_{min}(T) \leqslant \frac{2 \log d + \log T}{\log Q(O) + \log T} \; H(Z)\,, \quad (3.11)$$

and the result follows taking the lim for $T \to \infty$ in eq.s (3.10) and (3.11). Q.E.D.

Theorem 3.1 above constitutes in variable-to-length version of the direct part of Shannon's First Coding Theorem.

4. Block-to-Block Coding

One of the main features of this encoding technique is that, if it is to be efficient, some source sequences must be ambiguously encoded. These "ambiguous" source sequences should be associated to a single "ambiguous" codeword, and the problem is whether we can shorten the codeword length m significantly, while keeping the probability of the ambiguous codeword sufficiently low. Let us first give a precise description of the B-B encoding scheme.

Let $P(\mathbf{z}^k)$ be the probability that the k-length sequence $\mathbf{z}^k$ is generated by the source, and let A be a positive constant satisfying

$$\min_{\mathbf{z}^k} P(\mathbf{z}^k) < A^k < \max_{\mathbf{z}^k} P(\mathbf{z}^k)\,. \quad (4.1)$$

For every integer number k, fixing A determines the set $\mathcal{T}_k$ defined as follows:
Definition 4.1. Let $\mathcal{T}_k$ be the set of all source sequences $\mathbf{z}^k(z_i \in J_c;\ i = 1,2,\ldots,k)$ such that $P(\mathbf{z}^k) \geqslant A^k$, and let $\mathcal{T}_k^c$ be the complement of $\mathcal{T}_k$, When no confusion arises, th subscript k in the $\mathcal{T}_k$ and $\mathcal{T}_k^c$ will be omitted. We shall refer to $\mathcal{T}$ and to $\mathcal{T}^c$ as the unambiguous set and the ambiguous set,respectively.

Let T be the number of elements in $\mathcal{T}$(T obviously depends on k), and let m be such that

$$d^{m-1} < T + 1 \leqslant d^m \ . \tag{4.2}$$

It is then possible to select T + 1 distinct sequences, each of m symbols from the coding alphabet J_d and assign one to all the sequences of $\mathcal{T}^c$ and the remaining T to the sequences of $\mathcal{T}$ in a one-one way; this association law coincides with our code Φ. Thus:

$$\Phi(\mathbf{z}^k) = \mathbf{s}_T^m = s_1^{(T)} s_2^{(T)} \ldots s_m^{(T)} \quad \text{for all } \mathbf{z}^k \in \mathcal{T}^c .$$

$$\Phi(\mathbf{z}^k) \neq \Phi(\mathbf{z}_*^k) \quad \text{for} \quad \mathbf{z}^k, \mathbf{z}_*^k \in \mathcal{T},\ \mathbf{z}^k \neq \mathbf{z}_*^k \ .$$

For an illustration of this code, see figure 4.1. This code has the usual exponentially growing code book instrumentation.

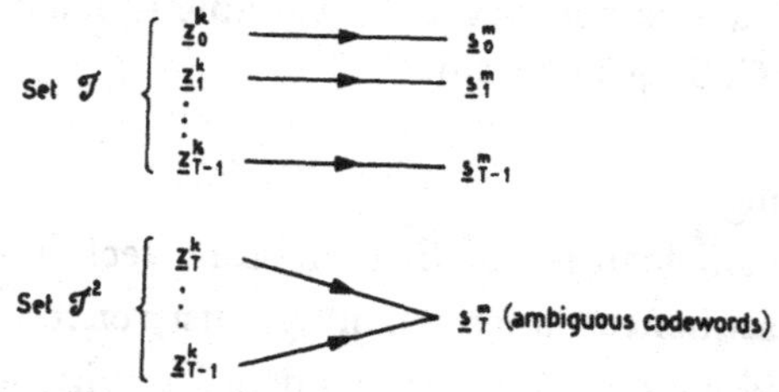

Fig. 4.1 **Diagram of the block-to-block coding method**

From the definition (4.2) of m, we get

$$m \sim \frac{\log(T+1)}{\log d},$$

and therefore the encoding rate R for this scheme is

$$R = \frac{\log(T+1)}{k} \sim \frac{\log T}{k}.$$

We are interested in he tradeoff between the rate R and the probability $P\{z^k \epsilon \mathscr{T}^c\}$ that the source puts out a sequence belonging to $\mathscr{T}^c$. Should it turn out that, in order for $P\{z^k \epsilon \mathscr{T}^c\}$ to approach zero as $k \to \infty$, the ratio $(T+1)/c^k$ must tend to 1, then the coding scheme would be useless and we should reject it.

Although the results that follow are straightforwardly generalizable to all ergodic Markovian sources, we shall restrict our discussion to the constant memoryless source (CMS), for which the output letters are independent, identically distributed random variables, and hence the probability of a sequence $z^k = z_1 \dots z_k$ is the product of the probabilities of its letters: $P\{z_1 \dots z_k\} = \prod_{i=1}^{k} Q(z_i)$.

For a CMS the probability of ambiguous encoding, P_a say, is given by (cf. definition 4.1):

$$P_a \triangleq P\{z^k \epsilon \mathscr{T}_c\} = P\{\log P(z^k) < k \log A\} =$$

$$= P\{\frac{1}{k} \sum_{i=1}^{k} \log Q(z_i) < \log A\}. \tag{4.3}$$

Obviously P_a is a function of k.

Since

$$E[\log Q(Z)] = \sum_{z \epsilon \mathscr{T}_c} Q(z) \log Q(z) = -H(Z), \tag{4.4}$$

from the following inequality satisfied by P_a:

$$P_a = P\{-\log A - H(Z) < \frac{1}{k} \sum_{i=1}^{k} (-\log Q(z_i)) - H(Z)\} \leqslant$$

$$\leqslant P\{-\log A - H(Z) < \left| \frac{1}{k} \sum_{i=1}^{k} -(\log Q(z_i)) - H(Z) \right|\},$$

by the Law of Large Numbers we can conclude that

$$\lim_{k\to\infty} P_a = 0,$$

provided the positive constant A satisfies $-\log A - H(Z) > 0$, i.e. provided

$$\log A < -H(Z) .$$

At the same time for the number T of elements of $\mathscr{T}$ the following upper bound can be obtained:

$$1 \geqslant P\{\mathbf{z}^k \in \mathscr{T}\} \triangleq \sum_{\mathbf{z}^k \in \mathscr{T}} P(\mathbf{z}^k) \geqslant T \min_{\mathbf{z}^k \in \mathscr{T}} P(\mathbf{z}^k) , \tag{4.5}$$

and, from the definition of $\mathscr{T}$ also

$$T \leqslant A^{-k}. \tag{4.6}$$

If we now select A to satisfy

$$\log A = -H(Z) - \epsilon ,$$

then $P_a \to 0$ as $k \to \infty$, while from (4.6)

$$T \leqslant 2^{-k \log A} = 2^{-k[-H(Z)-\epsilon]} = 2^{k[H(Z)+\epsilon]}$$

and therefore the rate $R = 1/k \log T$ satisfies

$$R \leqslant H(Z) + \epsilon .$$

What we have just achieved is the proof of the following theorem:

Theorem (Shannon [7]).

For any given $\epsilon > 0$ and $\delta > 0$, if k is large enough it is possible to encode k-length sequences of the source into $T+1$ codewords in such a way that the probability of the ambiguous set is upperbounded by

$$P_a < \delta , \tag{4.7}$$

while at the same time the inequality

$$\frac{\log T}{k} < H(Z) + \epsilon$$

holds.

Remark

We note that all the coding theorems we have given (for B-VL, VL-B, and B-B schemes) show that efficient coding is possible, in the limit as $k \to \infty$, provided the coding rate R is larger than the source entropy H(Z). A converse, whose proof we omit, shows that efficient encoding is impossible if $R < H(Z)$, see [1], [7], [8].

5. Instrumentable Encoding Based on the Elias Process [9]

As we noted earlier, although from a theoretical point of view the important thing about an encoding scheme is the asymptotic behaviour of the average codeword length and possibly of the probability of the ambiguous codeword, nevertheless from the practical viewpoint of instrumentation the **complexity** of the scheme plays a fundamental role. In this section we shall describe some coding algorithms that obey Shannon's theorem and whose complexity grows only **linearly** (and not exponentially) with the codeword length [1], [2]. A single principle, due to Elias, can be used to provide such algorithms for B-VL, VL-B, and B-B coding schemes. We first describe it for the B-VL case.

Consider the c^k source sequences of length k, and the corresponding probabilities $\{P(\mathbf{z}^k)\}$. Divide the interval [0,1] into c^k subintervals, whose lengths are $\{P(\mathbf{z}^k)\}$ and associate the sequence $\mathbf{z}^k$ to the interval whose length is $P(\mathbf{z}^k)$. See figure 5.1 for an illustration in the case $c = d = 2$. The points $\pi(\mathbf{z}^k)$ $(k = 1,2,3,...)$ are of interest to us (in figure 5.1 the points $\pi(\mathbf{z}^k)$ are given for $k = 1,2$, and 3), and they can be determined in the following way. If the empty sequence λ is associated to the point $\pi(\lambda) = 1$ then $\pi(\mathbf{z}^k)$ can be obtained recursively as follows. In the binary case:

$$\begin{aligned} \pi(\mathbf{z}^{\ell-1}\, 1) &= \pi(\mathbf{z}^{\ell-1}) \\ \pi(\mathbf{z}^{\ell-1}\, 0) &= \pi(\mathbf{z}^{\ell-1}) - P(\mathbf{z}^{\ell-1}\, 1)\,, \end{aligned} \tag{5.1}$$

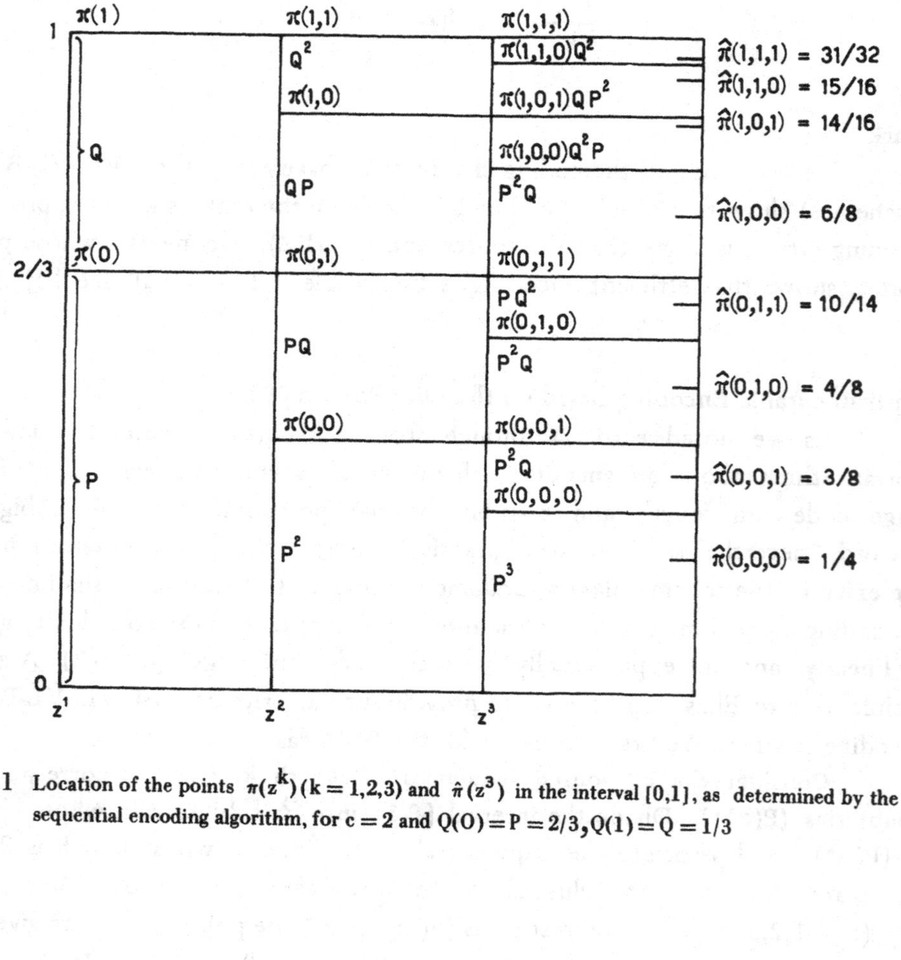

Fig. 5.1 Location of the points $\pi(z^k)\,(k=1,2,3)$ and $\hat{\pi}(z^3)$ in the interval $[0,1]$, as determined by the Elias sequential encoding algorithm, for $c=2$ and $Q(0)=P=2/3, Q(1)=Q=1/3$

and in the general c-nary case:

$$\pi(\mathbf{z}^{\ell-1}(c-1)) = \pi(\mathbf{z}^{\ell-1})$$
$$\pi(\mathbf{z}^{\ell-1}\, z) = \pi(\mathbf{z}^{\ell-1}) - P(\mathbf{z}^{\ell-1}) \cdot \sum_{j=z+1}^{c-1} Q(j) \quad (z = 0,1,\ldots, c-2). \tag{5.2}$$

See figure 5.2 which illustrates this for $c = 3$. The ordering of the points $\pi(\mathbf{z}^k)$ thus determined is the lexicographical ordering induced by the natural ordering in $\mathcal{T}_c$.

Now let $m(\mathbf{z}^k) = [1 - \log_d P(\mathbf{z}^k)]$ (integral part) and expand the numbers $\pi(\mathbf{z}^k)$ in d-nary basis:

$$\pi(\mathbf{z}^k) = s_1 d^{-1} + s_2 d^{-2} + \ldots + s_{m(\mathbf{z}^k)} d^{-m(\mathbf{z}^k)} + \ldots, \tag{5.3}$$

where $s_i \in J_d$. Define the truncated expansion

$$\hat{\pi}(\mathbf{z}^k) = s_1 d^{-1} + s_2 d^{-2} + \ldots + s_{m(\mathbf{z}^k)} d^{-m(\mathbf{z}^k)} \tag{5.4}$$

and the codeword $\Phi(\mathbf{z}^k)$ associated to $\mathbf{z}^k$:

$$\Phi(\mathbf{z}^k) = s_1 s_2 \ldots s_{m(\mathbf{z}^k)} \,. \tag{5.5}$$

We first note that the distance between the points $\pi(\mathbf{z}^k)$ and $\hat{\pi}(\mathbf{z}^k)$ satisfies

$$0 \leqslant \pi(\mathbf{z}^k) - \hat{\pi}(\mathbf{z}^k) < d^{-m(\mathbf{z}^k)} = d^{-[1-\log_d P(\mathbf{z}^k)]} < d^{\log_d P(\mathbf{z}^k)} = P(\mathbf{z}^k)\,,$$

where, by the above construction, $P(\mathbf{z}^k)$ can be seen to be the distance between $\pi(\mathbf{z}^k)$ and the next lower representative point (cf. eq. (5.2)). Therefore $\pi(\mathbf{z}^k)$ leads to a unique $\hat{\pi}$ point lying within the interval whose top point is $\pi(\mathbf{z}^k)$, hence $\mathbf{z}^k$ can be recovered without ambiguity from $\hat{\pi}(\mathbf{z}^k)$.

It is actually possible to recover $\mathbf{z}^k$ recursively, according to the following rule:

Rule. Given $\pi(\lambda) = 1$ and $\mathbf{z}^0 = \lambda$ (the empty sequence), then given $\mathbf{z}^\ell, \mathbf{z}^{\ell+1}$ is chosen as follows: if the inequality

$$\pi(\mathbf{z}^\ell) - P(\mathbf{z}^\ell) \sum_{j=z+1}^{c-1} Q(j) \geqslant \hat{\pi}(\mathbf{z}^k) \qquad (\ell = 0,1,\ldots, k-1) \tag{5.6}$$

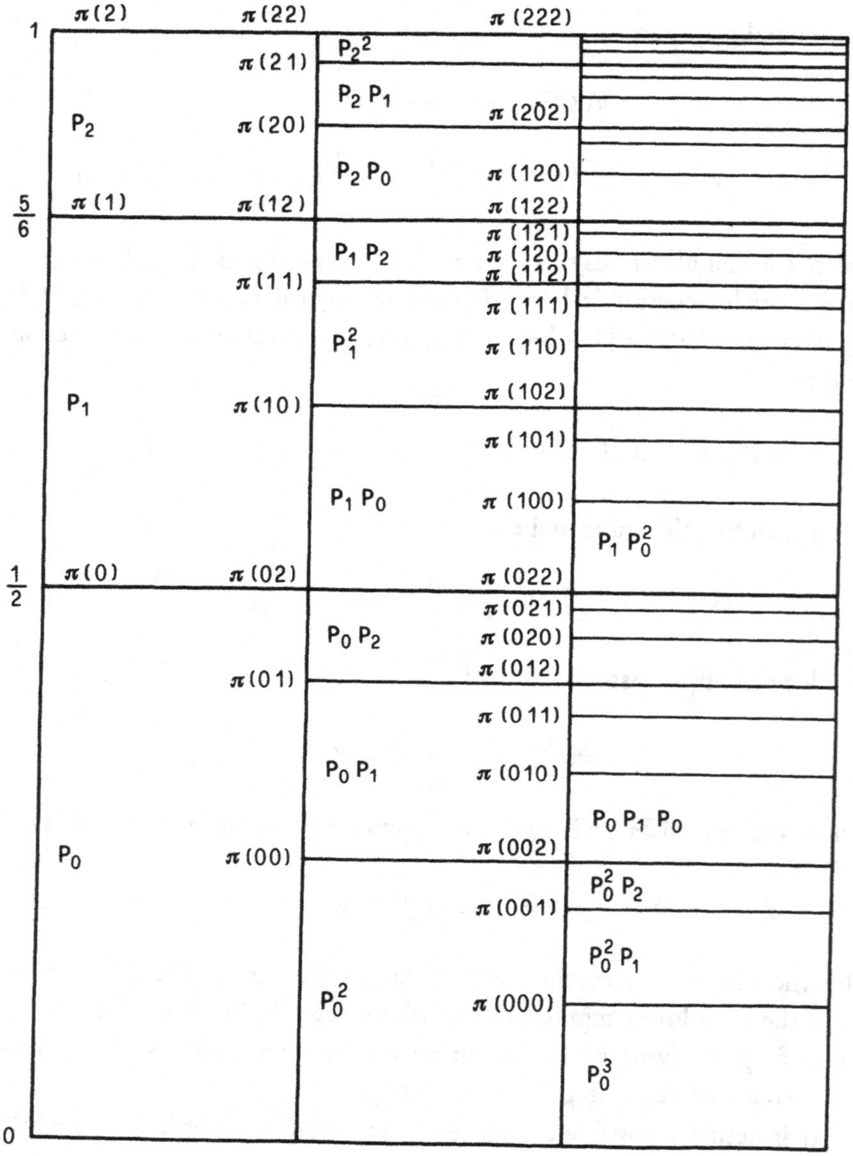

Fig. 5.2 **Location of the points $\pi(z^k)$ $(k = 1,2,3)$ in the interval [0,1], as determined by the Elias sequential encoding algorithm, for $c = 3$ and $Q(0) = P_o = 1/2$, $Q(1) = P_1 = 1/3$, and $Q(2) = P_2 = 1/6$**

is satisfied for some element of J_c, then put $\mathbf{z}^{\ell+1} = \mathbf{z}^{\ell} z$, where z is the smallest element of J_c satisfying (5.6). If, however no elėment of J_c satisfies (5.6), then put $\mathbf{z}^{\ell+1} = \mathbf{z}^{\ell}(c-1)$.

Now remark that if one has to decode a sequence of codewords $\{ \hat{\pi}(\mathbf{z}^k) \}$, it is necessary to know the length $m(\mathbf{z}^k)$ of each of them, otherwise it is not possible to know when one ends and the next one begins. This implies that the codeword set must be uniquely decipherable. But this is not the case in general. To make the code uniquely decipherable, every $\Phi(\mathbf{z}^k)$ is then prefixed with another codeword, which specifies its length $m(\mathbf{z}^k)$. The corresponding codeword set can be a block set of length ℓ, where ℓ is the smallest integer satisfying

$$d^{\ell} \geqslant \max m(\mathbf{z}^k) - \min m(\mathbf{z}^k) + 1 ,$$

i.e.

$$d^{\ell} > -\log_d Q^k_{min} + \log Q^k_{max} = k \log_d \frac{Q_{max}}{Q_{min}} ,$$

and therefore

$$\ell = \log_d k + \log_d \log_d \frac{Q_{max}}{Q_{min}} .$$

Now, since $m(\mathbf{z}^k) = [1 - \log_d P(\mathbf{z}^k)]$, the resulting compression ratio is upperbounded by

$$\frac{1}{k} [k\, H(Z) + 1 + \ell] = H(Z) + \frac{1}{k}(\text{const} + \log_d k) ,$$

which tends to H(Z) as $k \to \infty$. Therefore this encoding method is quasi-optimal (in the limit).

We next re-adjust the algorithm so as to have a block code. Given the sequence $\mathbf{z}^k$, the point $\pi(\mathbf{z}^k)$ is computed as before; $\hat{\pi}(\mathbf{z}^k)$ and $\Phi(\mathbf{z}^k)$, however, have the new values

$$\hat{\pi}(\mathbf{z}^k) = s_1 d^{-1} + \dots + s_m d^{-m}$$

$$\Phi(\mathbf{z}^k) = s_1 s_2 \dots s_m \quad (s_i \in J_d),$$

where m is the chosen codeword length. Note that

$$\pi(\mathbf{z}^k) \geq \hat{\pi}(\mathbf{z}^k) > \pi(\mathbf{z}^k) - d^{-m}$$

and consequently, since the interval corresponding to $\mathbf{z}^k$ has length $P(\mathbf{z}^k)$, $\pi(\mathbf{z}^k)$ is the smallest π-point above the computed $\hat{\pi}(\mathbf{z}^k)$ provided $P(\mathbf{z}^k) \geq d^{-m}$. If this is the case, $\pi(\mathbf{z}^k)$ and therefore $\mathbf{z}^k$ can be uniquely recovered from $\hat{\pi}(\mathbf{z}^k)$ by the previously described recursive algorithm. Therefore, the probability of error is upper-bounded by

$$P_k(e) \leq P\{\mathbf{z}^k :\ P(\mathbf{z}^k) < d^{-m}\} =$$

$$= P\{\mathbf{z}^k : -\frac{1}{k}\sum_{i=1}^{k} \log_d Q(z_i) - H_d(Z) > \frac{m}{k} - H_d(Z)\},$$

where $H_d(Z) = H(Z)/\log d$. By the Law of Large Numbers $P_k(e) \to 0$ as $k \to \infty$ as long as the condition

$$R \triangleq \log d \,\frac{m}{k} > H(Z)$$

is satisfied and so Shannon's theorem is again obeyed.

We now illustrate a further modification of the algorithm which will give us a linearly instrumentable VL-B encoding. This version is based on lemma 3.6 of the previous section, which states that if $Q(0) \leq Q(1) \leq \dots \leq Q(c-1)$, then for an optimal word set W(T).

$$(5.7) \qquad \max_{\mathbf{w} \in W(T)} P(\mathbf{w}) \leq \frac{1}{Q(O)} \min_{\mathbf{w} \in W(T)} P(\mathbf{w}).$$

Let m be the desired codeword length; then we decide that a sequence $\mathbf{w} = z_1 z_2 \dots z_{L-1} z_L$ will be a codeword provided

$$(5.8) \qquad P(z_1 \dots z_{L-1}) \geq \frac{1}{Q(O)}\, d^{-m},$$

$$P(z_1 \dots z_{L-1}\, z_L) < \frac{1}{Q(O)}\, d^{-m} . \tag{5.8'}$$

If this rule is obeyed, for each codeword **w**

$$P(\mathbf{w}) \geq d^{-m} , \tag{5.9}$$

since Q(0) is the smallest probability of the letters of J_c, and consequently the number of codewords will be less than or equal to d^m. Now (5.8) and (5.9) imply that for this class of codes (5.7) holds, i.e. lemma 3.6 is true, and so is true the corresponding coding theorem that is entirely based on that lemma.

The corresponding algorithm works recursively as follows:

1. Set $\pi(\lambda) = P(\lambda) = 1,\ L = 1;$
2. If $P(\mathbf{z}^L) < d^{-m}/Q(O)$, go to 4, else continue;
3. $\pi(\mathbf{z}^{L+1}) = \pi(\mathbf{z}^L) - P(\mathbf{z}^L) \sum_{j=z_{L+1}+1}^{c-1} Q(j)$ for $\mathbf{z}^{L+1} = \mathbf{z}^L\, z_{L+1}$,

   $P(\mathbf{z}^{L+1}) = P(\mathbf{z}^L)\, Q(z_{L+1})$,

   $L = L + 1$,

   go to 2 ;
4. $\mathbf{w} = \mathbf{z}^L ,\ \pi(\mathbf{w}) = \pi(\mathbf{z}^L) = s_1 d^{-1} + \dots + s_m d^{-m} + \dots ,$

   where $s_i \in J_d$;
5. $\hat{\pi}(\mathbf{w}) = s_1 d^{-1} + \dots + s_m d^{-m}$,

   $\Phi(\mathbf{w}) = s_1 s_2 \dots s_m$.

From what we said above, it is obvious that

$$\pi(\mathbf{w}) \geq \hat{\pi}(\mathbf{w}) > \pi(\mathbf{w}) - d^{-m}$$

and since $P(\mathbf{w}) \geq d^{-m}$, then $\pi(\mathbf{w})$ is the smallest π-point whose value exceeds $\hat{\pi}(\mathbf{w})$. The usual recovery algorithm can be used again for obtaining **w** from $\pi(\mathbf{w})$ with the additional stopping rule: $\mathbf{w} = \mathbf{z}^L$ where L is the least integer such that $P(\mathbf{z}^L) < 1/Q(O)\, d^{-m}$.

6. Schalkwijk Coding Schemes

In this section we shall outline B-VL and VL-B coding schemes due to Schalkwijk [10]. We shall limit ourselves to the binary source alphabet, but it is not difficult to generalize the argument to the c-nary case, along the same lines.

Among the 2^k binary sequences $\mathbf{z}^k = (z_1, z_2, \dots, z_k)$ of length k consider those having weight w, i.e., those for which $\sum_1^k z_i = w$. There are $\binom{k}{w}$ such sequences, for $w = 0,1,2,\dots,k$, and for each value of w they can be ordered in a natural way as follows. Put

$$I(\mathbf{z}^k) = \sum_{i=1}^{k} 2^{k-i} z_i , \tag{6.1}$$

i.e. let each sequence correspond to a number $I(\mathbf{z}^k)$ of which it is the set of coefficients in binary expansion. Now let $N(\mathbf{z}^k)$ be a numbering of the k-length sequences of weight w, with

$$N(\mathbf{z}^k) = 0,1,\dots,\binom{k}{w} - 1. \tag{6.2}$$

The association between these numbers and the sequences is done as follows: given any two sequences $\mathbf{z}^k$ and $\mathbf{z}_*^k$ of weight w, then

$$N(\mathbf{z}^k) > N(\mathbf{z}_*^k) \text{ if and only if } I(\mathbf{z}^k) > I(\mathbf{z}_*^k) . \tag{6.3}$$

Each of the $\binom{k}{w}$ numbers $N(\mathbf{z}^k)$ can be expanded in base d:

$$N(\mathbf{z}^k) = \sum_{i=1}^{m} s_i d^{m-i} , \quad (s_i \in J_d) , \tag{6.4}$$

where

$$m = \lceil \log_d \binom{k}{w} \rceil . \tag{6.5}$$

[In (6.5) $\lceil x \rceil$ denotes the smallest integer greater than or equal to x.] Thus the number $N(\mathbf{z}^k)$ can be uniquely encoded by means of the d-ary codeword $\Phi(\mathbf{z}^k) = s_1 \dots s_m$, whose length m depends on the weight w of $\mathbf{z}^k$.

Now, in turn, $w \in \{0,1,\dots,k\}$ can be specified by a d-ary codeword of length $\lceil \log_d (k+1) \rceil$, $\psi(w)$ say. The Schalkwijk B-VL coding scheme then associates the binary sequence $\mathbf{z}^k$ to the double codeword

$$\psi(w) \ \Phi(\mathbf{z}^k),$$

where $w = \sum_{i=1}^{k} z_i$ is the weight of $\mathbf{z}^k$. This code is clearly uniquely decipherable, since the $\psi(w)$ portion (of constant length) specifies the length m of the second portion, $\Phi(\mathbf{z}^k)$, which in turn uniquely specifies $\mathbf{z}^k$ according to the corresponding ordering $N(\mathbf{z}^k)$.

We now consider the expected codeword length E[L]. If p is the probability with which a 1 is output by the binary constant memoryless source, we have

$$\begin{aligned} E(L) &= \lceil \log_d (k+1) \rceil + \sum_{w=0}^{k} \binom{k}{w} p^w (1-p)^{k-w} \lceil \log_d \binom{k}{w} \rceil \\ &\tilde{<} 1 + \log_d (k+1) + \sum_{w=0}^{k} \binom{k}{w} p^w (1-p)^{k-w} \log \left[\left(\frac{w}{k}\right)^{-w} \left(1-\frac{w}{k}\right)^{-(k-w)} \right] \\ &\cong \log_d (k+1) + k\,H(p), \end{aligned}$$

where the approximate inequality is obtained using the Stirling approximation for n! to evaluate the binomial coefficient $\binom{k}{w}$ and the approximate equality holds for large values of k and is based on the law of large numbers, by which w/k tends to p with probability 1. Therefore

$$\lim_{k\to\infty} \frac{1}{k} E[L] = H(p),$$

which means that this coding scheme is optimal and conforms to Shannon's theorem.

We now consider the complexity of implementation of this scheme.

The complexity of obtaining $\psi(w)$ grows only linearly with k, and therefore the scheme can be considered as "good" if also the computation of $N(\mathbf{z}^k)$ involves a number of steps which grows only linearly with k. This we proceed to show.

Let w be the weight of the sequence $\mathbf{z}^k = z_1, z_2, \ldots, z_k$ and let the weight w_i of the sequence $z_1, z_{i+1}, \ldots, z_k$ (suffix of length $(k-i+1)$ of $\mathbf{z}^k$) be defined as

(6.6) $$w_i = \sum_{j=i}^{k} z_j \qquad (i = 1,2,\ldots,k).$$

Note that $w_1 = w$ is exactly the weight of $\mathbf{z}^k$.

Then the following nice result holds: the ordinal number $N(\mathbf{z}^k)$ of $\mathbf{z}^k \in S(k,w)$ is given by

(6.7) $$N(\mathbf{z}_k) = \sum_{i=1}^{k} z_i \binom{k-i}{w_i},$$

where $\binom{\ell}{w} = 0$ whenever $\ell < w$, and $0 \leqslant N(\mathbf{z}^k) \leqslant \binom{k}{w} - 1$.

The proof of (6.7) is by induction on the length k of $\mathbf{z}^k$. For $k = 1$ only two weights are permitted, i.e. $w = 0$, and $w = 1$ and (6.7) holds in both cases, since if $w = 0$, then $z_1 = 0$, yielding $N(0) = 0$, and if $w = 1$, then $\binom{0}{1} = 0$, again yielding $N(1) = 0$. Assume now that (6.7) is true for each $S(k,w)$, $0 \leqslant w \leqslant k$, with $k \leqslant M$; we wish to prove that (6.7) holds true for $S(M+1,w)$, with $0 \leqslant w \leqslant M+1$. The crucial point is to remark that $S(M+1,w) = S_0(M+1,w) \cup S_1(M+1,w)$, where $S_0(M+1,w)$ is that subset of $S(M+1,w)$ whose sequences have $z_1 = 0$, and $S_1(M+1,w)$ is the complementary set, i.e. contains the sequences of $S(M+1,w)$ for which $z_1 = 1$. Now $S_0(M+1,w)$ has $\binom{M}{w}$ elements and, by the inductive hypothesis, it can be ordered according to

$$N_0(\mathbf{z}^{M+1}) = \sum_{i=1}^{M} z_{i+1} \binom{M-i}{w_{i+1}} = \sum_{i=2}^{N+1} z_i \binom{M+1-i}{w_i},$$

whereas $S_1(M+1,w)$, again by the inductive hypothesis, can be ordered according to

$$N_1(\mathbf{z}^{M+1}) = \sum_{i=1}^{N} z_{i+1} \binom{M-i}{w_{i+1}} = \sum_{i=2}^{N+1} z_i \binom{M+1-i}{w_i}.$$

Now, by the rule (6.3), each element in S_1 follows all elements in S_0, and consequently the total set S(M + 1,w) can be ordered according to

$$N(z^{M+1}) = (1-z_1)\, N_0(z^k) + z_1 [\binom{M}{w} + N_1(z^k)]$$

$$= \sum_{i=1}^{M+1} z_i \binom{M+1-i}{w_i},$$

which is (6.7).

From (6.7) it is seen that the computation of $N(z^k)$ involves a linear amount of work if the coefficients $\binom{k-i}{w_i}$ are stored. Since the coefficients have the form $\binom{k-1}{w}, \binom{k-2}{w}, \ldots, \binom{0}{w}$ it is only necessary to store them when w does not exceed k−1, k−2,...,0 respectively. This requires $k + (k-1) + \ldots + 1 = 1/2 \cdot k(k+1)$ locations, and to make the algorithm truly linear we have to avoid this storage. Consider then the computation of the coefficients $\binom{k-i}{w_i}$.

First remark that w_{i+1} is the weight of the suffix of length k−i, and therefore either $w_{i+1} = w_i$ or $w_{i+1} = w_i - 1$. This implies that if $\binom{k-j}{w_i} = 0$ for some j, then $\binom{k-i}{w_i} = 0$ for all $i > j$. If i' is the largest index for which $\binom{k-i'}{w_{i'}} \neq 0$, then $\binom{k-i'}{w_{i'}} = 1$ and all coefficients with an index i smaller than i' can be obtained from this one by multiplication. Actually, since w_{i-1} is equal either to w_i or to $w_i + 1$, the coefficient $\binom{k-i+1}{w_{i-1}}$ is either $\binom{k-i+1}{w_i}$ or $\binom{k-i+1}{w_i+1}$ and these two numbers can be obtained from $\binom{k-i}{w_i}$ by one multiplication.

It is useful to explicitly illustrate the ordering of k-sequences specified by (6.7) by means of Pascal's triangle. Consider the sequence $z^6 = 010100$, which belongs to S(6,2); an application of (6.7) yields N(010100) = 8, and this can be obtained as follows: write Pascal's triangle down to the 6-th row (6 being the length of the sequence), and consider $\binom{6}{2} = 15$ (2 being the weight of the sequence).

The digits of z^6 determine a path from 15 up to the topmost 1 according to the following rule: consider the digits of z^6 from the leftmost one and take a step upward to the right for each 0 and a step upward to the left for each 1 (see figure 6.1). The entire path, for any sequence belonging to S(6,2), is contained in the rectangle shown in figure 6.1. Now whenever a step is taken in the 1 direction (upward left), consider the triangle entry which would have been reached if the step were taken in the 0 direction (in figure 6.1 such entries are encircled); their sum yields $N(z^k)$. In our example 6 + 2 = 8, which is N(010100).

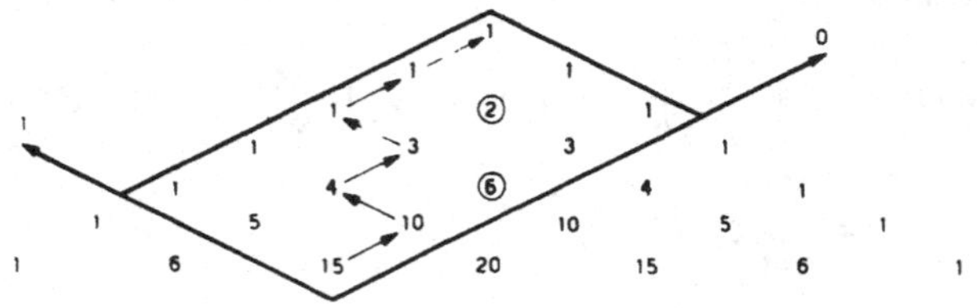

Fig. 6.1 **Pascal's triangle with the path corresponding to z = 010100**

An easy converse algorithm permits to obtain the sequence z^k corresponding to a given $N(z^k)$. The following table gives the complete set S(6,2) with the 15 sequences ordered according to (6.7).

0. 000011	5. 001100	10. 100001
1. 000101	6. 010001	11. 100010
2. 000110	7. 010010	12. 100100
3. 001001	8. 010100	13. 101000
4. 001010	9. 011000	14. 110000

We conclude these notes with an illustration of a VL–B encoding scheme due to Schalkwijk and closely connected with the previous B–VL scheme.

Suppose we fix a value w for the weight of the encoded sequences, and let $z_1\, z_2 \ldots z_t \ldots$ be the sequence of binary digits generated by the source. Assume that the probability that a 0 be generated is q and the probability that a 1 be generated is $p = 1-q$. Let now t be the smallest integer such that

$$\text{either} \qquad \text{i)} \quad \sum_{i=1}^{t} z_i = w$$

$$\text{or} \qquad \text{ii)} \quad \sum_{i=1}^{t} z_i = w - (k-t)\,.$$

In case i), heuristically, the source has output many ones in the first positions, and we consider the following sequence, which is completed with (k–t) dummy zeros:

$$z_1 z_2 \ldots z_t \; \overbrace{0\,0 \ldots 0}^{k-t}\,. \tag{6.8}$$

This sequence has length k and weight t. In case ii) the source has output many zeros in the first positions and we consider the following sequence, which is completed with (k–t) dummy 1's:

$$z_1 z_2 \dots z_t \; \overbrace{11 \dots 1}^{k-t} , \tag{6.9}$$

which again has length k and weight t. After this insertion of 0's and 1's, the next t source digits are considered, and the same procedure is applied.

Now the sequences (6.8) or (6.9) are encoded with the above-described procedure. Remark that in this case the weight of the sequence is always the same, and therefore there is no need of the ψ part of the codeword. The codewords are all of the same length, and therefore this is a VL–B encoding scheme.

The number of 1's in a binary sequence of length k tends in probability to pk, and it is interesting to consider the choice w = pk for the weight of the encoded sequences. It can be shown that in this case the average length $\bar{t}$ of the source sequences encoded into blocks of length $\lceil \log_2(k/pk) \rceil$ is

$$\bar{t} = k[1 - \binom{k}{kp} p^{pk} q^{qk}] . \tag{6.10}$$

Using Stirling's approximation in (6.10), we get

$$\bar{t} \approx k[1 - (2\pi pqk)^{-1/2}]$$

and therefore, for large k, the ratio $\lceil \log_2(k/pk) \rceil / \bar{t}$ approaches H(p), and the scheme obeys Shannon's theorem.

REFERENCES

[1] Jelinek, F.: *Probabilistic Information Theory*, McGraw-Hill, 1968.

[2] Jelinek, F.: "Buffer Overflow in Variable Length Coding of Fixed Rate Sources", IEEE Trans. Inform. Theory, IT-14(3), May 1968.

[3] Kraft, L.G.: "A Device for Quantizing, Grouping and Coding Amplitude Modulated Pulses", M.S. thesis, Dept. of Elec. Engin., M.I.T., 1949.

[4] Huffman, D.A.: "A Method of Construction of Minimum Redundancy Codes", Proc. IRE, 40(10), Sept. 1952.

[5] Gallager, R.G.: *Information Theory and Reliable Communication*, J. Wiley and Sons, 1968.

[6] Tunstall, A.: "Synthesis of Noiseless Compression Codes", Ph.D. dissertation, Georgia Inst. Technol., Atlanta, 1968.

[7] Shannon, C.E.: "A Mathematical Theory of Communication", Bell Systems Tech.J., 27, p. 379 and 623, 1948.

[8] Longo, G.: *"Source Coding Theory"*, CISM Courses and Lectures, No. 32, Springer, 1970.

[9] Elias, P.: unpublished result.

[10] Schalkwijk, J.P.: "An Algorithm for Source Coding", IEEE Trans. Inform. Theory, IT-18(3), May 1972.

LIST OF CONTRIBUTORS

BERGMANS, Patrick P.: State University of Ghent – Center for Computer Sciences – St. Pietersnieuwstraat 41 – B 9000 GENT (Belgium)

DE LUCA, Aldo: Laboratorio di Cibernetica CNR – Via Toiano 2 – 80072 ARCO FELICE (Napoli)

GOETHALS, Jean-Marie: MBLE Research Laboratory – Av. van Becelaere 2 – B-1170 BRUSSELS (Belgium)

JELINEK, F: IBM Thomas J. Watson Research Center – Yorktown Heights – N.Y., U.S.A.

LONGO, Giuseppe: Università di Trieste – Istituto di Elettronica – 34100 TRIESTE

MASSEY, James L.: University of Notre Dame – Department of Electrical Engineering – NOTRE DAME – Indiana 46556 (USA)

RUDOLPH, Luther D.: Syracuse University – Systems and Information Science – SYRACUSE – New York 13210 (USA)

SAVAGE, John E.: Division of Engineering – Brown University – Providence – Rhode Island

ZIABLOV, Victor: Institute for Problems of the Information Transmission – Aviamotornaya 8 Bild. 2 – Soviet Academy of Sciences – MOSCOW E-24 (USSR)

ZIGANGIROV, Kamil, Sh.: Institute for Problems of the Information Transmission – Aviamotornaya 8 Bild. 2 – Soviet Academy of Sciences – MOSCOW E-24 (USSR)

CISM SUMMER SCHOOL ON

"Coding and Complexity"

Udine, July 15–26, 1974

LIST OF PARTICIPANTS

BECK, Josef: Eötvös Lorand Tudomanyegyetem – Jozsef Attila u. 25 – 1212 BUDAPEST (Hungary)

BELLINI, Sandro: Politecnico di Milano – Istituto di Elettronica – Piazza L. da Vinci 32 – 20133 MILANO

BENELLI, Giuliano: Istituto di Onde Elettromagnetiche – Via Panciatichi 56 – 50127 FIRENZE

BRAVAR, Diego: Student – Via del Pratello 11 – 34135 TRIESTE

CHIRIACESCU, Sergiu: University of Brasov – B-dul Gh. Gheorghiu-Dej 29 – BRASOV (Rumania)

DELLA RICCIA, Giacomo: Ben Gurion University of the Negev – Department of Mathematics – BEER-SHEVA (Israel)

GALLO, Ferdinando: Student – Via Tommaso Pisano 1 – 56100 PISA

MICHELACCI, Giacomo: Via Fabio Severo 71 – 34127 TRIESTE

OSYCZKA, Andrzej: Instytut Technologii Maszyn – Politechnika Krakowska – ul. Warszawska 24 – KRAKOW (Poland)

SGARRO, Andrea: Via d'Isella 4 – 34144 TRIESTE

TARJAN, Tamas: Közgazdasagtudomonyi Int. MTA – Szentkiralyi u. 35 – 1088 BUDAPEST (Hungary)

TSOI, Ah Chung: Dept. of Electrical Engineering – Paisley College of Technology – High Street – PAISLEY – Renfrewshire (UK)

VETIER, Andras: Technical University of Budapest – Szentkiralyi u. 51 – 1088 BUDAPEST (Hungary)

Printed in the United States
By Bookmasters